DEVELOPMENTS IN REINFORCED PLASTICS—1

Resin Matrix Aspects

THE DEVELOPMENTS SERIES

Developments in many fields of science and technology occur at such a pace that frequently there is a long delay before information about them becomes available and usually it is inconveniently scattered among several journals.

Developments Series books overcome these disadvantages by bringing together within one cover papers dealing with the latest trends and developments in a specific field of study and publishing them within *six months* of their being written.

Many subjects are covered by the series including food science and technology, polymer science, civil and public health engineering, pressure vessels, composite materials, concrete, building science, petroleum technology, geology, etc.

Information on other titles in the series will gladly be sent on application to the publisher.

DEVELOPMENTS IN REINFORCED PLASTICS—1

Resin Matrix Aspects

Edited by

G. PRITCHARD

School of Chemical and Physical Sciences,
Kingston Polytechnic, Kingston upon Thames, Surrey, UK

APPLIED SCIENCE PUBLISHERS LTD
LONDON

APPLIED SCIENCE PUBLISHERS LTD
RIPPLE ROAD, BARKING, ESSEX, ENGLAND

British Library Cataloguing in Publication Data
Developments in reinforced plastics.—(Developments
series).
1: Resin matrix aspects
1. Reinforced plastics
I: Pritchard, Geoffrey   II. Series
668.4'94   TP1177
ISBN 978-94-009-8726-5   ISBN 978-94-009-8724-1 (eBook)
DOI 10.1007/978-94-009-8724-1

WITH 69 TABLES AND 87 ILLUSTRATIONS

FOREWORD

It needs only modest foresight to see that the future of plastics lies in their use as composites. The article made of plastics alone will, I believe, become a rarity. At the less critical end of the applications spectrum, users will 'dilute' plastics with cheap, solid fillers. Unstressed neutral axes will be foams, for air is a splendidly cheap filler. From here, it is a short step to get the diluent to reinforce, or alter, the material in some way. Reinforcement is needed because plastics tend to be weak, to lack stiffness and, above all, to creep. Glass fibre, especially, offers inherent cheapness, strength and stiffness and, although almost unusable by itself, it shows real advantages when married with a matrix. Greater use of filling and reinforcement will bring into prominence a relatively neglected group of polymers—the thermosets and chemisets. The great range of potential properties, obtained from these materials with their varied chemistry, makes them well suited for use in composites and they must surely move nearer to parity in tonnage with thermoplastics.

The key to all filling and reinforcement is to retain the one really outstanding property of plastics—their ease of fabrication. This is a property which is no less important for not being quantifiable. Consider the production of complex articles; these may be made in only one step using injection moulding, how many steps would it take with metal? Look at a GRP moulding made by hand lay-up; consider the complexity of making the same article in metal, in wood, or even in concrete. It is easy to go over the top, chipboard is an example. It has less than 10% polymer, and uses, otherwise, waste wood, but the real ease of fabrication is lost as it has to be cut, drilled and fastened like wood itself, not like plastic.

Plastics are complex materials. Attempts to use them without employing even our present inadequate knowledge of their science leads rapidly to disaster. Changes in morphology due to such factors as cooling, crystallisation, molecular orientation, or weld lines, can make one quite afraid to push melt into mould! The situation with composites is much more complex. To the existing factors is added the nature of the bond with the second material, i.e. the nature of the interface, the influence of another material on the polymer morphology, and the different mechanism of absorbing stress and of eventual failure.

The materials scientist, like other scientists, tends to attempt only that which he thinks has a good chance of success. He has tended to keep away from composites. Look at the relative amounts of literature on crack formation and propagation in unfilled and filled plastics! Consider our inadequate understanding of basic data on the match needed between matrix stiffness and fibre stiffness. If we have a thin interfacial layer with different properties to either fibre or matrix, then there is virtually no information on what those properties would best be.

This situation must change. The scientist, or the technologist, driven by what he has to do, must move to an intensive attack on the science of composites. Indeed, in the case of high performance materials for military applications, this is already happening.

There is a need to set out concisely, but precisely, where we start from. This series of volumes attempts to do just that, and is timely and welcome.

A. A. L. CHALLIS
Polymer Engineering Directorate,
Science Research Council,
London, UK

PREFACE

The subject of reinforced plastics is a fast-developing one. It is hoped that this book, and subsequent volumes, will help readers to keep up to date with some of the more important changes taking place. One of the benefits of a book such as this one may be that it encourages the consideration of reinforced plastics as a multidisciplinary field; some of the chapters are distinctly chemical, others discuss engineering properties, while some are a mixture. It should be beneficial if people engaged in research and development attempt to understand the relationships between synthesis, structure, properties and applications.

Perceptive readers may note one or two places where different authors give different opinions about the same matter. This is a fair reflection of the degree of uncertainty still persisting in our understanding of fibre–resin composites.

In his Foreword, Dr Challis stresses the importance of fabrication in determining the future of plastics. This subject is a large one and deserves a book to itself. Indeed it is hoped that a subsequent volume in this series will be devoted to that subject.

The editor would like to thank all the authors who have helped by contributing chapters, and also to acknowledge the part played by their companies and universities, etc., in facilitating publication.

G. PRITCHARD

CONTENTS

Foreword v

Preface vii

List of Contributors xi

1. Thermosetting Resins for Reinforced Plastics . . . 1
   G. PRITCHARD

2. Vinyl Ester Resins 29
   THOMAS F. ANDERSON and VIRGINIA B. MESSICK

3. Polyester Resin Chemistry 59
   T. HUNT

4. Phenol–Aralkyl and Related Polymers 87
   GLYN I. HARRIS

5. Initiator Systems for Unsaturated Polyester Resins . . 121
   V. R. KAMATH and R. B. GALLAGHER

6. High-Temperature Properties of Thermally Stable Resins . 145
   G. J. KNIGHT

7. Structure–Property Relationships and the Environmental
   Sensitivity of Epoxies 211
   ROGER J. MORGAN

8. Some Mechanical Properties of Crosslinked Polyester Resins   231
        W. E. DOUGLAS and G. PRITCHARD

9. Crack Propagation in Thermosetting Polymers   .   .   .   257
        ROBERT J. YOUNG

Index   285

LIST OF CONTRIBUTORS

THOMAS F. ANDERSON

    Senior Research Specialist Resins TS & D, Dow Chemical USA, Texas Division, Freeport, Texas 77541, USA

W. E. DOUGLAS

    School of Chemical and Physical Sciences, Kingston Polytechnic, Penrhyn Road Centre, Penrhyn Road, Kingston upon Thames, KT1 2EE, UK

R. B. GALLAGHER

    Project Leader—Applications, Lucidol Division, Pennwalt Corporation, PO Box 1048, Buffalo, New York 14240, USA

GLYN I. HARRIS

    Advanced Resins Ltd, Unit 8, Llandough Trading Estate, Penarth Road, Cardiff, South Glamorgan, UK

T. HUNT

    BP Chemicals Limited, Research and Development Department, South Wales Division, Sully, Penarth, South Glamorgan CF6 2YU, UK

V. R. KAMATH

    Lucidol Division, Pennwalt Corporation, PO Box 1048, Buffalo, New York 14240, USA

G. J. KNIGHT

Ministry of Defence PE, Materials Department, Royal Aircraft Establishment, Farnborough, Hants, UK

VIRGINIA B. MESSICK

Dow Chemical USA, Texas Division, Freeport, Texas 77541, USA

ROGER J. MORGAN

Lawrence Livermore Laboratory L-338, University of California, PO Box 808, Livermore, California 94550, USA

G. PRITCHARD

School of Chemical and Physical Sciences, Kingston Polytechnic, Penrhyn Road Centre, Penrhyn Road, Kingston upon Thames, KT1 2EE, UK

ROBERT J. YOUNG

Department of Materials, Queen Mary College, University of London, Mile End Road, London E1 4NS, UK

Chapter 1

THERMOSETTING RESINS FOR REINFORCED PLASTICS

G. Pritchard

Kingston Polytechnic, Surrey, UK

SUMMARY

The resins used in the reinforced plastics industry are mainly of the thermosetting variety. This means that processing and fabrication operations are accompanied by chemical crosslinking reactions, which make the production of moulded objects less straightforward than is the case for thermoplastics. However, many of the desirable physical and thermomechanical properties of thermosetting resins derive from their crosslinked structure.

Various classes of resin are discussed in brief outline. The common feature of these resins, their three-dimensional network structure, renders analysis and characterisation difficult, but new physical and chemical techniques are helping polymer scientists to achieve a better understanding of resin structure and properties. These techniques, together with other technical advances, for example in processing technology, will be important in determining the future of thermosetting resins.

1.1. INTRODUCTION

This book is mainly about resins—their synthesis, structure and properties. It is difficult to discuss resins without any reference at all to their practical uses, and without considering the fabrication procedures by which they are converted to marketable products. Nevertheless, it is going to be assumed that these important topics will receive much closer attention in later volumes, so the emphasis here is on chemical and physical or mechanical aspects.

It seems best to concentrate on thermosetting resins because, whatever the future may hold, thermoplastics constitute only about 10% of the present reinforced plastics matrix usage. Even this share is heavily dependent on a single market, i.e. automotive components.

Basic concepts and terminology will be introduced here as a preparation for the later, more specialised chapters.

1.2.   THE FUNCTION OF THE MATRIX

Every component of a composite material has its function. The reinforcement carries mechanical stresses, imparting resistance to creep, together with toughness, strength and stiffness. For example, the Young's modulus E_c of the composite, may be considered (as a first approximation) to be

$$E_c = KV_f E_f + V_m E_m$$

where V_f and V_m refer to the volume fraction of fibre and matrix respectively, and E_f and E_m refer to the moduli of fibre and matrix. K is an orientation factor, which is much greater for unidirectional composites, with alignment parallel to the principal stress axis, than for randomly oriented fibre arrays.

The reinforcement can operate properly only if there is good adhesion between it and the matrix, to facilitate load transfer. On the other hand very good adhesion reduces the ability of fibres to stop cracks. A balance must be found.

Fibrous reinforcements are vulnerable to mechanical damage, and it is a function of the resin to protect the fibres as well as to bind them together. Sometimes fibres also require protection against short-term chemical attack (although most resins available at present are as much affected by common chemical environments, especially organic solvents, as the fibres; but there are exceptions).

1.3.   SUPRAMOLECULAR STRUCTURE

Thermosetting resins are chemically reactive substances which undergo hardening to produce insoluble, infusible products. The three-dimensional network structures resulting are much more dense than those obtained during the vulcanisation of rubber. But it is still uncertain whether the crosslink densities of common crosslinked resins are (i) approximately uniform (as in Fig. 1.1(a)), or (ii) non-uniform but with interconnecting chain segments (Fig. 1.1(b)). The widely held view is that the latter is more likely. A third possibility is that the network consists of densely

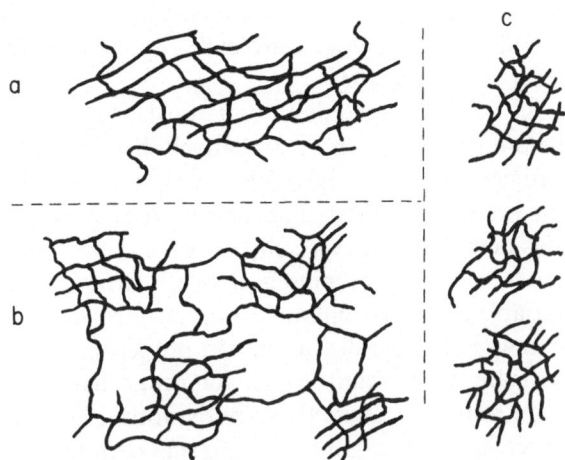

FIG. 1.1. Three models representing the structure of network polymers. (a) Continuous network of uniform density, (b) continuous network of non-uniform density, (c) discontinuous network.

crosslinked regions dispersed in non-bound matter. Figure 1.1(c) shows the dense micelles without the non-bound component. This third model has been suggested for epoxy resins;[1] the evidence against the first model comes from electron microscopy,[2] and from direct visual observation of swelling processes.[3] (See also Chapter 7.)

The structure of thermosetting resins is amorphous, and the behaviour of most resin matrices is that of rather brittle organic glasses. (Brittleness increases with crosslink density.)

1.4. PROPERTIES OF THERMOSETTING RESINS

Prior to crosslinking, thermosetting resins may be viscous liquids (in practice, they may be handled as dilute solutions); or they may be glassy solids which have been powdered and, perhaps, mixed with additives.

The change from the linear, uncrosslinked, soluble, fusible state to the true thermohardened network polymer occurs at a point called gelation.

Liquid resins or resin solutions need to undergo flow before gelation occurs, so as to adopt the mould shape. Viscosity is, therefore, of prime importance and many studies have been made of viscosity–temperature relationships. There is typically a departure from the classical behaviour of Newtonian fluids, especially at high shear rates.[4] The crosslinking or curing process is not necessarily carried out at high temperatures, and if a low-temperature cure is envisaged, the viscosity of the resin must initially be low.

After gelation, the resin becomes a soft, weak gel and later, slightly rubber-like, before finally becoming a relatively hard solid. Heating the final product sometimes shows up a relaxation, after which rubber-like behaviour again becomes apparent, but chemical decomposition can mask the onset of this development, especially with the more dense networks.

Whatever model is taken to represent network structures, a small 'sol' fraction is invariably found, which can be estimated by solvent extraction.

It would be unwise to generalise about the properties of thermosetting resins, but most are insufficiently tough for rough handling, and their practical uses are restricted unless they are reinforced in some way. One important respect in which thermosetting resins perform relatively well is in hardness; another is their maximum working temperature. In both

FIG. 1.2. Scanning electron micrograph of the surface of a pigmented, decorative laminate with a melamine–formaldehyde resin veneer, scratched with a diamond.

these cases they are rather better than thermoplastics, and these two advantages are utilised in, for example, the domestic kitchen worktop. Nevertheless, there is room for improvement in both respects. Figure 1.2 shows a scanning electron micrograph of a scratched piece of white melamine–formaldehyde paper-based decorative laminate.

The properties of some unreinforced thermosetting resins are given in Table 1.1. It will be seen that there is a very wide range of properties

TABLE 1.1
PROPERTIES OF UNFILLED, CROSSLINKED THERMOSETTING RESINS

Property	Units	Unsaturated polyester	Glycidyl ether epoxide	Phenol formaldehyde
Tensile strength	$MN\,m^{-2}$	25–80	30–100	20–65
Compressive strength	$MN\,m^{-2}$	60–160	60–190	45–115
Tensile modulus	$GN\,m^{-2}$	2·5–3·5	2·5–6·0	2·0–6·5
Elongation at break	%	1·3–10·0	1·1–7·5	1·0–3·5
Flexural strength	$MN\,m^{-2}$	70–140	60–180	45–95
Flexural modulus	$GN\,m^{-2}$	2·5–3·5	1·8–3·3	2·5–6·5
Poisson's ratio	—	c. 0·35	0·16–0·25[a]	—
Specific gravity	—	1·11–1·15	1·15–1·25	1·31
Water absorption (24 h)	mg	10–30	7–20	15–30
d.c. volume resistivity	ohm–cm	10^{12}–10^{14}	10^{10}–10^{18}	10^9–10^{14}

[a] Calculated.

for a given resin, because of differences in precise formulation and structure. Even greater variations are found in the data sheets of commercial manufacturers, because these refer mostly to filled or reinforced compositions, and depend on the nature, orientation and volume fraction of the second phase, as well as on interfacial bound strength.

Table 1.1 is inadequate in that it gives no indication of how easily one specified property can be reconciled with another. In fact it is commonly found that tensile modulus decreases as the % elongation at break increases, so the highest values quoted for these two properties cannot actually be combined in the same material. It is usually found that high heat distortion temperature, hardness and modulus are combined with increased brittleness, i.e. lower impact strength.

Modification of resin properties is achieved partly by alteration of chemical formulation and partly by the use of particulate, fibrous or laminar reinforcement (see Table 1.2). There is such a variety of reinforcement forms that quoted values can differ greatly even when the materials

TABLE 1.2
PROPERTIES OF FILLED THERMOSETTING RESINS

Property	Units	Material				
		Mineral filled alkyd	Glass fibre filled alkyd	Woodflour phenolic	Cellulose filled phenolic	Paper filled melamine
Tensile strength	MN m^{-2}	20–35	25–75	24–35	30–40	45–75
Compressive strength	MN m^{-2}	65–90	100–130	55–70	130–180	140–300
Flexural strength	MN m^{-2}	45–70	100–150	24–60	55–75	65–95
Specific gravity	—	1·8–2·5	1·8–2·2	1·3	1·4	1·4–1·6
Volume resistivity	ohm–cm	10^{13}–10^{15}	10^{13}–10^{15}	10^{12}–10^{13}	10^{11}–10^{13}	10^{8}–10^{11}
Power factor (1 MHz)	—	0·012–0·016	0·019–0·021	0·02–0·05	0·035–0·05	0·02–0·04
Dielectric constant (1 MHz)	—	4–5	4	6·0	4·8–5·4	6–8
Dielectric strength	volts/mil	180–320	200–300	50–250	70–120	60–200
Water absorption (24 h)	mg	7–75	10–25	30–65	30–60	10–50

are ostensibly similar. However, certain limitations are apparent in all the common thermosetting resins. There is a need for improved heat resistance (measured not only by the onset of chemical decomposition but also by the temperature at which mechanical properties are adversely affected). The present generation of thermosetting resins is also inadequate in resistance to combustion, to plasticisation or softening by solvents, liquids or water vapour, and to oxidation. But major advantages have been demonstrated by reinforced thermosets in their resistance to dilute aqueous inorganic fluids, and some other chemicals, at moderate temperatures—resulting in their extensive use for chemical process equipment.[5,6]

1.5. PHENOL–FORMALDEHYDE RESINS

Phenol condenses with aldehydes to give products which can be either linear and soluble (provided that phenol is in excess, and the pH is low) or branched, sometimes crosslinked, and insoluble (if formaldehyde is in excess). The linear, soluble products are called *novolaks*, while the multifunctional products from formaldehyde-rich formulations result in *resoles*. The novolaks can be safely heated while remaining soluble, and they can only be crosslinked by the addition of further aldehyde. So the novolak moulding powder will contain a hardening agent such as hexamethylene tetramine:

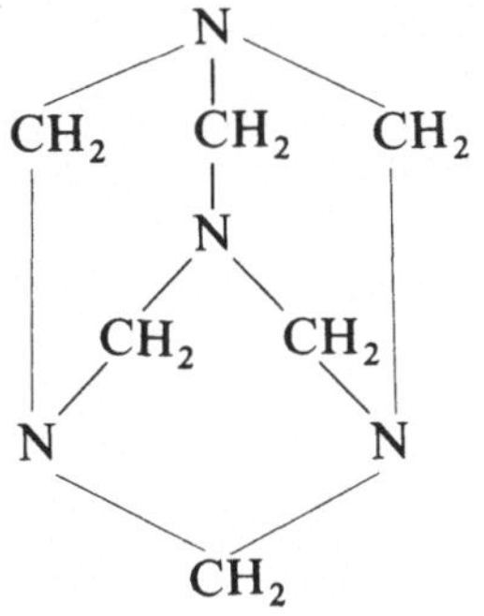

$$
\begin{array}{ccc}
& N & \\
\diagup & | & \diagdown \\
CH_2 & CH_2 & CH_2 \\
| & | & | \\
& N & \\
& \diagup \quad \diagdown & \\
& CH_2 \quad CH_2 & \\
\diagup & & \diagdown \\
N & & N \\
\diagdown & & \diagup \\
& CH_2 &
\end{array}
$$

This compound decomposes on heating to generate ammonia and formaldehyde.

The first stage in novolak formation is the slow reaction between phenol and (usually) formaldehyde, to give *ortho* and *para* methylol phenols:

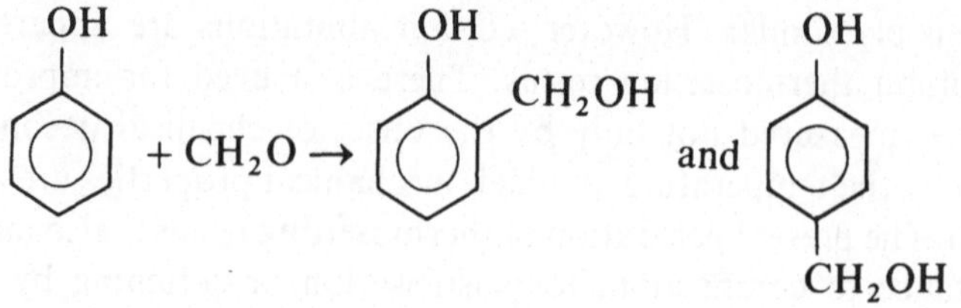

Further addition to the ring does not occur, because condensation with phenol is more rapid:

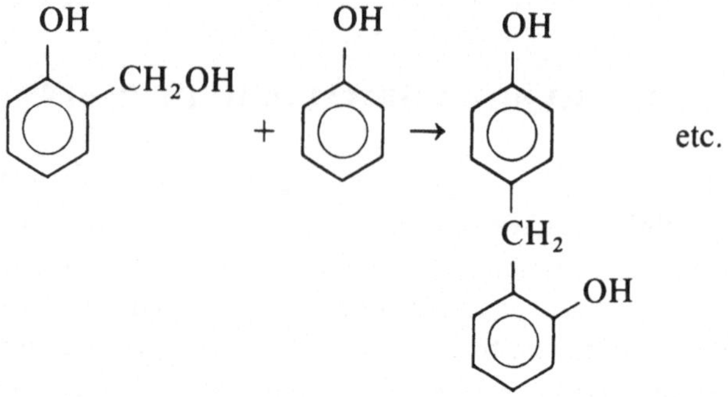

Several two-ring products can form, but the major products are:

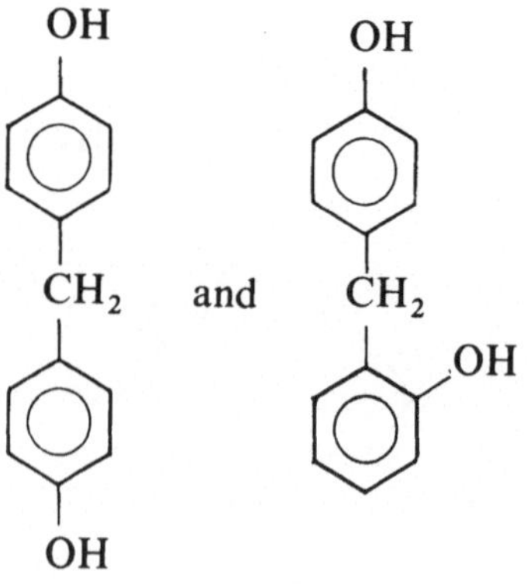

These intermediates undergo further reaction with formaldehyde and phenol alternately, eventually reaching six or seven rings in length, but without crosslinking. A typical novolak has a structure such as:

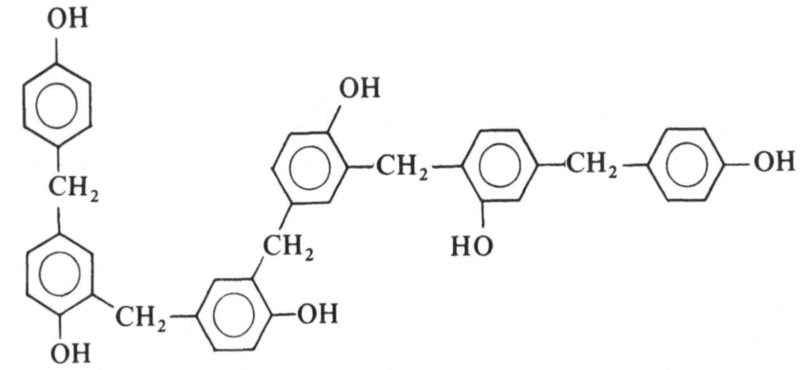

Crosslinking with hexamethylene tetramine brings about the formation of additional —CH_2OH groups and their mutual condensation, eliminating water, to give complex network structures with —CH_2— and —CH_2OCH_2— bridges between aromatic rings.

Resole formation can lead to crosslinked products, because addition of aldehyde to phenol occurs rapidly, and results in reactive polyfunctional intermediates such as:

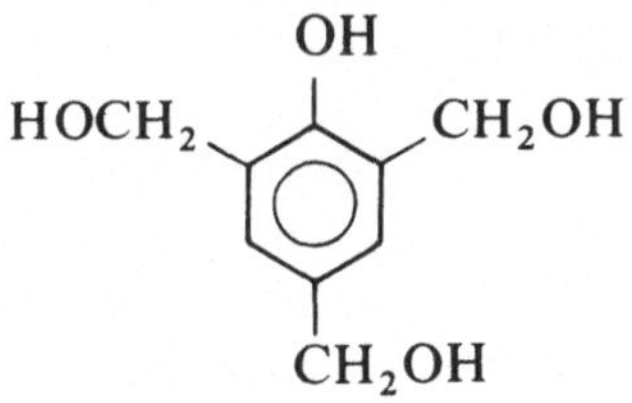

These molecules undergo self-condensation to form insoluble, infusible networks. A strong acid catalyst will bring about the curing of a thin film of resole at ambient temperatures.

Cold curing is used to produce castings, which can be made water-white. It is found that hot-cured phenolic resins are generally dark, and this affects their application for domestic use. The dark colour is believed to be the result of a reaction between the oxygen of —CH_2—O—CH_2 bridges and two neighbouring hydrogen atoms of phenolic OH groups:

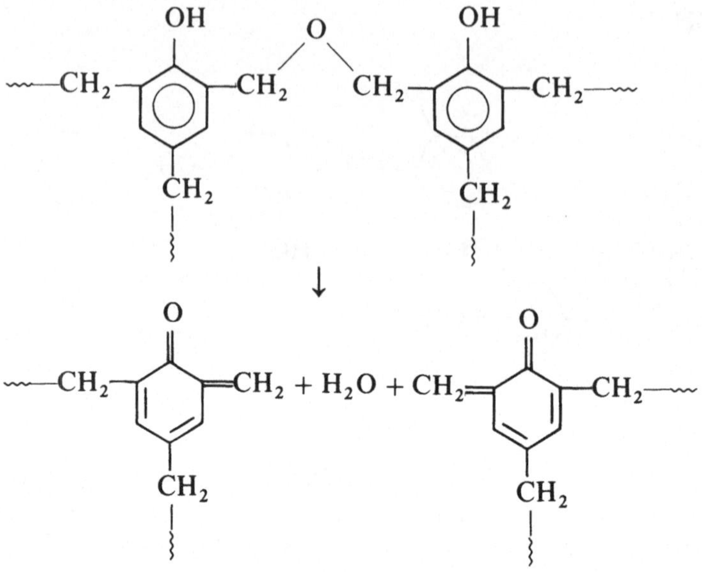

This reaction takes place when the temperature reaches 160 °C and becomes rapid at 180 °C.

Phenol–formaldehyde resins are used, with fillers, as moulding compounds (sand is used for foundry moulds, alumina for grinding wheels, asbestos for friction linings, glass and mineral wool for insulation, wood-flour for general purpose applications, cotton flock for impact resistance). A typical formulation might be:

Resin	35 to 45	parts by weight
Hardener	6	parts by weight
Filler	45 to 55	parts by weight
Magnesium oxide	1 to 2	parts by weight
A stearate	1 to 2	parts by weight
A plasticiser	1 to 2	parts by weight
A dark pigment	1 to 2	parts by weight

Alternatively resins can be dissolved in solvent so as to impregnate cloth or paper for the production of decorative laminates (using a melamine–formaldehyde veneer) or of industrial laminates (such as printed circuit boards).

The last decade has seen a great advance in the injection moulding of phenolic resins. This is more economical than compression moulding because of sharply reduced cycle times, and lower labour costs. About

60 % of phenolic moulding powder consumed in some European countries is already injection moulded, although in the United Kingdom the figure is nearer 35 % (in 1979). There are several problems; the resins are not easily obtained in convenient granular form, and tend to suffer from inconsistent properties. Long barrel life is desirable, but almost incompatible with fast cure cycles unless there is a cure 'trigger' temperature of around 120 °C. Injection moulding imparts fibre orientation, but the use of 'injection–compression'—i.e. injection into a slightly opened compression mould—allows for the recovery of anisotropy. There is still a need for compounding techniques which do not shorten fibre length too much.

Space has been devoted to phenolic resins because (a) they have undergone surprising adaptation in the past decade, and (b) they have certain promising advantages, which could maintain these oldest of the true synthetic plastics in a competitive position while other materials are adversely affected (see Section 1.17). They are now marketed for hand lay-up, spray-up, and filament-winding as well as for foams. Their excellent fire-retardant properties and low smoke emission offer great advantages over many competitive plastics. On the debit side, the high concentration of polar groups results in moisture uptake, and rather poor electrical property retention.

1.6. AMINO RESINS

The reaction of aldehydes (again, usually formaldehyde) with amines or amides gives rise to crosslinked products. Several amines and amides have been tried, but only two starting materials are of commercial importance. These are *urea* (Structure (I)) and *melamine*, i.e. 1,3,5-tri-amino-2,4,6-triazine (Structure (II)). Both can be obtained from non-petroleum sources.

(I)   (II)

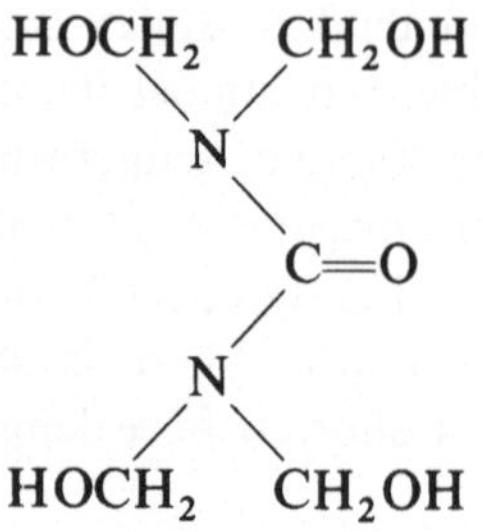

$$\begin{array}{ccc} HOCH_2 & & CH_2OH \\ & \diagdown N \diagup & \\ & | & \\ & C & \\ & \diagup \diagdown & \\ N & & N \\ | & & \| \\ C & & C \quad CH_2OH \\ \diagup \diagdown & & \diagdown \diagup \\ N \quad N & & N \\ HOCH_2 \quad CH_2OH & & CH_2OH \end{array}$$

$$\begin{array}{cc} HOCH_2 \quad CH_2OH \\ \diagdown N \diagup \\ | \\ C=O \\ | \\ N \\ \diagup \diagdown \\ HOCH_2 \quad CH_2OH \\ \text{(III)} \end{array} \qquad \begin{array}{c} \text{(IV)} \end{array}$$

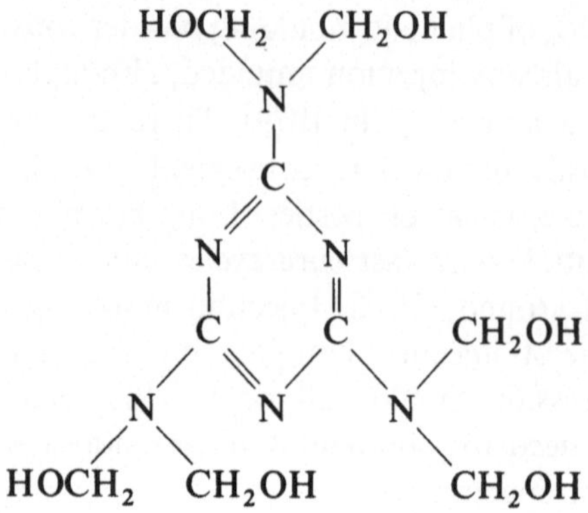

Amino resin synthesis is similar to that for phenolic resins. Excess formaldehyde can lead to reactive intermediates with up to four methylol groups in urea (Structure (III)) or six in melamine (Structure (IV)). Acidic conditions favour the formation of polyfunctional intermediates. Subsequent crosslinking leads to network structures with bridges such as:

(a) $\sim\sim$—NH—CH$_2$—NH—$\sim\sim$

(b)
$$\sim\sim—N\begin{array}{c} \diagup CH_2—\sim\sim \\ \diagdown CH_2—\sim\sim \end{array}$$

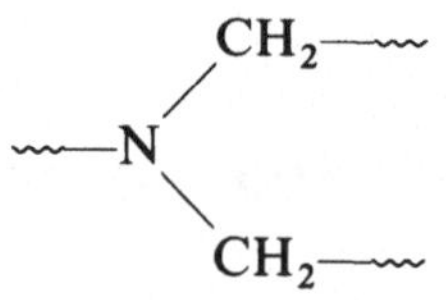

(c)
$$\sim\sim—N\begin{array}{c} \diagup CH_2—\sim\sim \\ \diagdown CH_2 \end{array}$$
$$\sim\sim—N\begin{array}{c} \diagup \\ \diagdown CH_2—\sim\sim \end{array}$$

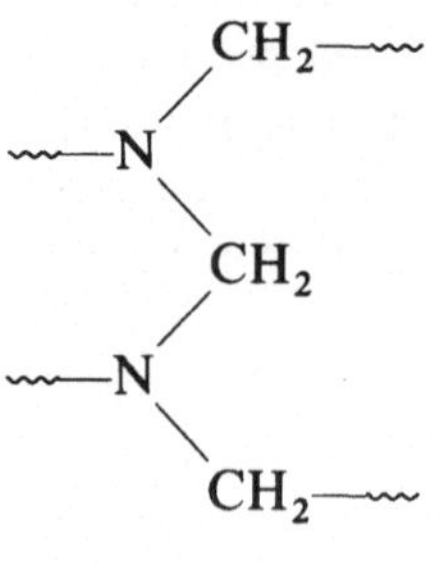

Both urea and formaldehyde resins are used as moulding powders, with formulations which include a filler and a hardener. The number of fillers used with urea resins is limited, because of problems in achieving a wide range of colours, whereas melamine resins can be produced with various colours and also have the advantages of superior hardness, heat resistance, water absorption characteristics and stain resistance.

Urea resins are widely used for adhesives, in chipboard manufacture, and as foams. These applications are mainly outside the field of reinforced plastics as it is generally understood. On the other hand, melamine resins are important in the production of electrical grade laminates. Together, the amino resins constitute one of the world's major categories of thermosetting resin.

1.7. SILICONE RESINS

These are the only resins discussed in this chapter which are not carbon-based. Despite the availability of silica, they are expensive and therefore very limited in their present use. The preparation of silicone resins is by condensation reactions, as illustrated below:

(1) Monochlorosilanes give hexamethyldisiloxane:

$$2(CH_3)_3SiCl + 2H_2O \xrightarrow{\text{hydrolysis}} 2(CH_3)_3SiOH + 2HCl$$

$$\xrightarrow{\text{condensation}} (CH_3)_3Si-O-Si(CH_3)_3 + H_2O$$

(2) Dichlorosilanes give linear polymers:

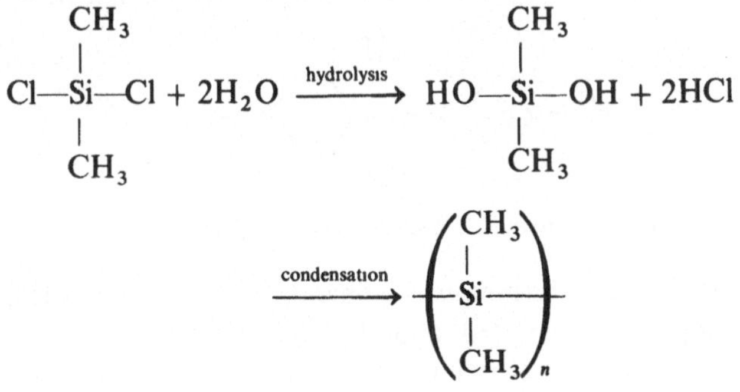

(3) Trichlorosilanes give network polymers:

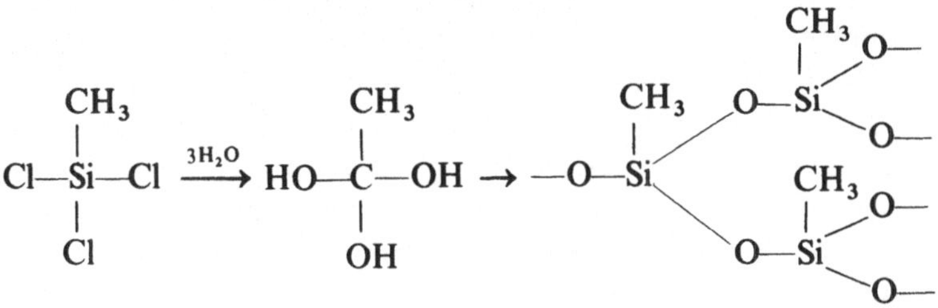

Because of their Si—O and Si—CH₃ bonds, the silicone resins have the advantage of good heat resistance, but their mechanical properties are poor. This is believed to be partly because crosslinking has to compete with cyclisation reactions, so the final network is not comparable with those of organic carbon-based resins. They have very good water resistance, and electrical properties, and can be blended with organic resins, provided that they (the silicones) contain sufficient phenyl substituents. As a result of their electrical properties, silicone–glass laminates are used for printed circuit board manufacture, electric motor components and transformer formers.

It is interesting to note that their poor mechanical properties are also reflected in the laminates. This demonstrates the importance of the matrix in composite materials.

1.8. UNSATURATED POLYESTER RESINS

Unsaturated polyester (UP) resins constitute an important and growing sector of the thermosetting resin industry. The resins are produced by condensation of dibasic acids (including one unsaturated acid or anhydride) with diols, followed by crosslinking with a reactive diluent, usually styrene. Most UP resins use phthalic anhydride, maleic anhydride

(4,4′ dihydroxy, 2,2′ diphenylpropane) (bisphenol A)

and propane-1,2-diol, although 'bisphenol A' and its hydrogenated, alicyclic analogue are also used, and aliphatic dibasic acids or long-chain diols may be added. The crosslinking reaction produces no volatile products, provided that the reaction mixture does not become hot enough to evaporate the styrene. It is, therefore, possible to produce mouldings at ambient temperatures, without pressure, i.e. by 'contact laminating'. This is one of the reasons for the adoption of UP resins in very large structures such as swimming pools and large hulls. However, low-temperature cure is slow, and high-temperature, rapid cures require pressure. The unfilled resins are brittle and most applications are for glass laminates. Two of the most successful fields are the marine market and the chemical process equipment sector.

Premix compounds have been developed for the compression, transfer and, recently, injection moulding of intricate parts. These are called dough or bulk moulding compounds. Sheet moulding compound is similar, but contains a thickening agent, and is used for shallow-profile mouldings such as vehicle body parts, furniture and trays.

UP resins are not fire resistant unless specially formulated. These formulations are effective but expensive, and they also give uncertain weathering resistance. Some of the attempts to optimise fire retardancy have been summarised by Wilson.[7] Other advances in UP chemistry are described more fully in Chapter 3, and are also discussed by Bruins.[8]

1.9. EPOXY RESINS

Epoxy resins[9] are like polyester resins in having no volatile reaction (cure) products, although there may be a volatile solvent present. They have superior mechanical and electrical properties, but are more expensive than polyester resins and present greater fabrication problems, at least in the field of large glass laminate products.

Epoxy resins are compounds containing two or more epoxy groups, the commonest being the reaction product of epichlorhydrin with 4,4'-dihydroxy-2,2'-diphenyl propane:

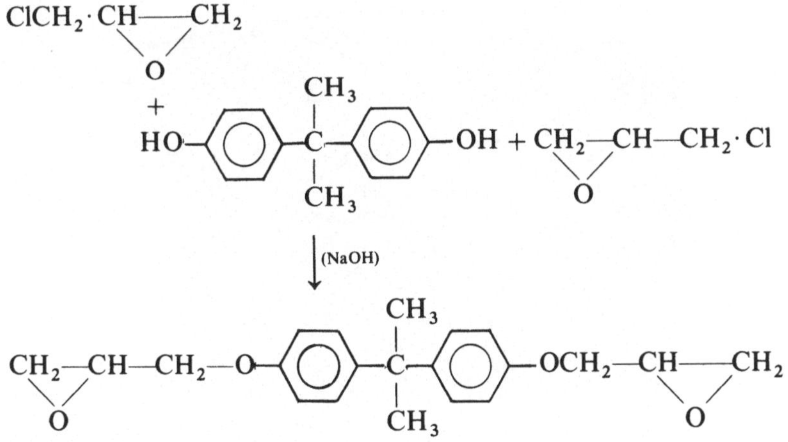

Larger molecules may be produced, but the average molecular weight is low. Hardening occurs by additive reactions, opening the epoxide ring, using primary or secondary amines, amides or anhydrides; or catalytically, e.g. by tertiary amines, or boron trifluoride.

FIG. 1.3.   Heavy duty filled epoxy resin screed, able to withstand tracked military vehicles. (Courtesy of Structoplast Ltd, Leatherhead, England.)

FIG. 1.4.   Low viscosity epoxy adhesive, specially formulated for the repair of cracks in concrete structures. (Courtesy of Structoplast Ltd, Leatherhead, England.)

Sometimes substantial quantities of hardener are required, and the nature of the hardener has an important effect on resin properties. Some hardeners are effective only at elevated temperatures.

Other epoxide resins include epoxy-novolaks; cycloaliphatic epoxy resins (formed by epoxidising cycloaliphatic intermediates such as dicyclopentadiene); acyclic (chain type) aliphatic epoxides; epoxidised polybutadiene, and epoxidised drying oils.

The range of possible properties of epoxy resins is wide. Apart from the base resin itself, additives such as diluents, flexibilisers and reactive rubbers are added.

Epoxide resins are widely used for high performance laminates for first class mechanical and electrical applications. They also have a range of other uses: adhesives, coatings, abrasion-resistant floorings and road surfaces. Figure 1.3 shows a specially formulated, highly filled epoxy resin screed, able to withstand not only ordinary traffic but tracked military vehicles. There has been a growth in the use of epoxy resins for the repair of concrete structures; this too requires special formulations (see Fig. 1.4).

1.10. VINYL ESTER RESINS

These resins have chemical structures intermediate between those of epoxides and polyesters. They are similar to epoxides at the prepolymer stage, except for the presence of terminal unsaturation. This unsaturation facilitates polyester-type crosslinking with styrene, thus making hand lay-up fabrication practicable. These resins are discussed fully in Chapter 2.

1.11. FURAN RESINS

Furan resins have excellent chemical resistance. This has led to their useful employment as anti-corrosion linings in chemical plant. Unfortunately, they have fabrication difficulties, and have met with little success as laminates or moulding compounds. Recently, spraying equipment has been successfully adapted for use with furan resins, and this may lead to increased growth in the future.

The starting material is furfural, obtained from oat husks and corn cobs by digestion with sulphuric acid and steam. It is then hydrogenated to furfuryl alcohol (Structure (V)).

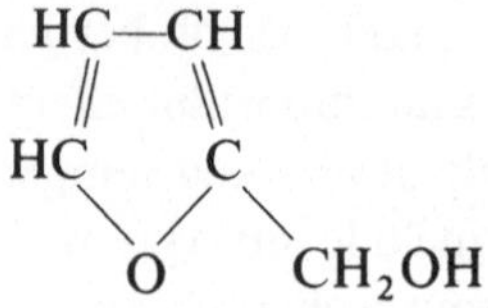

$$HC\text{---}CH$$
$$HC \quad C$$
$$O \quad CH_2OH$$

(V)

which self-condenses on heating with acid, giving Structure (VI):

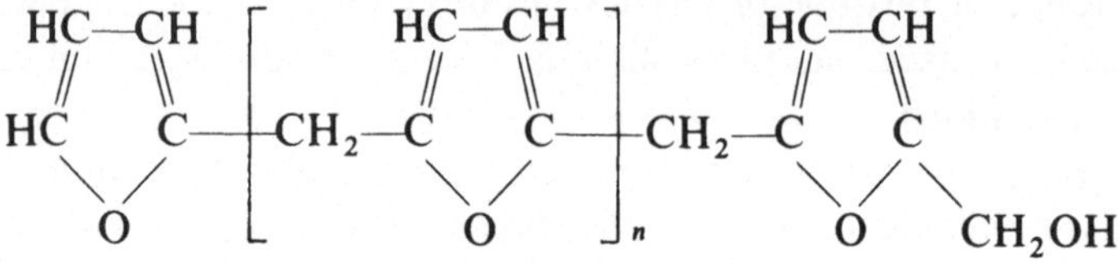

$$HC\text{---}CH \quad \left[\quad HC\text{---}CH \quad\right] \quad HC\text{---}CH$$
$$HC \quad C\text{---}CH_2\text{---}C \quad C\text{---}CH_2\text{---}C \quad C$$
$$O \quad \left[\quad O \quad\right]_n \quad O \quad CH_2OH$$

(VI)

The resins tend to crosslink in bulk unless exothermic heat is removed, and normal crosslinking is achieved by addition of 4% p-toluene sulphuric acid, causing a loss of some of the unsaturation; the final product can be represented by Structure (VII):

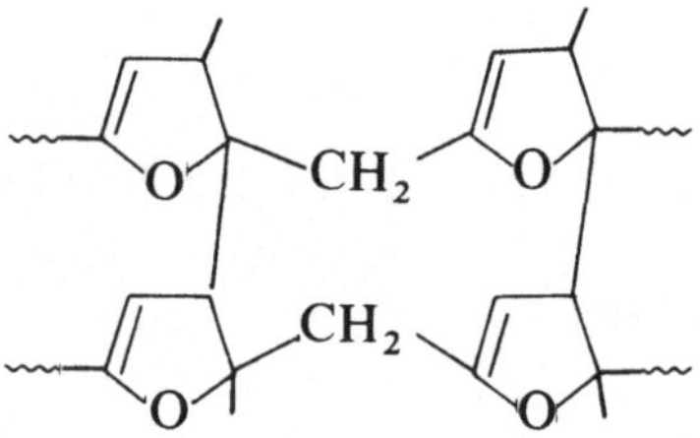

(VII)

The furan resins not only have chemical resistance, but also withstand heat well, and burn only with difficulty at high temperatures, producing no smoke problem.

1.12. POLYIMIDES

Polyimides have been developed for high temperature applications where cost is of secondary importance. In supersonic aircraft, for example, the temperature of the outer walls is proportional to the square of the velocity, and at Mach 3, the maximum temperature would exceed 300 °C. Cyclic-chain structures such as Structure (VIII) (where R may be aliphatic or aromatic) possess satisfactory physical and thermomechanical properties. But not all polyimides are genuinely thermosetting.

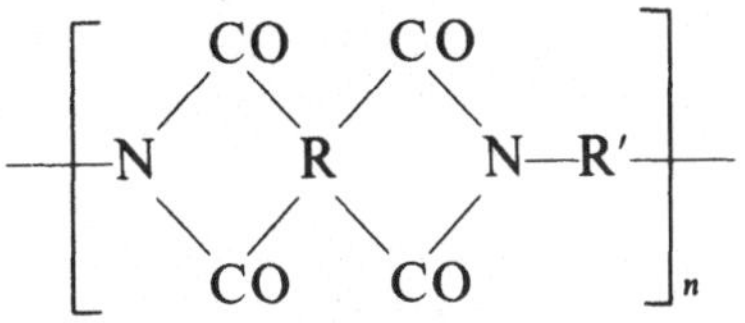

$$\left[-N \underset{CO}{\overset{CO}{\diamond}} R \underset{CO}{\overset{CO}{\diamond}} N-R'- \right]_n$$

(VIII)

First produced fairly early in the twentieth century, polyimides were developed commercially by Du Pont Co. in the USA in the late 1950s.

Reaction of a diamine with the dianhydride of a tetracarboxylic acid, in a polar solvent, gives a polyamic acid, e.g. Structure (IX),

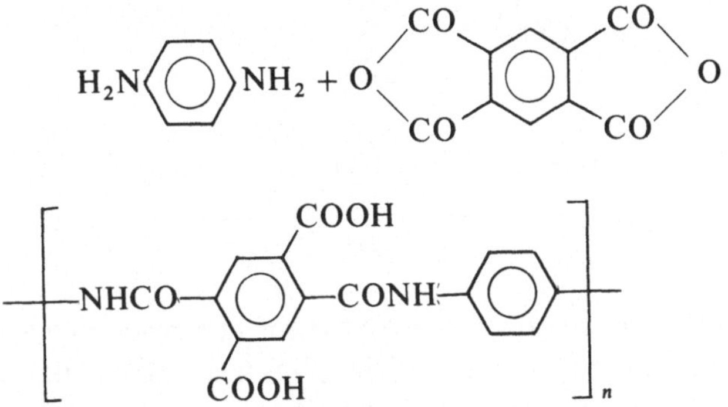

$$H_2N \langle \bigcirc \rangle NH_2 + O \overset{CO}{\underset{CO}{\diamond}} \bigcirc \overset{CO}{\underset{CO}{\diamond}} O$$

$$\left[-NHCO - \bigcirc \overset{COOH}{\underset{COOH}{}} - CONH - \bigcirc - \right]_n$$

(IX)

The formation of the polyimide is then achieved by a further reaction to remove water from the polyamic acid, either by heating or by a dehydrating reaction, to give:

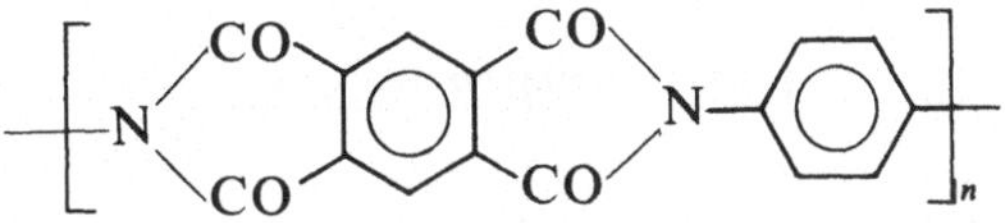

$$\left[-N \overset{CO}{\underset{CO}{\diamond}} \bigcirc \overset{CO}{\underset{CO}{\diamond}} N - \bigcirc - \right]_n$$

Polyimides may be cast as films while still in the intermediate form, and subsequently baked to produce the final product. But the polyamic acid (Structure (IX)) is hydrolytically unstable and it melts at a temperature very close to that needed for conversion to the polyimide; so fabrication is not easy. If the melt impregnation is obviated by the use of solvents, another difficulty arises; the most suitable solvents are polar, difficult to remove, and responsible for void formation. Voids cause a serious deterioration in the mechanical properties.

Crosslinking by radical reactions has been developed. This avoids the problem of volatiles and requires a crosslinking agent such as Structure (X) together with the intermediate product from the reaction of Structure (XI) with Structure (XII):

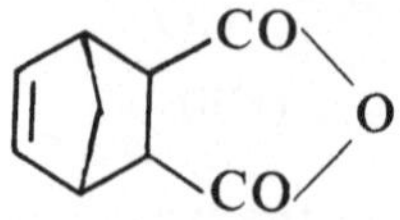

(X)

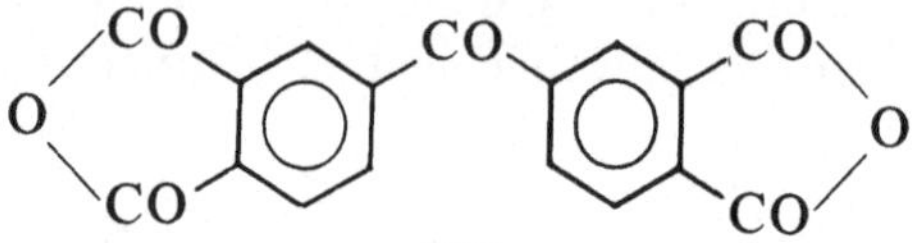

(XI)

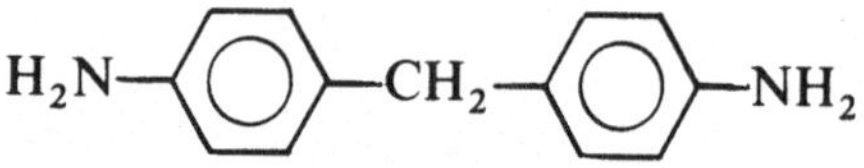

$$H_2N-\!\!\bigcirc\!\!-CH_2-\!\!\bigcirc\!\!-NH_2$$

(XII)

Polymerisation *in situ* (with the reinforcement already immersed in the original reagents) has been tried. This gives water elimination problems. Several other routes are still being developed.

The addition polyimides suffer from, as yet, inadequate toughness and thermo-oxidative stability, and from a susceptibility to micro-cracking during cure. The thermoplastic varieties require a high temperature and pressure for fabrication, and suffer from creep at high temperatures.

1.13. OTHER RESINS

Many exotic high-temperature polymers have been developed.[10] The more practical ones are discussed in Chapter 6, and one particular class of resin having moderately good thermal stability is discussed in Chapter 4.

All the examples of laminating resins mentioned so far include several polar groups. Apart from the silicones, none have really low water absorption capacity. This is a limitation not only for their electrical properties, but also for their retention of mechanical properties under hot, humid conditions.

The bis-diene resins, developed in the 1960s, represented an attempt to produce crosslinked, non-polar resins.[11] They were made by heating

oligomeric bis-cyclo-pentadienyl compounds to produce monocyclopenta-
dienyl radicals, which then reacted with the remaining oligomer to give
a crosslinked product.

1.14.   IMPROVEMENTS IN RESIN PROPERTIES

As implied in the previous paragraph, there is a need for resins with
improved resistance to hygrothermal ageing. Attempts are also being
made to achieve a higher extensibility without sacrificing modulus, and
to obtain improved hardness and heat distortion temperature without
loss of toughness.

Attempts have been made to toughen epoxide and polyester resins
by addition of elastomers, notably, carboxyl-tipped butadiene–acrylonitrile rubbers,[12] and urethane rubber.[13] Polyethersulphones have also
been tried. The problem is to achieve the correct particle size distribution
and the right degree of compatibility between phases. Success has been
very limited indeed with polyesters, and although the toughness of epoxy
resins has certainly been increased by carboxyl tipped nitrile addition,
the viscosity–pressure characteristics leave room for improvement. Moreover, improvements in matrix toughness are not always reflected in
improved composite properties. The addition of any second phase affects
weathering, UV absorption, and other aspects of durability.

Perhaps the most important single improvement sought in the past
decade for thermosetting resins has been the non-burning polyester resin.
It is unfortunate that the material most suited for the convenient fabrication of fire risk products such as boats, caravans, car bodies, building
panels, electrical components and chemical process equipment is
combustible. Polyesters are also prone to smoke generation.

Halogen-containing derivatives are increasingly used as starting
materials for the synthesis of polyesters, and additives such as antimony
oxide are employed to react with the halogen during combustion. The
choice of halogenated derivatives is restricted to halogenated acids or
diols capable of being esterified without loss of halogen during the
synthesis. Some possible candidates have to be eliminated on the grounds
of cost, and others do not provide sufficient halogenation (chloromaleic
acid, for instance). Two acids, tetrachlorophthalic anhydride, (Structure (XIII)) and 'HET' acid (chlorendic anhydride, Structure (XIV)) have
become established choices.

Diols prepared from decachlorodiphenyl and similar phenyl compounds

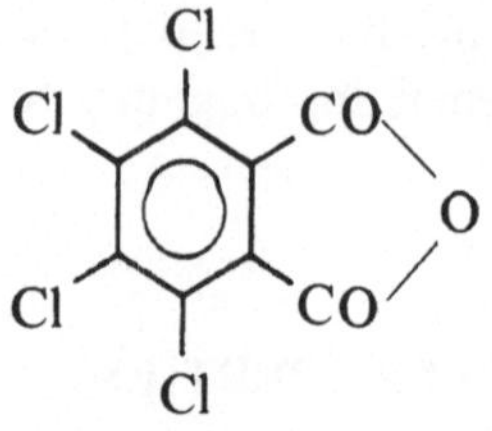

(XIII) 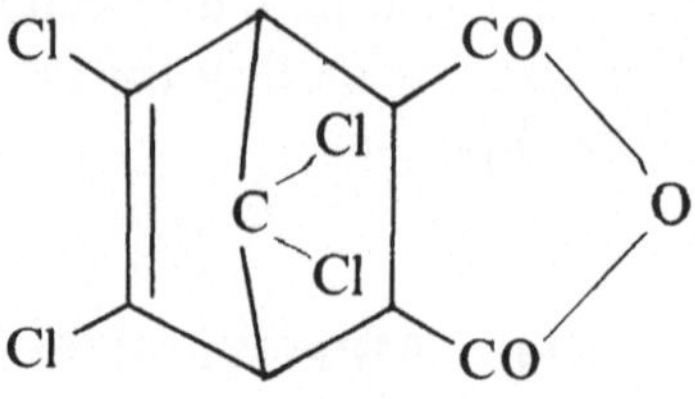   (XIV)

result in polyesters of exceptionally high chlorine content.[14] (See also Chapter 3.)

Disadvantages of some fire-retardant systems can be seen in the increased rate of photodegradation, causing yellowing and eventual darkening in sunlight. Addition of halogenated additives, e.g. chlorinated paraffin wax, also affects long term mechanical properties. The incorporation of aluminium trihydrate has a very beneficial effect on fire resistance. Protection of glass–polyester laminates can also be achieved by intumescent coatings.

Smoke generation is very undesirable, increasing the number of fatalities in fires. Polyester resins produce smoke from the styrene and phthalic components, both of which are virtually essential constituents. Efforts have been made to reduce smoke emission by:

(1) substituting non-aromatic monomers for styrene, at considerable cost,
(2) inducing char formation or intumescence.

The fire retardant properties of urea–formaldehyde (U–F) and some other thermosetting resins offer great advantages in the market for insulating foam, despite the sometimes inferior thermal insulation capability in comparison with expanded polystyrene and polyurethane foam.

1.15. HANDLING AND PROCESSING OF RESINS

It has already been noted that an adequate account of this subject is beyond the scope of this book. Nevertheless, the chemistry of thermosetting resins is inseparable from their processability.

Considerable effort has been directed to controlling the hardening reaction, so that long shelf lives are achieved along with rapid gelation and cure at the right time. One-component epoxy resins are now available which rely on heat-curing systems and give rapid cure at 150°C, but still allow long storage life without refrigeration.

The search for more convenient methods of curing unsaturated polyester resins has led to further investigation of ultra-violet initiation[15] and visible light curing.[16] Savings in materials, cleaner operation, and reduced styrene evaporation have been claimed for visible light cured polyesters. On the other hand special light sources are required, and the method is restricted to open mould fabrication methods with thin sections.

Bulk handling of polyester resins is economic for high throughputs. One method is to pump resin directly from a thermostatted, walled-in storage tank, while another requires the use of smaller, mobile containers filled from road tankers.

Injection moulding of dough moulding compounds and of phenolic resins has become widespread. Polyester sheet moulding compounds have been developed to achieve better product consistency and improved surface quality. There is a better understanding of how the resin structure affects the thickening process and the final SMC properties.[17]

Recycling of scrap thermoset material has been advocated and practised.[18] This is because the increasing cost of raw materials has made the policy seem attractive economically. Up to 15 % of reground phenolic, amino or alkyd is said to leave the mechanical properties of compression-, transfer-, and injection-moulded parts unchanged. Above this point, impact and flexural strength are severely affected. Addition of scrap acts like inert filler in reducing mould shrinkage.

However, the use of reground thermosetting material has been severely criticised because it could lead to wide variations in properties and occasional product failure. Contamination with metal dust from the grinding process could seriously affect electrical properties. Regrinding also increases the production of undesirable dust.

1.16. NEW METHODS OF RESIN QUALITY CONTROL

May et al.[19] have pointed out a growing realisation of the need for quality control in the manufacture of high performance resins. The new instrumental techniques for chemical analysis and for physical examination of materials, many of which were developed in the 1950s and 1960s, have now begun to demonstrate their usefulness in the study of thermosetting resins.

Variability between batches of epoxy resin affects prepreg processing by causing changes in tack, flow and gel times. The effect of batch variability on composite durability has yet to be determined.

Infra-red spectroscopy can determine whether a reactive diluent containing a major functional group (such as carbonyl) has been accidentally omitted from a batch. It can also determine quantitatively the percentage of some curing agents such as dicyandiamide or diaminodiphenylsulphone.

Gel permeation chromatography[20] determines the molecular weight distribution of a resin. Figure 1.5 shows a GPC trace obtained from an unsaturated polyester resin in styrene solution. The GPC indicates

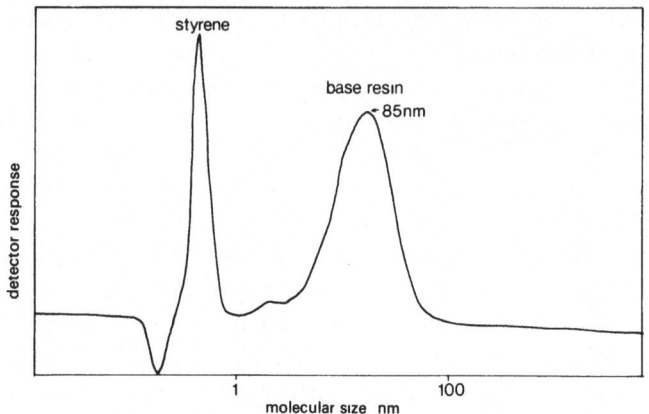

FIG. 1.5.   Gel permeation chromatograph of an unsaturated polyester resin in styrene solution.

whether too much or too little epoxy diluent has been used, detects the presence of two or more resin types in a mixture and measures free phenol in phenolic resins.

Thin layer chromatography facilitates separation of the resin components on a silica gel plate. Differential scanning calorimetry or measurement of electrical loss tangent, can indicate the degree of ageing of a resin after prolonged storage.

More recent innovations which may prove useful to the resin technologist include laser–Raman spectroscopy,[21] ^{13}C NMR neutron scattering, and ESCA ^{13}C NMR has already been used to study the curing mechanisms of acetylene-terminated imide monomers when reacted to form polyimide resins. Tomita and Hatono[22] used the same technique to characterise random urea–formaldehyde structures. Proton NMR has been used for the analysis of phenol–formaldehyde[23] and polyester[24] resins. (See Fig. 1.6.)

Pyrolysis gas chromatography has been a standard method for studying

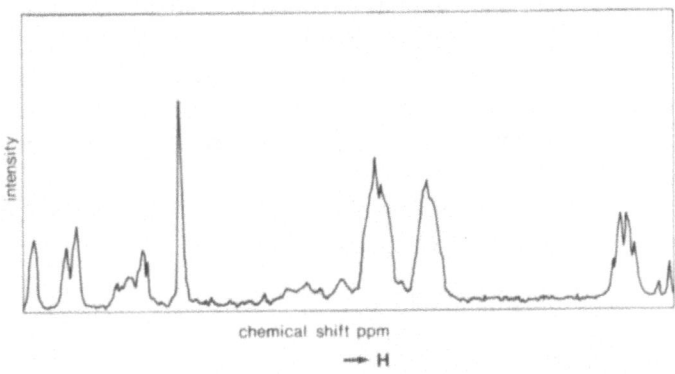

chemical shift ppm

→ H

FIG. 1.6.   NMR spectrum of an unsaturated polyester resin.

crosslinked rubbers for many years. It has some potential for studying crosslinked polymers in general.[25]

There are still many problems. It is not easy to determine quantitatively the hydroxyl content of thermosetting resins. Hase and Hase[26] proposed a silylation method based on NMR spectroscopy designed to give a nine-fold enhancement of the proton signal. This technique was applied to polyester resins. Fritz *et al.*[27] substituted the hydroxyl group of a polyethylene glycol by a chromophoric siloxy group, purified the silylated polymer, and determined the chromophoric concentration photometrically. This method was claimed to be a thousand times more sensitive than the acetylation procedure.

The quality control of finished products can be still more difficult. Attention has been directed to developing non-destructive methods for the determination of void content and distribution[28,29] of resin, fibre ratio[30] and of the magnitude and location of built-in stresses. Barrett and Predecki[31] have described how stresses in epoxy resins can be measured by high angle X-ray diffraction. Voids and other defects can be detected and assessed by ultrasonic scanning[29] and by radiotracer methods,[32] but there is a need for further development of practical non-destructive techniques.

1.17. THE FUTURE PROSPECTS FOR THERMOSETTING RESINS

It has been estimated that about 24% of the world's plastics moulding materials are of the thermosetting variety. The share has declined in the last thirty years, because thermoplastics based on cheap oil and easy processing had many economic advantages. In the future, the ability of thermosetting formulators to increase the inorganic proportion of their moulding compositions may be very important. Also the development of injection moulded thermosets has reduced their processing disadvantage.

About 90 % of the major organic chemicals for the chemical industry are derived from just six feedstocks:

(1) synthesis gas from coal or natural gas
(2) ethylene
(3) propylene
(4) butadiene
(5) benzene
(6) *para*-xylene

The cost of plastics, and of metals too, depends on oil prices, but in different ways. Plastics are mostly made from oil with alternative sources being coal, fermentation for ethanol, and biomass for glycerol, ethanol and furfural.

The total energy requirement for phenolic resin production, including both feedstock and conversion, is about $60 \, kJ/dm^3$. This is less than for engineering thermoplastics (about 100 to $300 \, kJ/dm^3$) and also less than for metals, (up to $400 \, kJ/dm^3$). This is partly responsible for the relatively low cost of phenolic and melamine resin materials. But the cost of fabricating finished articles can sometimes be higher for thermosets than for thermoplastics.

The rise in material prices since 1973 has been dramatic for both thermosetting and thermoplastic materials. (See Table 1.3.)

Attempts will be made to increase the volume fraction of fibres, inorganic fillers, hollow microspheres and foam material to cushion the effect of oil price rises. Styrene monomer (used in polyester resins as well as polystyrene) increased in price from 18 US cents/lb to 31 US cents/lb in the twelve months up to mid 1979. This compares with 9 cents/lb in 1972.

Reinforced thermosets offer certain energy advantages. They save energy in various ways: in their transportation and in their marine, chemical process, building and aerospace applications. Low weight, low maintenance costs and good thermal insulation characteristics will be useful characteristics in the future. Taking these factors together, there are good prospects for structural foam composites.

TABLE 1.3
UK PRICE RISES FOR MOULDING MATERIALS
(Units: pence per kg, tonne lots)

	September 1973	December 1979	% Increase
Phenol–formaldehyde	20·2	62·5	209
Urea–formaldehyde	24·0	56·4	135
Melamine–formaldehyde	38·5	85·0	121
Polyester DMC	26·7	75·0	181
Polypropylene	23·6	55·0	133
Nylon 6	61·5	171·0	178
Acetal	69·4	144·0	107
General purpose polystyrene	28·3	63·0	123

Source: Plastics and Rubber Weekly.

The materials producers and the moulders will need to maintain and improve their standards of pollution control and health and safety. Dust, fumes and skin-sensitising substances are common in the reinforced plastics industry, and Sweden has already legislated to improve practices.[33,34] Increasing research into the toxicological properties of chemicals in general, and the inevitable dissemination of over-simplified accounts of the results, are facts of life.

Fundamental research into the science of crosslinked polymers has not been so extensive as that directed towards more tractable linear polymers. It is nevertheless increasing and may be expected to result in improved commercial products. Some examples of the research and development recently carried out in the thermosetting field are given in the remaining chapters of this book.

REFERENCES

1. MORGAN, R. J. and O'NEAL, J. E. J. Macromol. Sci., Phys., 1978, **B15,** 139.
2. NELSON, B. E. and TURNER, D. T. J. Polym. Sci., Polym. Phys. Ed., 1972, **10,** 2461.
3. FUNKE, W. J. Polym. Sci., 1967, Part C3, 1497.
4. ALEMAN, J. V. Polym. Eng. Sci., 1978, **18,** 1160.
5. FOWLE, D. J. Chem. Ind., 1978, 361.
6. DESAI, R. R. Composites, 1974, **5,** 16.
7. WILSON, E. L. in Flame Retardancy of Polymeric Materials, Vol. 3, ed. Kuryla, W. C. and Papa, A. J., 1975, Marcel Dekker, New York, p. 254.

8. BRUINS, P. F., ed., *Unsaturated Polyester Technology*, 1976, Gordon and Breach, New York.
9. POTTER, W. G. *Epoxide Resins*, 1970, Iliffe, London.
10. IDRIS JONES, J. *Chem. Brit.*, 1970, **6**, 251.
11. JUDD, N. C. W. and WRIGHT, W. W. *Reinforced Plastics*, 1978, **22**, 39.
12. RINDE, J. A., MONES, E. T., MOORE, R. L. and NEWEY, H. A. *34th SPI Reinforced Plastics/Composites Conf., New Orleans, La., USA*, 1979, Paper 17-A.
13. HANCOX, N. L. and WELLS, H. *32nd SPI Reinforced Plastics/Composites Conf., Washington D.C., USA*, 1977, Paper 9-C.
14. ANON. *Plastics and Rubber Weekly*, 1979, Sept. 21, pp. 24–6.
15. SCHICK, J. P. *Plastica*, 1978, **31**, 4.
16. LONGENECKER, D. M. and GRETH, G. G. *Plastics Eng.*, 1977, **33**, 52.
17. BURNS, R., LYNSKEY, B. M., GANDHI, K. S. and HANKIN, A. G. *Plastics and Polymers*, 1975, **43**, 228.
18. BAUER, S. H. *Plastics Eng.*, 1977, **33**, 44.
19. MAY, C. A., HADDAD, D. K. and BROWNING, C. E. *33rd SPI Reinforced Plastics/Composites Conf., Washington, D.C., USA*, 1978, Paper 15-D.
20. LEE, W. Y. *J. Appl. Polym. Sci.*, 1978, **22**, 3343.
21. KOENIG, J. L. and SHIH, P. T. K. *J. Polym. Sci.*, 1972, **10**, Part A2, 721.
22. TOMITA, B. and HATONO, S. *J. Polym. Sci., Polym. Chem. Ed.*, 1978, **16**, 2509.
23. POSPISIL, L. and NAVRATIL, M. *Chem. Prum.*, 1979, **29**, 34.
24. BIRLEY, A. W., DAWKINS, J. V. and KYRIACOS, D. *Polymer*, 1978, **19**, 1433.
25. HAEUSLER, K. G., SCHROEDER, E., GROSSKREUZ, G. and HUBE, H. *Plastic u. Kaut*, 1978, **5**, 691.
26. HASE, A. and HASE, T. *Analyst*, 1972, **97**, 998.
27. FRITZ, D. F., SAHIL, A., KELLER, H. P. and KOVATS, E. S. *Analyst, Chem.*, 1979, **51**, 7.
28. STONE, D. E. W. and CLARKE, B. Technical Report No. 74162 (Dec. 1974) Royal Aircraft Establishment, Farnborough, England.
29. DEAN, G. Characterization of fibre composites using ultrasonics. *Proceedings of Conference 'Composites—Standards, Testing and Design'*, 1974, IPC Science and Technology Press, Guildford, England, p. 126.
30. TORP, S., FØRLI, O. and MALMO, J. *32nd SPI Reinforced Plastics/Composites Conf., Washington, D.C., USA*, 1977, Paper 9-A.
31. BARRETT, C. S. and PREDECKI, P. *Polym. Eng. Sci.*, 1976, **16**, 602.
32. JOINER, J. C. *The determination of voids in carbon fibre composites*, Report AQD/NM 00296, (July 1973), Ministry of Defence, Aircraft Quality Directorate, Woolwich, England.
33. ANON. *Reinforced Plastics*, 1975, **19**, 148.
34. BRIGHTON, C. A., PRITCHARD, G. and SKINNER, G. A. *Styrene Polymers: Technology and Environmental Aspects*, Chapters 5 and 7, 1979, Applied Science Publishers Ltd, London.

Chapter 2

VINYL ESTER RESINS

Thomas F. Anderson and Virginia B. Messick

Dow Chemical USA,† Texas, USA

SUMMARY

Vinyl ester resins are relatively recent additions to the thermosetting family. They have some features in common with unsaturated polyesters, and most have structural features similar to those of epoxides. They are notable for their high-temperature properties, their chemical resistance, their high elongation and their convenient processing characteristics.

This chapter describes the history, the synthesis and typical structures of vinyl esters, and relates these structures to the properties of cast resins and laminates.

Applications in corrosion-resistant equipment, land transport, electrical insulation, marine, and several other fields are described. The toxicological properties of vinyl ester resins are summarised.

2.1. INTRODUCTION

Vinyl ester resins are resins produced by the addition of an ethylenically unsaturated monocarboxylic acid to a backbone (usually epoxy) producing terminal unsaturation and which can be cured with vinyl monomers similar to those used for crosslinking polyesters. Vinyl ester resins combine the excellent thermal and mechanical properties of epoxy resins with the ease of processing and rapid curing of polyester resins. They have opened up broad new applications for thermoset resins since they (1) can be

† Dow CRI No. B-600-024-80.

29

cured rapidly with relatively nontoxic catalysts, as can polyester resins, (2) have excellent wetting and bond to glass fibres, as epoxy resins do, (3) retain high elongation at moderate and high heat distortion temperatures due to controlled crosslinking structure and (4) have excellent high-temperature and heat ageing properties when novolac epoxy resins are a part of the basic structure.

2.2. EARLY HISTORY

The chemistry of the vinyl ester resins was developed in the late 1950s and early 1960s by several people, each having a different base. Bowen, working for the US Commerce Department, was looking for a way to get greater toughness and improved bonding to teeth in the acrylic polymers used in dental applications. He reacted glycidyl acrylate and glycidyl methacrylate with bisphenol and used the resulting difunctional resin he obtained to give crosslinked polymers.[1,2] These resins were so reactive that they did not have a useful working life and it was several years before stable resins for the dental market were commercialised.

Fekete et al. developed resins primarily for use in electrical insulation and corrosion resistant equipment.[3,4] These resins were homopolymerised through their acrylic end groups and copolymerised with monomers. One development with these resins was the appearance of very fast curing moulding compounds made with a blend of the vinyl ester resin with a polyester resin.[5] These compounds were unique in that they cured faster than compounds made with either of the other resins by itself.

Bearden worked on polyalkyl acrylate polymers and reacted several different vinyl unsaturated compounds with diepoxides to give di- and tetra-functional unsaturated resins.[6] These resins were developed to be cut in styrene monomer and designed for use in matched metal die moulding. Bearden, in early work, discovered methods of stabilising these tremendously reactive resins so that they could be shipped to moulders without polymerising in the drum, and yet would cure rapidly and completely when properly catalysed. These stabilised systems were the key that allowed commercialisation of these resins in manufacturing structural and electrical matched metal die moulded laminate structures in 1964. Electrical laminates made with these resins were found to have exceptional retention of physical and electrical properties when heat aged continuously at temperatures of 180° to 200°C, and in special cases at 220°C. Yet they could be processed easily and rapidly like polyester resins. As a result, insulation based on these resins made possible the development of more compact (higher operating temperature) motors, generators, and dry type transformers.

Bearden and coworkers, Jernigan, Najvar and Hargis, recognised that the heat resistance, high reactivity activity and toughness of these resins would offer real advantages in the corrosion resistance field. They modified one of the resins by substituting methacrylic acid for acrylic acid on the ends of the molecule. The resultant shielding of the ester linkage gave a resin with excellent resistance to hydrolysis and broad general corrosion resistance. This group recognised the need for fire retardant formulations and thickenable formulations and developed such resins.[7-9] The fast glassfibre wet-out and rapid, complete curing of these resins was used to develop highly automated filament winding technology.[10]

2.3. SYNTHESIS

Vinyl ester resins are based on the reaction product of an epoxy resin and an ethylenically unsaturated carboxylic acid which results in terminal unsaturation. Various epoxy resins are used, including the diglycidyl ether of bisphenol A, or higher homologues thereof, the diglycidyl ether of tetrabromo bisphenol A, epoxylated phenol-formaldehyde novolac and polypropylene oxide diepoxide. The most commonly used acids are acrylic and methacrylic acids, although use of other unsaturated acids such as cinnamic and crotonic acids has been reported in the literature.[11] The acid–epoxide reaction is straightforward and is catalysed by tertiary amines, phosphines, alkalis or onium salts.[12] The acid–epoxide reaction results in pendant hydroxyl groups which provide adhesion and/or reactive sites for further modification with compounds such as anhydrides or isocyanates.[13] Vinyl ester resins are diluted with a reactive monomer such as styrene, vinyl toluene or dicyclopentadiene acrylate.

2.4. STRUCTURE-IMPARTED CHARACTERISTICS

The generalised structure of a vinyl ester resin is shown below.

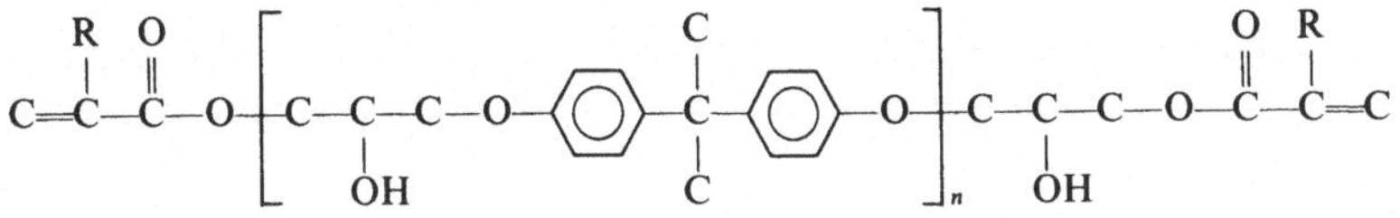

R = H or CH$_3$

2.4.1. Toughness

Toughness is imparted to vinyl ester resins by the epoxy resin backbone[14] (the portion enclosed in brackets in Structure (I)). The molecular weight of the epoxy resin portion can be controlled by reacting the diglycidyl ether of bisphenol A with specific amounts of bisphenol A. Physical properties such as tensile strength, tensile elongation and heat distortion temperature are relatable to the molecular weight of the epoxy resin portion of the vinyl ester resin. Thus, vinyl ester resins can be 'tailor made' to meet the requirements of specific applications.

2.4.2. Corrosion Resistance

Corrosion resistance is obtained first from the phenyl ether linkage of the epoxy resin backbone. This linkage is quite stable in many chemical environments. The ester linkage appears to be shielded somewhat by the pendant methyl group. Furthermore, the ester linkages are formed only at the ends of the molecule, so the number of ester groups present are minimal. This is in contrast to polyesters where ester linkages form part of the repeating unit.

2.5.   STRUCTURES

The following structures illustrate the diversification of vinyl ester resins that has taken place over the last 10 years to meet specific end-use needs.

2.5.1. Basic Vinyl Ester Resin (Bisphenol A-Epoxy)

Example: Derakane® 411 vinyl ester resin.[15]

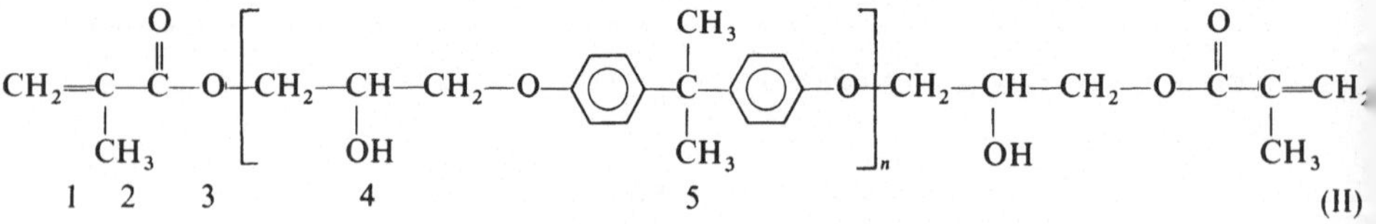

$$CH_2=\underset{\underset{CH_3}{|}}{C}-\overset{\overset{O}{\|}}{C}-O\left[CH_2-\underset{\underset{OH}{|}}{CH}-CH_2-O-\bigcirc-\underset{\underset{CH_3}{|}}{\overset{\overset{CH_3}{|}}{C}}-\bigcirc-O\right]_n CH_2-\underset{\underset{OH}{|}}{CH}-CH_2-O-\overset{\overset{O}{\|}}{C}-\underset{\underset{CH_3}{|}}{C}=CH_2$$

1   2   3   4   5   (II)

1.   Terminal vinyl unsaturation—located at the ends of the molecule where they are very reactive, causing the vinyl ester resin to cure rapidly to give fast green strength, and enable the vinyl ester resin to homopolymerise or copolymerise to give polymers with high corrosion resistance.

® Trademark of The Dow Chemical Company.

2. Methyl group—shields the ester linkage increasing resistance to hydrolysis.
3. Ester groups—vinyl ester resins have 35 to 50 % fewer ester groups per unit of molecular weight than corrosion resistant polyesters. This contributes to their resistance to hydrolysis by alkaline solutions.
4. Secondary hydroxyl—the interaction of the chain secondary hydroxyl groups with the hydroxyl groups on the surface of glass fibre gives improved wetting and bonding. This is one of the factors responsible for the higher strengths obtained with vinyl ester resin laminates.
5. Epoxy resin backbone—imparts toughness and allows controlled molecular weight for low viscosity. The epoxy ether bond provides superior acid resistance.

2.5.2. SMC Resins

Example: Derakane 786 vinyl ester resin.

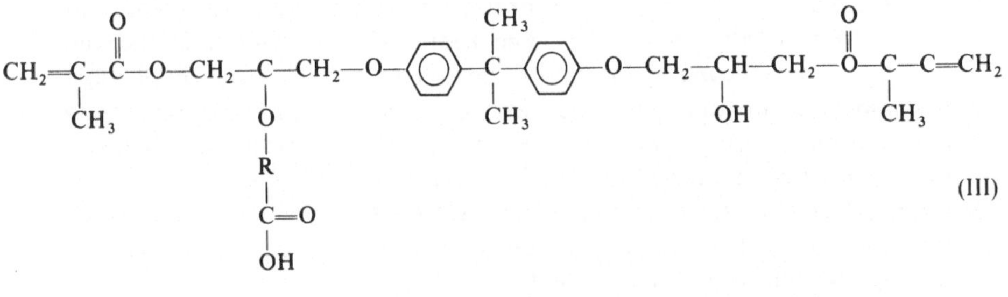

$$
CH_2{=}C{-}C{-}O{-}CH_2{-}C{-}CH_2{-}O{-}\langle O\rangle{-}C{-}\langle O\rangle{-}O{-}CH_2{-}CH{-}CH_2{-}O{-}C{-}C{=}CH_2
$$

(III)

A B-stageable vinyl ester resin was developed to overcome problem areas in preimpregnated moulding mats or sheet moulding compound (SMC).[16] For a resin to be appropriate for use in SMC technology, it must be chemically thickenable (using divalent metallic oxides and/or hydroxides) while retaining the molecular unsaturation which is polymerised during the moulding operation. This is accomplished by introducing acid functionality on the vinyl ester resin molecule.[9] This resin has a more rapid B-staging rate, good glass wettability, and better rheology (allowing lower moulding pressures) than polyesters.

The SMC resins have found application in structural parts, particularly in the automotive industry, where weight reduction is an important factor for improved fuel consumption.[17,18,19] Addition of a thermoplastic material[20] has greatly improved the surface characteristics and shrinkage.

2.5.3. Epoxy Novolac Vinyl Ester Resins for High-Temperature Applications
Example: Derakane 470 vinyl ester resin.

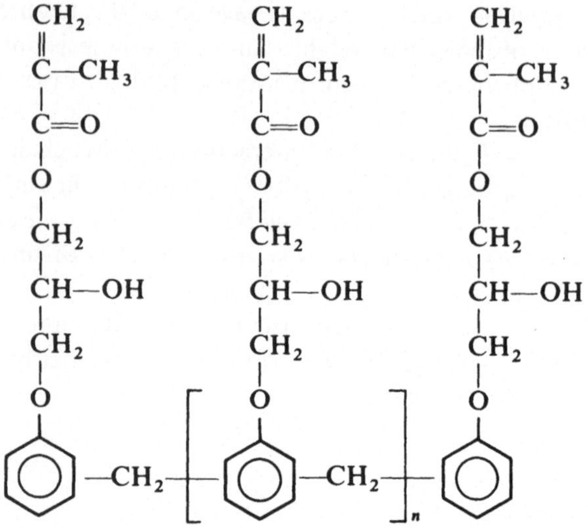

(IV)

In 1972 a new high-temperature organic resistant vinyl ester resin was made commercially available.[21] The heat resistance and high thermal stability were achieved by incorporating an epoxy resin based on phenol–formaldehyde novolac into the vinyl ester resin backbone, increasing the crosslink density when the resin is cured. This resin, which has a heat distortion temperature of 270–300°F (132–149°C) significantly extended the useful operating temperature of vinyl ester resins while retaining excellent corrosion resistance, especially in environments containing chlorine or organic solvents.

2.5.4. Flame Retardant Resins
Example: Derakane 510-A-40 vinyl ester resin.

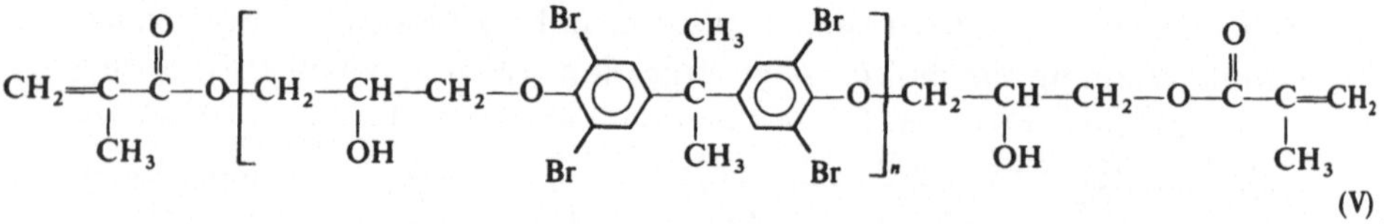

(V)

Brominated vinyl ester resins were developed to meet the industry's need for reduced flammability characteristics while retaining corrosion resistance. This flame retardant vinyl ester resin is used extensively in RP

TABLE 2.1

THE PERFORMANCE OF A FLAME RETARDANT RESIN IN FIRE RESISTANCE TESTS[a]

Test type	Test designation	Derakane 510A resin
60 s Burning test[b]	ASTM D757	0·4 in/min (10·2 mm/min)
Intermittent exposure test[c]	HLT 15	100
Tunnel test flame spread		
Rating test[c]	ASTM E84	
unfilled		30
with 5% Sb_2O_3		10
Limiting oxygen index test[c]	ASTM D2863	
unfilled		29·7% oxygen
with 5% Sb_2O_3		40·8% oxygen

[a] The results shown in this table were obtained from controlled and/or small scale bench tests. They are not necessarily predictive of behaviour in a real fire situation. Vinyl ester resins are organic materials, and the resins and products made therefrom will burn under the right conditions of heat and oxygen supply.
[b] 60% glass, press moulded at 60 psi (0·41 MN m^{-2}); benzoyl peroxide cure.
[c] 25% glass, 0·125 inch (3·2 mm), hand lay-up with Methylethylketone peroxide (MEKP) and Cobalt Naphthenate cure. Glass sequence: 10 mil. C-veil (V), three 1·05 oz ft^{-2} (457 gm^{-2}) mats (M), 10 mil. veil.

hoods, ductwork and stack applications where decreased flammability is desirable. The performance of this flame retardant resin in fire resistance tests is shown in Table 2.1.

2.5.5. Radiation Curable Resins
Example: Dow XD-9002 experimental vinyl ester resin.

(VI)

Radiation curable resins were developed to meet needs in several areas of application including coatings and printing inks. The ability to cure when exposed to ultraviolet or electron beam radiation is obtained by replacing the terminal methacrylate groups of vinyl ester resins with acrylate end groups. Photoinitiators such as benzophenone or benzoin ethers are used in ultraviolet curing to absorb the UV energy and transfer it to the resin system so as to effect vinyl polymerisation. The resins have a low viscosity in the undiluted state, or reactive diluents such as 2-ethyl hexyl acrylate or 2-hydroxypropyl acrylate can be used to further reduce viscosity. Cure times are quite rapid, and are measured in seconds rather than minutes, as for conventional coatings.

2.5.6. Bisphenol A Fumaric Acid Condensation Polyester
Example: Atlac® 382.

$$
\text{HO}-\overset{\overset{\text{O}}{\|}}{\text{C}}-\text{CH}=\text{CH}-\overset{\overset{\text{O}}{\|}}{\text{C}}-\text{O}\left[\overset{}{\underset{\underset{\text{CH}_3}{|}}{\text{CH}}}-\text{CH}_2-\text{R}-\text{CH}_2-\overset{}{\underset{\underset{\text{CH}_3}{|}}{\text{CH}}}-\text{O}-\overset{\overset{\text{O}}{\|}}{\text{C}}-\text{CH}=\text{CH}-\overset{\overset{\text{O}}{\|}}{\text{C}}-\text{O}\right]_n
$$

$$
\overset{}{\underset{\underset{\text{CH}_3}{|}}{\text{CH}}}-\text{CH}_2-\text{R}-\text{CH}_2-\overset{}{\underset{\underset{\text{CH}_3}{|}}{\text{CH}}}-\text{OH}
$$

R = Bisphenol A

(VII)

The structure of a bisphenol A fumarate polyester has been included for comparison. The physical properties are shown in Table 2.2.

2.5.7. Urethane-Based Vinyl Ester Resin
Example: Atlac 580.

$$
\text{H}_2\text{C}=\overset{\overset{\text{R}_2}{|}}{\text{C}}-\text{U}\left[\text{O}-\text{R}_1-\text{O}-\overset{\overset{\text{O}}{\|}}{\text{C}}-\text{CH}=\text{CH}-\overset{\overset{\text{O}}{\|}}{\text{C}}-\text{O}-\text{R}_1-\text{O}\right]_n-\text{U}-\overset{\overset{\text{R}_2}{|}}{\text{C}}=\text{CH}_2
$$

(VIII)

R$_1$ = Bisphenol A as shown in ICI literature
R$_2$ = Alkyl group or hydrogen
U = Urethane Connecting Group

A urethane-based vinyl ester resin was introduced in 1975.[13] It is reported to combine the best properties of polymers containing both internal and terminal unsaturation. There is some controversy as to whether or not this resin can be properly termed a 'vinyl ester resin' due to a large amount of internal unsaturation. The urethane modification is said to improve glass wettability and adhesion.

® ICI United States Inc.

TABLE 2.2
TYPICAL PHYSICAL PROPERTIES OF VINYL ESTER RESINS

	II Bis A epoxy vinyl ester resin	III SMC vinyl ester resin + thermoplastic additive	IV Epoxy novolac based vinyl ester resin	V Brominated vinyl ester resin	VI Radiation cured vinyl ester resin	VII Bisphenol A-fumarate polyester	VIII Urethane-based vinyl ester resin[f]	IX Rubber-modified vinyl ester resin[f]
Liquid Properties								
Viscosity kinematic, 77°F (25°C), cs	450	2 250	200	300	4 700[c]	—	—	1 000
Specific gravity	1 04	1 02	1 07	1 22	1 15	—	—	1 07
77°F (25°C) Gel time (min)[a]	28	—	12	24	—	—	—	—
180°F (82°C) Gel time (min)[b]	12	30 0	13	10	—	—	—	—
% Styrene	45	44	36	40	neat	50	—	45
Clear Casting Properties[d]								
Tensile strength, psi (MN m^{-2})	12 000 (82 7)	6 000[c] (41 4)	11 000 (75 8)	10 600 (76 1)	—	10 000 (69 0)	13 100 (90 3)	10 000 (69 0)
% Elongation	5	6	3	5	—	1 5	4 2	10
Flexural strength, psi (MN m^{-2})	18 000 (124 0)	11 000 (75 8)	20 000 (137 9)	18 000 (124 0)	—	13 500 (93 1)	22 600 (155 8)	17 000 (117 2)
Flexural modulus psi × 10⁵ (GN m^{-2})	4 5 (3 1)	3 4 (2 3)	5 5 (3 8)	5 2 (3 6)	—	—	4 9 (3 4)	4 4 (3 0)
HDT, °F (°C)	215 (102)	220 (104)	290–300 (143 9)	230 (110)	—	270 (132)	221 (105)	170 (77)
Barcol hardness	35	—	40	40	—	37	37	30

[a] 1 5% MEKP solution (60% MEKP in dimethyl phthalate) and 0 5% cobalt naphthenate solution (6% Co in mineral spirits)
[b] 1% Benzoyl peroxide
[c] 60°C
[d] 1 0% MEKP solution and 0 3% cobalt naphthenate solution, cured 16h at 77°F (25°C) and 2h at 311°F (155°C)
[e] 1% Benzoyl peroxide and 0 1% N,N-dimethyl aniline, cured 16h at 77°F (25°C) and 2h at 311°F (155°C)
[f] Data taken from 'Introducing Atlac® 580 Vinyl Ester Resin', ICI United States Inc , 1975

2.5.8. Rubber-Modified Vinyl Ester Resin

Example: Dow XD-8084 experimental vinyl ester resin. To meet the
need for improved toughness for applications where severe mechanical
abuses are encountered, a rubber-modified vinyl ester resin was developed.
Rubber is incorporated into the vinyl ester molecule. In addition to

TABLE 2.3

TYPICAL REVERSE IMPACT DATA ON $\frac{5}{16}$ INCH (7·9 mm) HAND
LAY-UP LAMINATES

Resin	First crack
Bisphenol A-fumarate polyester	16 in lb (1·77 J)
Epoxy novolac based vinyl ester resin	28 in lb (3·04 J)
Basic vinyl ester resin	57 in lb (6·38 J)
Rubber-modified vinyl ester resin	207 in lb (22·96 J)

improved toughness, the rubber-modified resin has improved adhesion,
lower exotherm temperatures and less shrinkage than the basic vinyl
ester resin.

Tables 2.3 and 2.4[22] illustrate the improved impact toughness and
adhesion of the rubber-modified vinyl ester resin compared to the basic
vinyl ester resin.

TABLE 2.4

TYPICAL ADHESIVE STRENGTH OF RUBBER-MODIFIED VINYL ESTER RESIN COMPARED TO
THE BASIC VINYL ESTER RESIN AND BISPHENOL A FUMARATE POLYESTER

Test	Result
t-Peel (ASTM D-1876-61T) on 2024 T3 Aluminium	
Rubber-modified vinyl ester resin	4·5 lb/linear in (779 N m^{-1})
Basic vinyl ester resin	2·9 lb/linear in (502 N m^{-1})
Bisphenol A-fumarate polyester	0·9 lb/linear in (156 N m^{-1})
Lap Shear (ASTM D-1002) on 2024 T3 Aluminium	
Rubber-modified vinyl ester resin	> 1 000 psi (> 6·9 MN m^{-2})
Basic vinyl ester resin	500 psi (3·5 MN m^{-2})
Bisphenol A-fumarate polyester	500 psi (3·5 MN m^{-2})

2.6. CURING

The choice of peroxide catalysts is determined by the particular resin in question and the temperature at which it is to be cured. Generally, methyl ethyl ketone peroxide (MEKP) is used for room temperature curing, and benzoyl peroxide (BPO) or *t*-butyl perbenzoate is used for elevated temperature curing. BPO with N,N-dimethyl aniline as an accelerator may also be used for curing vinyl ester resins at room temperature. Cobalt causes ketone peroxides to dissociate into free radicals. Thus the room temperature cure can be effected without the application of external heat.[23] N,N-dimethyl aniline or other aromatic tertiary amines may also be used to further accelerate the MEKP curing system. MEKP with a high dimer content is more reactive than low dimer MEKP in curing vinyl ester resins, while the opposite is true with polyester resins.[24,25]

Criteria for determining the catalyst system and judging the degree of cure include exotherm temperature, residual monomer, physical properties, working time and development of hardness or tack-free state. The exotherm temperature should be high enough to cure the resin but not so high as to cause cracking of the resin. This is particularly important in cast parts of large mass and thickness where the heat dissipation will be slower. Residual monomer will deleteriously affect physical properties and corrosion resistance.[26]

Oxygen in the air will inhibit the complete cure of an exposed vinyl ester resin surface. This inhibited surface may vary in depth, and may result in reduced weatherability, poor chemical resistance and/or premature failure. This problem is overcome by preventing or reducing contact of the curing surface with air.

2.7. INHIBITORS

Phenolic inhibitors such as hydroquinone or the monomethyl ether of hydroquinone are used during the synthesis of vinyl ester resins to prevent polymerisation during processing. An inhibitor such as one of the phenolics may also be added at the completion of synthesis to extend the shelf-life. Sulphur-containing compounds have also been claimed to be effective in extending shelf-life.[27] The effectiveness of an inhibitor can be determined by following its consumption using gas or liquid chromatography or by uncatalysed shelf-life studies under controlled conditions.

The ideal inhibitor would give infinite shelf-life while not interfering with, or retarding, the peroxide-catalysed cure of the vinyl ester resin. Frequently the catalyst system and the curing temperature for a particular application will determine the choice of inhibitors as with polyester resins. Periodic aeration of vinyl ester resins has also been found to be helpful in extending the shelf-life.

2.8. EFFECTS OF CAST RESIN HIGH TENSILE ELONGATION ON PERFORMANCE OF LAMINATED STRUCTURES

The effect of the very high tensile elongation of basic vinyl ester resin castings shows up in the performance of laminated structures in areas such as, (1) the strain level at first resin cracks, (2) impact load at first reverse side cracking and (3) hydraulic pressure in pipe at first weeping.

(1)   In Table 2.5 it is shown that there is direct linear relationship between cast resin tensile elongation and the % tensile strain at first visible cracks in SPI laminates made with the same resins. The laminates were made from: veil (V)—Owens–Corning Fibreglass M514 Treatment 236; mat (M)—Owens–Corning Fibreglass M-711 1–1·5 oz chopped strand mat (457 g m^{-2}). Construction was V–M–M–V. Mould surface was Mylar® film on both sides. The first fine cracks were visible in the chopped strand layers only at the values shown in the table. These fine cracks were

TABLE 2.5

PERCENT STRAIN AT FIRST VISIBLE RESIN CRACKS IN CHOPPED STRAND MAT LAMINATE V. TENSILE ELONGATION OF RESIN CASTINGS

Resin type	% Tensile elongation of cast resin	% Tensile strain in laminate to first visible resin cracks
Basic vinyl ester	5–7	1–1·2
Epoxy novolac vinyl ester	3–4	0·6–0·7
Polyester (rigid)	1·5–2	0·3–0·5
Polyester (very rigid)	<1	0·2–0·3

® Trademark of the E. I. Du Pont Company.

TABLE 2.6

REVERSE IMPACT ON $\frac{5}{16}$ INCH (7·9 mm) THICK HAND LAY-UP
LAMINATES

Resin type	% Tensile elongation of cast resin	Impact at first crack
Polyester (rigid)	1·5–2	16 in lb (1·77 J)
Vinyl ester (epoxy novolac)	3–4	28 in lb (3·04 J)
Vinyl ester (basic)	5–7	57 in lb (6·38 J)

visible when the samples were illuminated with a parallel light beam
at a 45° angle to the laminate surface. They disappeared when the
stress was removed.

(2) Table 2.6 shows the impact load to first reverse surface cracking
on $\frac{5}{16}$ in (7·9 mm) thick laminate v. resin tensile elongation. In
this case we see that there is a direct relationship between resin
tensile elongation and impact resistance. In the resins tested here
the relationship is also linear. However, the linear relationship dis-
appears with high elongation rubber modified resins, which show
stress marks, but do not reverse crack at impact loads of almost
twice that which would be obtained by linear extrapolation of the
points shown in Table 2.6. This information is previously given in
Table 2.3 in the discussion on rubber modified resin.

(3) Filament wound pipe having a 2 in (50 mm) inside pipe diameter
was made using a 0·010 in (0·254 mm) C-veil liner and nominal
0·080 in (2·0 mm) wall from two of the polyester resins and the basic
vinyl resin used in the previous tests. The results are shown in
Table 2.7. Note that the pipes made from the 'very rigid' and 'rigid'
polyester resins started to weep at less than half and slightly more
than half of the weep pressure of the high elongation vinyl ester
resin although their 'Split ring Tensile' strength is equal or almost
equal to that of the vinyl ester resin. The burst pressure of the polyester pipes could not be determined because the weep rate became
so high that the pump capacity was reached not very far above the
first weep pressure. The epoxy novolac type of vinyl ester resin was
not run in this test. In tests on 4 in (102 mm) diameter pipe made
from this 3–4% tensile elongation resin it started to weep at
approximately 75% of the weep pressure of pipe made from the
basic vinyl ester resin.

TABLE 2.7

WEEP AND BURST PRESSURE AND HOOP STRENGTH OF 2 INCH (50 mm) FILAMENT
WOUND PIPE V. TENSILE ELONGATION OF CAST RESIN 0·080 INCH (2·0 mm) WALL
THICKNESS, 0·010 INCH (0·254 mm) C-VEIL LINER

Resin type	% Tensile elongation of cast resin	Pressure to weep psi (MNm^{-2})	Burst pressure × 10^3 psi (GNm^{-2})	Hoop tensile^a × 10^3 psi (MNm^{-2})
Polyester (very rigid)	<1	960 (6·6)	—^{b}	30 (207)
Polyester (rigid)	1·5–2	1 280 (8·8)	—^{b}	36 (248)
Vinyl ester (basic)	5–7	2 180 (15·0)	2 310 (15·9)	36 (248)

^{a} Split ring tensile.
^{b} Rate of weeping exceeded pump capacity so that pipes could not burst.

The examples given above show the effects of cast resin elongation on performance of laminate in three different types of tests. From the practical side, this shows up dramatically in performance in applications that range from chemical equipment to ballistic armour. This is covered in Section 2.11 (applications).

2.9. MONOMERS

While styrene is the most common monomer used to dilute vinyl ester resins, other monomers which have been used include divinyl benzene,[28] vinyl toluene, chlorostyrene,[29] α-methyl styrene, t-butyl styrene and dicyclopentadiene acrylate.

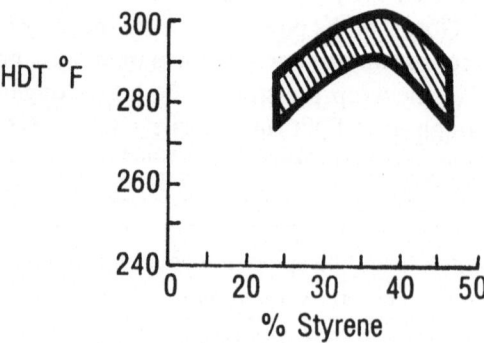

FIG. 2.1.   The effect of styrene concentration on the heat distortion temperature (HDT) of Derakane 470.

TABLE 2.8

MECHANICAL PROPERTIES OF DERAKANE 411 RESIN AT 45% MONOMER[a]

Monomer	Tensile strength (psi (MN m^{-2}))	% Elongation	Flexural strength (psi (MN m^{-2}))	Flexural modulus (psi × 10^5 (GN m^{-2}))	HTD (°F (°C))	Barcol hardness
Chlorostyrene	12 600 (87)	3·80	24 100 (166)	5·66 (3·9)	231 (111)	45
Dicyclopentadiene acrylate (70% monomer)	7 800 (54)	1·36	20 400 (141)	5·39 (3·7)	191 (88)	46
Divinyl benzene	3 600 (25)	0·94	12 200 (84)	5·76 (4·0)	226 (107)	53
Styrene	12 000 (83)	6·29	22 400 (154)	5·68 (3·9)	222 (105)	41
Vinyl toluene	11 500 (79)	4·15	21 300 (147)	5·44 (3·7)	205 (96)	44

[a] Cured with 0·3% Cobalt naphthenate solution (6% Co in mineral spirits) and 1·0% MEKP solution (60% MEKP in dimethyl phthalate) 16 h at room temperature and 2 h at 311 °F (155°C).

Styrene monomer offers advantages of low cost, desirable reactivity and many years of formulating experience in the field. The generalisation may be made that for a given vinyl ester resin, there is an optimum styrene concentration with respect to physical properties. Thus with Derakane 470, epoxy novolac vinyl ester resin, the maximum heat distortion temperature (HDT) is achieved with a 36% styrene dilution as shown in Fig. 2.1.

The effect of the styrene content on the corrosion resistance of vinyl ester resins has been reported.[30] In contrast to polyester resins, vinyl ester resins have good corrosion resistance at low monomer levels.

Vinyl toluene diluted vinyl ester resins were developed for use in electrical applications. These resins were shown to have less water absorption, higher flexural strength and electrical strength retention v. heat ageing than a conventional polyester resin diluted in styrene. Vinyl toluene diluted vinyl ester resins offer the advantages of easier processability (especially in matched metal die moulding) and lower cost than epoxy resins in electrical applications. Vinyl toluene is less volatile and has a higher flash point than styrene. The former property is important in view of the push for lower monomer emissions during fabrication and processing.

Dicyclopentadiene acrylate-diluted vinyl ester resins have the advantage of longer shelf-life than conventional styrene-diluted vinyl ester resins. Dicyclopentadiene acrylate is more compatible with vinyl ester resins than styrene, which is advantageous since higher levels of dicyclopentadiene acrylate are required to produce comparable viscosities. A disadvantage of dicyclopentadiene acrylate monomer is that it yields resins which are more brittle.

Table 2.8 shows the mechanical properties of a vinyl ester resin in several monomers.

The low tensile strengths and elongations of the resin diluted in dicyclopentadiene acrylate and divinyl benzene are indicative of brittleness.

2.10. VINYL ESTER RESIN LAMINATE PROPERTIES

Although laminates have been made and tested from vinyl ester resins using graphite or carbon fibres, Nexus® polyamide fibres, and a wide variety of polyester fibres, glass fibres are commercially by far the most important reinforcements. Since this book is primarily about resins, only a few basic glass fibre laminates will be discussed. *Hand lay-up, filament*

® Trademark of Burlington Industries.

TABLE 2.9
TYPICAL PROPERTIES OF HAND LAY-UP LAMINATE ($\frac{1}{4}$" (6·4 mm) THICK)

Property	PS 15-69 requirement	Structure II basic resin	Structure IV epoxy novolac resin	Structure V flame retardant resin
% Monomeric styrene	—	45	36	40
Cast resin HDT °F (°C)	—	210–215°	290–305	225–235
Flexural strength, psi (MN m^{-2})				
Room temperature	19 000 (131)	29 600 (204)	24 000 (165)	23 800 (164)
150°F (66°C)		28 500 (196)	—	23 800 (164)
200°F (93°C)		27 400 (189)	24 500 (169)	24 000 (165)
225°F (107°C)		14 700 (101)	—	21 000 (145)
250°F (121°C)		5 000 (34)	24 100 (166)	12 000 (83)
300°F (149°C)		3 200 (22)	21 000 (145)	—
325°F (163°C)		—	12 000 (83)	—
350°F (177°C)		—	8 000 (55)	—
Flexural modulus, × 10^5 psi (GN m^{-2})				
Room temperature	8 (5·5)	10·3 (7·1)	12·5 (8·6)	11 (7·6)
150°F (66°C)		10·1 (7·0)	—	11 (7·6)
200°F (93°C)		8·5 (5·9)	11·8 (8·1)	9·7 (6·7)
225°F (107°C)		4·9 (3·4)	—	8·2 (5·7)
250°F (121°C)		2·3 (1·6)	10·6 (7·3)	5·8 (4·0)
300°F (149°C)		2·3 (1·6)	8·3 (5·7)	—
325°F (163°C)		—	6·1 (4·2)	—
350°F (177°C)		—	5·2 (3·5)	—

TABLE 2.9—contd.

Property	PS 15-69 requirement	Structure II basic resin	Structure IV epoxy novolac resin	Structure V flame retardant resin
Tensile strength, psi (MN m^{-2})				
Room temperature	12 000 (83)	20 700 (143)	18 000 (124)	16 400 (113)
150°F (66°C)		25 100 (173)	—	18 300 (126)
200°F (93°C)		21 800 (150)	18 600 (128)	19 500 (134)
225°F (107°C)		18 200 (126)	—	18 500 (128)
250°F (121°C)		11 700 (81)	18 800 (130)	17 000 (117)
300°F (149°C)		7 700 (53)	17 000 (117)	—
325°F (163°C)		—	14 400 (100)	—
350°F (177°C)		—	11 000 (76)	—
Tensile modulus, $\times 10^5$ psi (GN m^{-2})				
Room temperature	—	17·4 (12·0)	16·5 (11·4)	15 (10·3)
150°F (66°C)		18·1 (12·5)	—	17 (11·7)
200°F (93°C)		14·9 (10·3)	17·1 (11·8)	13 (9·0)
225°F (107°C)		11·1 (7·7)	—	12·6 (8·7)
250°F (121°C)		7·6 (5·2)	17·1 (11·8)	12 (8·3)
300°F (149°C)		—	10·4 (7·2)	—
325°F (163°C)		—	9·1 (6·3)	—
350°F (177°C)		—	7·3 (5·0)	—

Wr = Woven roving glass.
Glass Content: 40%.
Laminate Construction: V–M–M–Wr–M–M–Wr–M.
V = Std. 10 mil. corrosion-grade C-glass veil.
M = Chopped strand mat of 1·5 oz ft^{-2} (457 g m^{-2}).
Courtesy of The Dow Chemical Company.

wound and *matched metal die moulded laminates* make up the bulk of the commercial laminates being sold.

The biggest use of vinyl ester resin *hand lay-up laminate* is in making chemical resistant equipment. Table 2.9 shows the typical properties of $\frac{1}{4}$ in (6·4 mm) thick, 40% glass fibre laminate, made with three different vinyl ester resins. Their properties are compared with the US Bureau of Standards' requirement PS 15-69 which was developed for polyester corrosion resistant equipment. Note the high physical properties and excellent retention of physical properties at elevated temperatures. This typical data has been used extensively in the design of reinforced plastics equipment. Four pressure vessels designed using this data were pressure tested up to 1·5 times operating pressure. Only one of more than 100 strain gauges used showed a significantly higher strain level than was predicted.

The properties of *filament wound laminate* vary with the wind angle of the glass fibre. In pipe wound at an angle of 54·5° to the axis, having a nominal 65% glass content, the tensile stress at break is 37 000–43 000 psi (255–296 MN m^{-2}).

TABLE 2.10

PHYSICAL PROPERTIES OF MATCHED DIE MOULD VINYL ESTER SMC AND HSMC

Property	SMC	HSMC	HSMC
Composition			
% glass	40	50	65
% resin	30	25	35
% filler	30	25	0
Tensile Strength psi (MN m^{-2})	19 000 (131)	26 000 (179)	32 000 (221)
Flexural Strength psi (MN m^{-2})	43 000 (296)	50 000 (345)	62 000 (428)

Matched die laminates of vinyl ester resins have slightly better static physical properties and significantly better dynamic properties than similar laminates made with polyester resins. One of the sizeable use areas is automotive parts moulded from SMC and HSMC (high strength sheet moulded compound) vinyl ester resins. Table 2.10 gives some of the physical properties of automotive type moulded laminate.

2.11. VINYL ESTER RESIN APPLICATIONS

Because vinyl ester resins are made from higher cost raw materials and often use extended process times, they have higher costs than do standard polyester resins. For this reason they are only used commercially in applications where their superior properties are needed.

The major uses of vinyl ester resins will be discussed by markets. The markets are:

> corrosion resistant,
> land transportation,
> electrical insulation,
> marine and radiation curing,
> others.

2.11.1. Corrosion Resistant Applications

By far the largest use of vinyl ester resins is in making corrosion resistant equipment. When the basic vinyl ester resin is made with methacrylic acid end caps (Structure (II)) the resultant shielding of the ester linkages gives a resin with exceptional resistance to a broad range of acids and alkalis. The 'E' type glass fibres used as reinforcements in making corrosion resistant equipment have high strength but poor resistance to acids and alkalis. The pendant hydroxyl groups on the vinyl ester chain assist in wetting and bonding to glass fibres. Superior protection of the glass fibre results. These characteristics, coupled with the high tensile elongation

FIG. 2.2.   Pulp mill recovery boiler scrubber demister.

(toughness) of the resins, have been responsible for their wide use in this market. Unstressed corrosion resistance testing in the laboratory shows the vinyl ester resins to be very similar to the bisphenol A fumarate and chlorendic anhydride polyesters in many chemicals. In plant service, however, the higher toughness of the resin decreases the dangers of mechanical damage to the corrosion liner during fabrication, shipping, installation, operation and maintenance. The resulting superior performance has led to wide use of the vinyl ester resins in tanks, piping, absorption

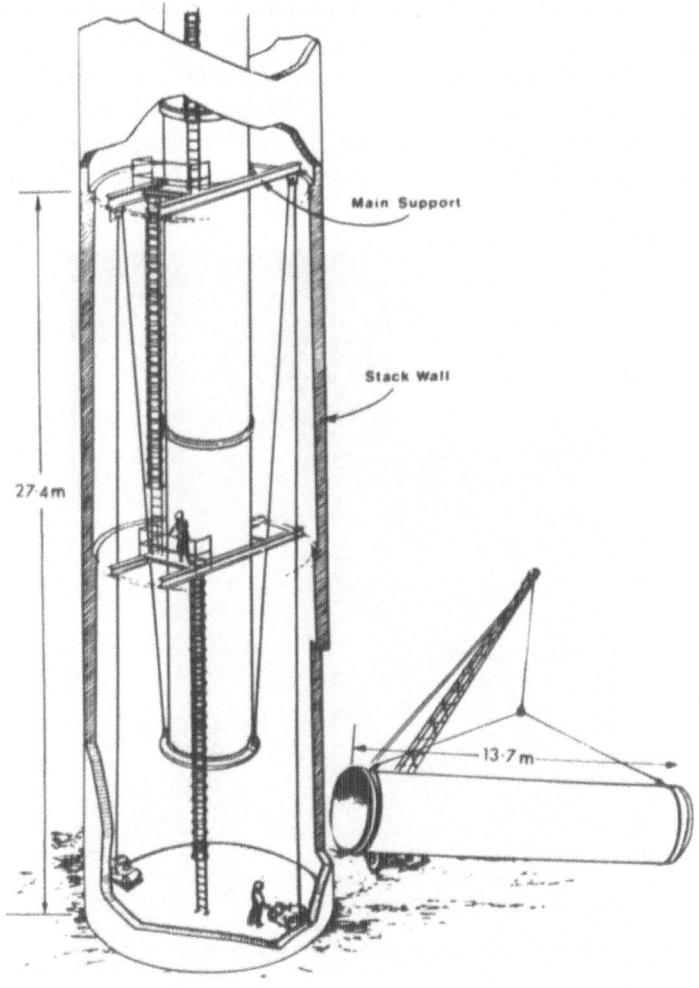

FIG. 2.3.   400 ft (122 m) chimney liner installation (schematic).

towers, process vessels, hoods, ducts, scrubbers and stacks. The trio of scrubbers, demisters and stacks shown in Fig. 2.2 are an excellent example of a cost effective use of the basic vinyl ester resin. These units, 13 ft 8 in (4.17 m) in diameter at the base and 100 ft (30·5 m) tall, scrub sulphur dioxide gases at 180 °F (82 °C) from a pulp mill's two recovery boilers. These large structures were transported hundreds of miles to the mill site and erected. They had been in service over six years at the time that an inspection showed them to be essentially in original condition.

Some idea of the importance of toughness in resins for corrosion applications can be gained by looking at the possibilities for rough mechanical overstressing shown in Figs. 2.3 and 2.4. These show an 11 ft 3 in (3·43 m) diameter stack liner that was made from the fire retardant vinyl ester resin (Structure (V)) in 45 ft (13·7 m) long sections and installed in a 400 ft (122 m) tall concrete chimney. Figure 2.3 shows schematically

FIG. 2.4.   Inserting chimney section through breaching.

the method of installation, while Fig. 2.4 shows one of the 45 ft (13·7 m) long sections being inserted through the breaching in the concrete chimney. No cracks were found in this 400 ft (122 m) tall stack liner when it was originally inspected after installation, nor when inspected after seven years of service handling smelter gases.

The vinyl ester resins made with a novolac backbone (Structure (IV)) retain good general corrosion resistance, and much of the resistance to solvents and high temperatures that the phenolic resins, used in their synthesis, display. Equipment made with these resins has been in service over 10 years at 340 °F (172 °C) handling chlorine and hydrogen chloride wet vapour. These resins are widely used to make tanks and piping that handle organic contaminated waste water and by-product hydrochloric acid that is contaminated with organic and chlorinated organic compounds. Figure 2.5 shows a tank trailer, made with the epoxy novolac vinyl ester on the inside with the fire retardant resin on the exterior, that hauls by-product hydrochloric acid one direction and 50% caustic on the return.

FIG. 2.5.   Reinforced plastic tank trailer.

2.11.2. Land Transportation Applications

The second largest use of vinyl ester resins is in making body and structural parts for land transportation equipment. Much of this equipment, e.g. automobiles, trucks, buses, snowmobiles and tractors, is very competitive and is assembled in large numbers on assembly lines from painted, press moulded parts. These parts must have a high heat distortion temperature to withstand the paint ovens, while being tough enough to withstand rough assembly line handling and the vibration, impact and other mechanical stresses of the particular service.

In production of the body for the Chevrolet Corvette, there were problems with cracked polyester SMC (sheet moulding compound) exterior parts at the end of the assembly line. When vinyl ester SMC parts replaced the polyester parts, cracks due to assembly line handling disappeared. With the emphasis on improved fuel economy and weight reduction in automobiles, many development programmes for structural parts are moving rapidly. The excellent fatigue life of parts made from vinyl ester resins makes them the logical choice for structural parts. One of the most promising is the development of automobile wheels from vinyl ester HSMC (Fig. 2.6). The glass fibre, high strength sheet mouldings

FIG. 2.6.   Vinyl ester HSMC wheel on test car.

weigh only 50–60 % as much as the steel wheel.[31,32] They have, to date, passed all the specifications for steel wheels, while wheels made from other resins (polyester and epoxy) have not passed all the required tests.

Railroad hopper car covers made from vinyl ester HSMC pass all the operation requirements, while being light enough for one man to open and close easily.

2.11.3. Electrical Insulation Applications

Large motors, generators, transformers and other heavy electrical equipment require insulation that is very tough and does not shrink or crack under continuous operation at 160 to 200 °C. Special vinyl ester resins are used to make rigid insulation that retains adequate electrical strength and dimensional stability for more than 10 years at 200 °C.[33]

2.11.4. Marine Applications
Use of vinyl ester resins in marine applications has been limited to high performance craft such as kyaks, canoes, speed boats and special fishing rigs. In many cases Kevlar® fibre is used as all or part of the reinforcement to give lighter weight boats with more puncture resistant hulls.

2.11.5. Radiation Curing Applications
Resins of the Structure (VI) type are used in can coatings, printing inks, panel coatings and many other areas where rapid cures and short finishing lines are required.[34,35]

2.11.6. Other Applications
1. *The aircraft industry* has made relatively little use of vinyl ester resins. Developed in the late 1960s, the Cessna AG truck crop spraying aeroplane shown in Fig. 2.7 has been a continuing success. The spray tank shown in Fig. 2.8, made with the basic vinyl ester resin, is built in between the pilot and the engine as part of the aircraft structure. The resin was selected because of good resistance to severe mechanical stresses coupled with a broad resistance to agricultural spray chemicals.
2. *Dental filling* material consistently uses a small amount of vinyl ester resins. The resin's toughness, good resistance to food staining,

FIG. 2.7. Cessna AG truck crop duster.

® Trademark of the E. I. Du Pont Company.

FIG. 2.8.   Spray tank for Cessna AG truck.

good bonding chracteristics and ability to gel and cure rapidly and completely in the presence of moisture all contribute to its performance.

3.   *Ballistic armour* of very light weight has been developed by combining the very high tenacity of Kevlar® fibres with the toughness characteristics of the vinyl ester resins. This armour is light enough to be used for personnel and vital equipment protection on helicopters. It is used in armouring cars and, when made with fire retardant vinyl ester resin, is used aboard ships.

2.12.   NEW DEVELOPMENTS

Any time a new resin is developed by combining the best properties of two other resins, there is a strong effort to make the new resin better in all respects. The vinyl ester resins did combine the ease of processing

and curing of polyester resins with the toughness and bonding characteristics of epoxy resins.

By chemical structure the vinyl ester resins should, and in many cases
did, cure better than polyester resins. Total processibility as good as,
or better, than that obtained with polyester resins evolved over a period
of many years. The better bonding chracteristics obtained from the epoxy
backbone caused many parts to stick in the mould. The high reactivity
of the end of chain double bonds made stabilising the resins a very
delicate procedure; the problem was to prevent polymerisation during
shipping and storage without interfering with catalysed cure. The first
difficulty has been solved, but new resin producers still struggle with
resin stabilisation.

While the vinyl ester resins gave better bonding than polyester resins,
they did not bond as well as the epoxy resins to materials like stainless
steel and aluminium. The rubber modified vinyl ester resins were developed
to give improved bonding to a wide range of substrates (Table 2.4) and
to provide resins with vastly superior impact resistance.[22] Urethane
modified resins[13] (Structure (VIII)) were developed that gave improved
bonding.

New vinyl ester SMC resins have recently been developed that are
thickened through a network produced by reacting a polyisocyanate with
a polyol instead of the widely used reaction of a group II oxide or
hydroxide with acid end or pendant groups. Very fast SMC thickening
times and high laminate strengths are reported for these new resins.[36]

A brominated, novolac-based vinyl ester resin was recently developed
to provide a fire retardant resin with superior organic resistance and
superior heat ageing resistance. The improved organic resistance was
developed to handle the requirements of the process of solvent extraction of
uranium from crude phosphoric acid and from copper mine tailings.[37] The
improved heat ageing resistance was developed to provide a matrix resin
for high temperature ductwork and stacks for coal fired power plants,
smelters and related processing plants.

2.13. TOXICOLOGICAL ASPECTS OF VINYL ESTER RESINS

Vinyl ester resins in the uncured state are not highly toxic by ingestion,
but are capable of causing significant eye, skin and respiratory disorders.
Eye contact may cause slight irritation or transient corneal injury. Skin
contact may cause slight to moderate irritation on prolonged contact.

56 THOMAS F. ANDERSON AND VIRGINIA B. MESSICK

Inhalation of concentrated vapours may cause irritation of respiratory passages and slight anaesthesia. It is recommended that eye, skin and vapour contact with uncured vinyl ester resins be avoided. Normal laboratory and plant standards for good housekeeping and personal hygiene should be followed.

Vinyl ester resins that are completely polymerised (cured) are considered to be toxicologically inert and should present no health problems from handling. The cured resins may, however, present a health problem from inhalation of dust generated during machining or grinding, especially if they contain glass, silica powders, asbestos or metal powders.

Users should consult with the manufacturers of initiators and promoters or other materials used in formulating with vinyl ester resins for safe handling procedures.

REFERENCES

1. BOWEN, R. L., US Patent No. 3,066,112 issued to the US government, *Dental filling material comprising vinyl silane fused silica and a binder consisting of the reaction product of bisphenol and glycidyl acrylate*, Nov. 27, 1962.
2. BOWEN, R. L., US Patent No. 3,179,623 issued to the US government, *Method of preparing a monomer having phenoxy and methacrylate groups linked by hydroxy glyceryl groups*, April 20, 1965.
3. FEKETE, F., KEENAN, P. J. and PLANT, W. J., US Patent No. 3,221,043 issued to H. H. Robertson Co., *Ethylenically unsaturated dihydroxy diesters*, Nov. 30, 1965.
4. FEKETE, F., KEENAN, P. J. and PLANT, W. J., US Patent No. 3,256,226 issued to H. H. Robertson Co., *Hydroxy polyether polyesters having terminal ethylenically unsaturated groups*, June 14, 1966.
5. SZOBODA, G. R., SINGLETON, F. and ESHLEMAN, L., US Patent No. 3,621,093 issued to H. H. Robertson Co., *Process for making reinforced thermoset articles*, Nov. 16, 1971.
6. BEARDEN, C., US Patent No. 3,367,992 issued to The Dow Chemical Company, *2-Hydroxy alkyl acrylate and methacrylate dicarboxylic acid partial esters and the oxyalkylated derivatives thereof*, Feb. 6, 1968.
7. NAJVAR, D. J., US Patent No. 3,524,901 issued to The Dow Chemical Company, *Flame retardant vinyl esters containing acrylic or methacrylic phosphate esters*, Aug. 18, 1970.
8. HARGIS, S. R., JR, US Patent No. 3,524,903 issued to The Dow Chemical Company, *Flame retardant vinyl ester containing alkyl hydrogen phosphate resin and a halogenated epoxide resin*, Aug. 18, 1970.
9. SWISHER, D. H. and GARMS, D. C., US Patent No. 3,564,074 issued to The Dow Chemical Company, *Thermosetting vinyl resins reacted with dicarboxylic acid anhydrides*, Feb. 16, 1971.

10. PENNINGTON, D. W. and NORTON, F. E., US Patent No. 3,594,247 issued to The Dow Chemical Company, *Filament winding process*, July 20, 1971.
11. YOUNG, R. E. in *Unsaturated Polyester Technology*, ed. Bruins, P. F., 1976, Gordon and Breach Publishers, New York, p. 318.
12. YOUNG, R. E. in *Unsaturated Polyester Technology*, ed. Bruins, P. F., 1976, Gordon and Breach Publishers, New York, p. 316.
13. LEWANDOWSKI, R. J., FORD, E. C., JR, LONGENEEKER, D. M., RESTAINO, A. J. and BURNS, J. P. *30th SPI Reinforced Plastics/Composites Conf.*, 1975, New high performance corrosion resistant resin, Paper 6-B.
14. NIESSE, J. and FENNER, O. Effect of glass content and laminate structure on the mechanical properties of FRP structures, *Nat. Association of Corrosion Engineers, Corrosion 78*, March, 1978, Paper 105.
15. LINOW, W. H., BEARDEN, C. R. and NEUENDORF, W. R. *21st SPI Reinforced Plastics/Composites Conf.*, 1966, The DERAKANE resins—a valuable addition to thermosetting technology, Paper 1-D.
16. JERNIGAN, J. W., BEARDEN, C. R. and PENNINGTON, D. W. *22nd SPI Reinforced Plastics/Composites Conf.*, 1967, High speed 'B-staging' with vinyl ester resins —a route to the automation of reinforced plastics moulding, Paper 8-D.
17. STAVINOHA, R. F. and MACRAE, J. D. *27th SPI Reinforced Plastics/Composites Conf.*, 1972, DERAKANE vinyl ester resins—unique chemistry for unique SMC opportunities, Paper 2-E.
18. CUTSHALL, J. E. and PENNINGTON, D. W. *27th SPI Reinforced Plastics/Composites Conf.*, 1972, Vinyl ester resin for automotive SMC, Paper 15-A.
19. THOMAS, R. E. and ENOS, J. H. *33rd SPI Reinforced Plastics/Composites Conf.*, 1978, Very high strength SMC in automotive structures, Paper 4-E.
20. NOWAK, R. M. and GINTER, T. O., US Patent No. 3,674,893 issued to The Dow Chemical Company, *Thermosettable resins containing vinyl ester resin, polydiene rubbers and vinyl monomers*, July 4, 1972.
21. CRAVENS, T. E. *27th SPI Reinforced Plastics/Composites Conf.*, 1972, DERAKANE 470-45, a new high temperature corrosion resistant resin, Paper 3-B.
22. HAWTHORNE, K., STAVINOHA, R. and CRAIGIE, L. *32nd SPI Reinforced Plastics/Composites Conf.*, 1977, High bond—super tough—CR resin, Paper 5-E.
23. BRINKMAN, W. H., DAMEN, L. W. and SALVATORE, M. *23rd SPI Reinforced Plastics/Composites Conf.*, 1968, Accelerators for the organic peroxide curing of polyesters and factors influencing their behaviour, Paper 19-D.
24. CASSONI, J. P., HARPELL, G. A., WANG, P. C. and ZUPA, A. H. *32nd SPI Reinforced Plastics/Composites Conf.*, 1977, Use of ketone peroxides for room temperature cure of thermoset resins, Paper 3-E.
25. THOMAS, A., JACYSZN, O., SCHMITT, W. and KOLCZYNSKI, J. *32nd SPI Reinforced Plastics/Composites Conf.*, 1977, Methyl ethyl ketone peroxides, relationship of reactivity to chemical structure, Paper 3-B.
26. VARCO, P. *30th SPI Reinforced Plastics/Composites Conf.*, 1975, Important cure criteria for chemical resistance and food handling applicatons of reinforced plastics, Paper 6-C.
27. AWAJI, T. and ATOBE, D., US Patent No. 4,129,609 issued to Nippon Shokubai Kagaku Kogyo Co., Ltd, *Method for improving storage stability of epoxy ester thermosetting resins with thiuram compounds*, Dec. 12, 1978.

28. YOUNG, R. E. in *Unsaturated Polyester Technology*, ed. Bruins, P. F., 1976, Gordon and Breach Publishers, New York, p. 336.
29. RUBENS, L. C., THOMPSON, C. F. and NOWAK, R. M. *20th SPI Reinforced Plastics/Composites Conf.*, 1965, The advantage of chlorostyrene diluted polyesters in production of reinforced plastics, Paper 2-C.
30. SAMS, L. L., PIEGSA, J. G. and ANDERSON, T. F. *28th SPI Reinforced Plastics/Composites Conf.*, 1973, Monomer content vs. corrosion resistance of vinyl ester resins, Paper 13-A.
31. WOELFEL, J. A., O'NEIL, K. B., BREDE, A. and SMITH, R. W. *34th SPI Reinforced Plastics/Composites Conf.*, 1979, Reinforced composite wheels: light-wheel of the future, Paper 21-A.
32. STANLEY, A. *34th SPI Reinforced Plastics/Composites Conf.*, 1979, Wheel wells of structural SMC—for transportation industry, Paper 21-C.
33. WHARTON, W. H., ESCH, F. P. and ANDERSON, T. F. *Proceedings of the 10th Electrical Insulation Conf.*, Sept. 1971, Improving price–performance ratio of high temperature laminates, IEEE Pub. No. 71-C38-E1, p. 40.
34. CRARY, E., US Patent No. 3,661,576 issued to the W. H. Brady Company, *Photopolymerizable compositions and articles*, May 1, 1972.
35. SHUR, E. and DABAL, R., US Patent No. 3,772,062 issued to the Inmoat Corporation, *Ultraviolet curable coating compositions*, Nov. 13, 1973.
36. FERRANINI, J., MAGRANS, J. J. and REITZ, J. A., III. *34th SPI Reinforced Plastics/Composites Conf.*, 1979, New resins for high strength SMC, Paper 2-G.
37. CRAIGIE, L. J., RUSSELL, D. L. and HARTLESS, A. L. *34th SPI Reinforced Plastics/Composites Conf.*, 1979, A new brominated novolac-base vinyl ester resin, Paper 8-B.

Chapter 3

POLYESTER RESIN CHEMISTRY

T. HUNT

BP Chemicals Ltd, Penarth, South Glamorgan, UK

SUMMARY

This chapter reviews recent developments in the processing and formulation of unsaturated polyester resins. Considering the condensation polymer first, refinement of existing production techniques is discussed and the alkylene oxide route to the preparation of polyester resins is compared with the conventional diol based process. New, or relatively new, building blocks which can be used to prepare resins with improved properties are also considered.

Turning to the many additives which can be incorporated to modify the properties of polyester resins, flame retardant formulations are reviewed and efforts to improve the maturation, shrinkage and toughness of moulding compounds are summarised.

A novel polyester cement, curable by the addition of water, an ingenious new foaming system and various formulations that emit less styrene vapour during fabrication than standard grades are further examples of the valuable properties which can be introduced via additives.

Future trends in polyester synthesis are considered.

3.1. INTRODUCTION

A high proportion of the total sales of unsaturated polyester resins is made up of general purpose grades. These are mainly styrene solutions of polyesters derived from maleic anhydride, phthalic anhydride and

1,2-propane diol (propylene glycol). A wide variety of performance specifications can be satisfied by varying the proportions of these three basic
ingredients. Viscosity is varied by adjusting the styrene content; reactivity
is largely dependent on the molar ratio of maleic anhydride to phthalic
anhydride. Increased flexibility, if required, is introduced by partial
replacement of propane diol with a longer chain molecule such as
diethylene glycol. Alternatively phthalic anhydride can be partially
replaced with a linear dicarboxylic acid such as adipic acid.

Speciality resins which, when cured, have improved resistance to
hydrolysis and better mechanical properties are obtained by using isophthalic acid instead of phthalic anhydride.

Polyesters made with a halogenated building block such as tetrachlor-
phthalic anhydride instead of phthalic anhydride are the basis of formulations which give cured products with a reduced tendency to burn.

Further variations can be produced by the use of additives. Dispersion
of a small quantity (0·5–3 %) of fumed silica introduces thixotropy. A
small paraffin wax addition (c. 0·05–0·1 %) reduces residual surface tack
after cure. Tertiary aromatic amines boost the cure rate of resins polymerised with a cobalt salt/ketone hydroperoxide system. Addition of an
organohalogen compound plus antimony trioxide is an efficient means
of reducing the tendency to burn. Refractive index can be reduced, in
order to produce clearer fibreglass laminates, by substituting methyl methacrylate for some of the styrene. This also improves the weathering properties of the cured resin. These and many other variations have been described
elsewhere.[1,2]

In this chapter we shall be considering recent developments in polyester
chemistry and improvements/modifications in performance which can
be achieved by the use of various additives. It is in this latter area
that the main development effort has been concentrated in recent years.
Resin development has been largely confined to refinement of existing
processing techniques and modifications of standard formulations to suit
new application areas or improved methods of fabrication. Some building
blocks new to the unsaturated polyester resin field, which yield products
with improved properties, are also being used.

3.2. POLYESTER MANUFACTURING PROCESSES

3.2.1. The Alkylene Oxide Route to Polyesters
Increased interest is being shown in this process. Unsaturated polyester
resins can be prepared by reacting an alkylene oxide, especially propylene

oxide, with maleic anhydride and optionally phthalic anhydride in the presence of an hydroxyl-containing initiator (water, alcohols, carboxylic acids). The reactions taking place are shown below.

(a) Alcohol initiated

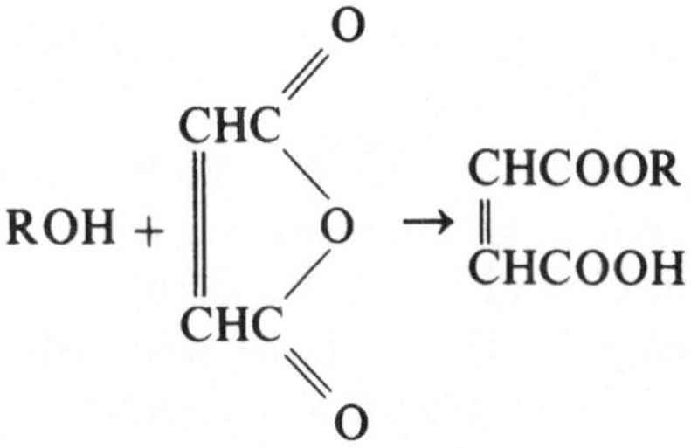

$$\text{ROH} + \begin{array}{c} \text{CHC} \\ \| \\ \text{CHC} \end{array} \!\! \begin{array}{c} O \\ \diagup \\ O \\ \diagdown \\ O \end{array} \rightarrow \begin{array}{c} \text{CHCOOR} \\ \| \\ \text{CHCOOH} \end{array} \qquad (1)$$

$$\begin{array}{c} \text{CHCOOR} \\ \| \\ \text{CHCOOH} \end{array} + R_1\text{CH} \!\!-\!\! \text{CH}_2 \ \ \overset{O}{\triangle} \rightarrow \begin{array}{c} \text{CHCOOR} \\ \| \\ \text{CHCOOCH}_2\text{CH(OH)}R_1 \end{array} \quad (2)$$

reacts with more anhydride, and so on

(b) Acid initiated

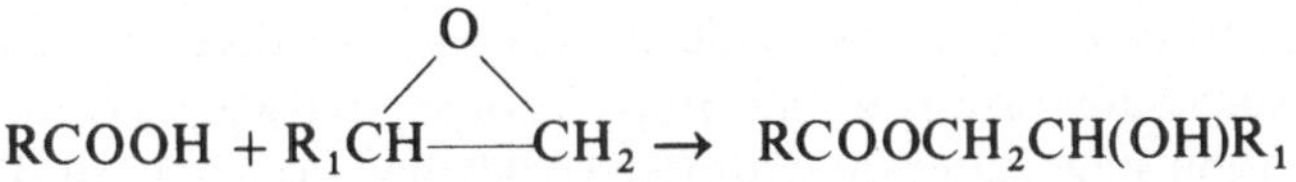

$$\text{RCOOH} + R_1\text{CH} \!\!-\!\! \text{CH}_2 \ \overset{O}{\triangle} \rightarrow \text{RCOOCH}_2\text{CH(OH)}R_1 \qquad (3)$$

$$\text{RCOOCH}_2\text{CH(OH)}R_1 \xrightarrow{\quad \overset{\displaystyle \begin{array}{c} O \\ \text{CHC} \diagup \\ \diagdown O \\ \text{CHC} \diagdown \\ O \end{array}}{} \quad} \begin{array}{c} R_1 \\ | \\ \text{RCOOCH}_2\text{CH} \\ | \\ \text{OCOCH} \\ \| \\ \text{HOOC} \cdot \text{CH} \end{array} \qquad (4)$$

reacts with more oxide

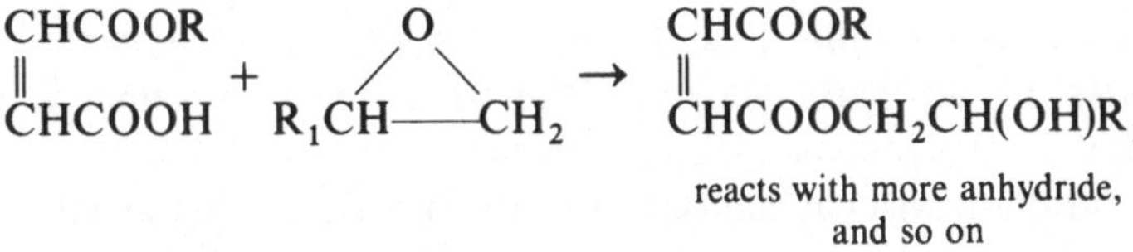

(c) Water initiated

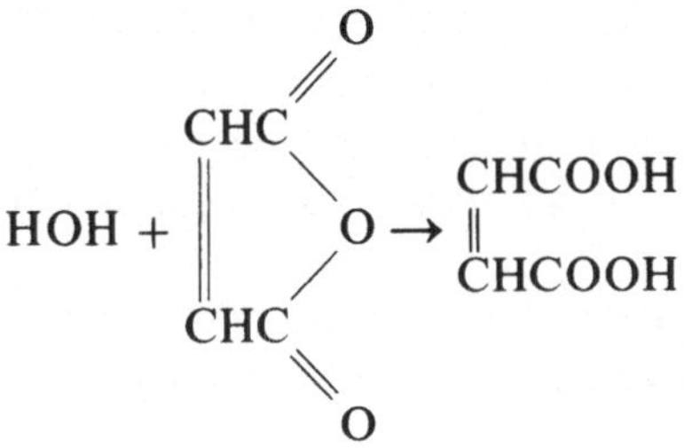

$$\text{HOH} + \begin{array}{c} \text{CHC} \\ \| \\ \text{CHC} \end{array} \!\! \begin{array}{c} O \\ \diagup \\ O \\ \diagdown \\ O \end{array} \rightarrow \begin{array}{c} \text{CHCOOH} \\ \| \\ \text{CHCOOH} \end{array} \qquad (5)$$

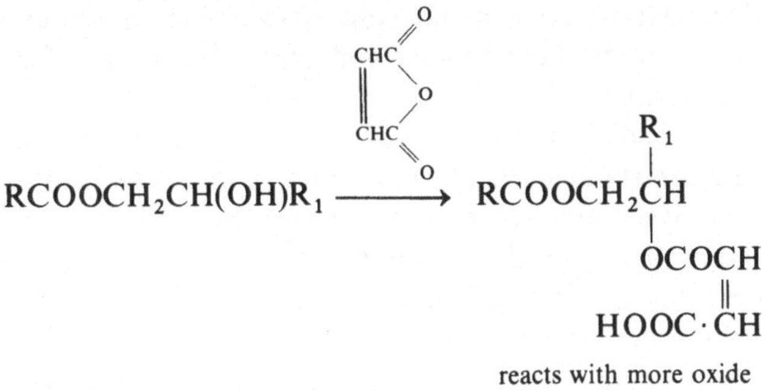

$$\begin{array}{c} \mathrm{CHCOOH} \\ \| \\ \mathrm{CHCOOH} \end{array} + \begin{array}{c} \mathrm{O} \\ \diagup \; \diagdown \\ \mathrm{R_1CH \!-\!-\! CH_2} \end{array} \rightarrow \begin{array}{c} \mathrm{CHCOOH} \\ \| \\ \mathrm{CHCOOCH_2CH(OH)R_1} \end{array} \qquad (6)$$

$$\mathrm{HOH} + \begin{array}{c} \mathrm{O} \\ \diagup \; \diagdown \\ \mathrm{R_1CH \!-\!-\! CH_2} \end{array} \rightarrow \mathrm{R_1CH(OH)CH_2OH} \qquad (7)$$

In all these systems, incoming oxide can react with both hydroxyl and carboxyl terminated fragments. It is important, therefore, to use a selective catalyst which promotes esterification rather than etherification if resins of similar structure to diol based polyesters are to be produced. Alternatively, reaction conditions can be adjusted so that esterification is the dominant activity, without using a catalyst. Both options are described in patents. Oxides and salts of magnesium, zinc and lead were preferred catalysts in an early patent.[3] Preferred reaction temperature was 120–130°C in order to get light coloured products and the propylene oxide addition was carried out over about 5 hours. It was later realised that maleate to fumarate isomerisation did not take place efficiently at these low reaction temperatures which explained the somewhat inferior cure properties of products made in this way, relative to normal diol based resins. Later processes, therefore, incorporated a higher temperature isomerisation stage, after the propylene oxide had been reacted. Preferred catalysts then included alkali metal hydroxides and salts,[4] lithium chloride or a quaternary ammonium halide,[5] and organic acid salts of zirconium, titanium or cerium.[6]

Addition of a small quantity of phosphoric acid before commencing the isomerisation stage at 200°C was found to prevent discoloration.

In an uncatalysed process[7] the alkylene oxide is added to a vigorously stirred mixture of anhydrides and initiator, in the most turbulent zone, the mixture being maintained within the temperature range 130–220°C. Pale coloured products are claimed to result.

The alkylene oxide route to unsaturated polyester resins has a number of advantages over the traditional glycol based process. No aqueous distillate is produced and the preferred propylene oxide is slightly cheaper than propylene glycol. The oxide ring opening is exothermic, so heat input is reduced. Fast cycle times are possible, provided that the addition rate and dispersion of the oxide are optimised. Also, a continuous process could be based on this route.

Disadvantages include the relatively limited range of formulations which can be made in this way, i.e. without a normal esterification stage in

POLYESTER RESIN CHEMISTRY

which water is evolved and the higher hazard ratings of the oxides relative to the corresponding glycols.

At the present time, the majority of resin manufacturers use the glycol route. When new plants are being planned, however, the oxide process will be a serious contender, particularly when manufacture of general purpose grades is the objective. Half the polyester output of one manufacturer is now made using propylene oxide.[8]

3.2.2. Refinement of Existing Processes

A recent innovation is the use of molten maleic and phthalic anhydrides to speed up charging and reduce heat-up times. The molten anhydrides are stored in heated tanks at 60 °C for maleic anhydride and 155 °C for phthalic anhydride.

Isophthalic acid based unsaturated polyester resins continue to be made by a two stage process, because of the low solubility and reactivity of this acid relative to maleic anhydride. In the first stage, isophthalic acid and the full glycol charge are reacted together until the hot mix is clear. Maleic anhydride is then added, and polyesterification is completed. If the glycol is propylene glycol, the stage-one temperature is limited to about 180 °C at atmospheric pressure, and overall process times are considerably longer than those for the corresponding orthophthalates. Use of catalysts, such as butyl stannoic acid, or higher temperature processing under pressure are now being recommended to reduce stage-one process times.

The kinetics and mechanism of the polyesterification process are also being examined in efforts to reduce production scale process times and improve product consistency.[9]

3.3. TEREPHTHALIC ACID

Polyester resins in which the main saturated dicarboxylic acid component is terephthalic acid are now available commercially. From published data[2] the main advantages gained by the use of terephthalic acid relative to equivalent isophthalic acid formulations appear to be increased heat distortion temperature and some reduction in shrinkage during cure. There is said to be little, if any, improvement in chemical resistance. Nevertheless, the terephthalates have gained a foothold, mainly for applications requiring chemical resistance.

The disadvantages associated with the use of this acid are considerable. It is the slowest reacting of the three phthalic acids and catalysts or pressure are required in the first stage of the two stage process (as already described for isophthalic acid in Section 3.2.2) to achieve acceptable process times. There are also more constraints on the formulation of resins than is the case with isophthalic acid or phthalic anhydride. These derive from its symmetrical structure and its tendency to give products with reduced solubility in styrene when reacted with symmetrical glycols such as neopentyl glycol.

3.4. DICYCLOPENTADIENE (DCPD)

DCPD used to be available in relatively small quantities from coal tar operations. Large quantities are now recoverable as a by-product of the steam cracking of naphtha for ethylene production at a price below those of the basic polyester raw materials.

The following reactions make DCPD an interesting raw material for the preparation of high performance unsaturated polyester resins.

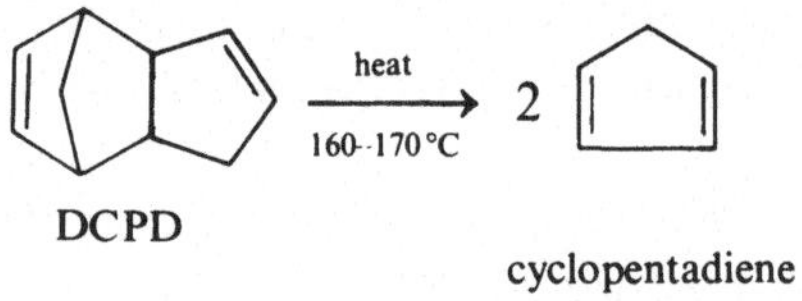

DCPD $\xrightarrow[160-170\,°C]{heat}$ 2 cyclopentadiene   (8)

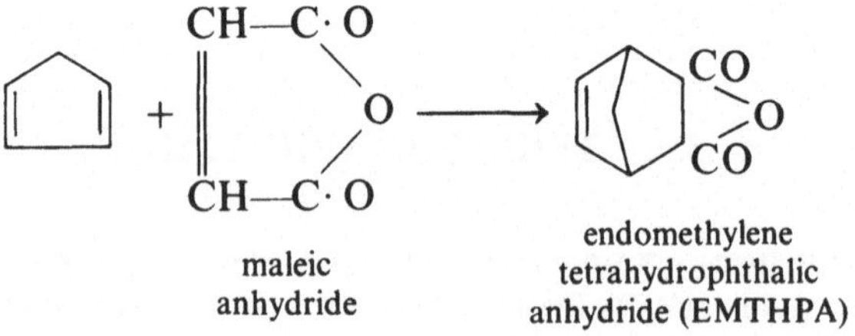

cyclopentadiene + maleic anhydride ⟶ endomethylene tetrahydrophthalic anhydride (EMTHPA)   (9)

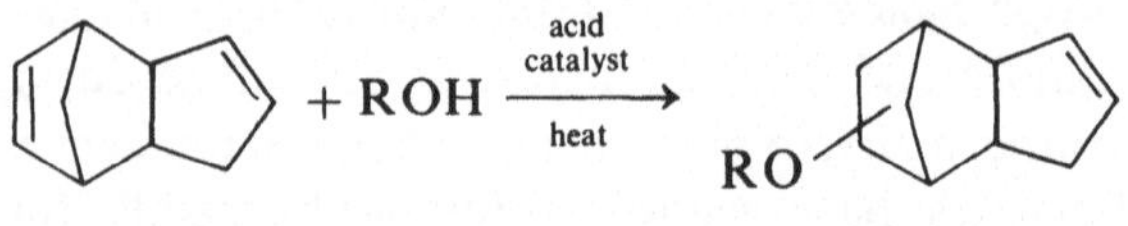

$+ ROH \xrightarrow[heat]{acid\ catalyst}$ (10)

ROH may be water, an alcohol or a carboxylic acid

Until recently, only the Diels–Alder addition of cyclopentadiene (eqn. (9)) has been utilised commercially in order to obtain unsaturated polyester resins containing endomethylene tetrahydrophthalate groupings.

There are two main methods of achieving this Diels–Alder addition. In the first, DCPD (1 mole) is heated with maleic anhydride (2 moles) under reflux in an atmosphere of nitrogen at 170–190 °C. The resulting EMTHPA is then reacted into an unsaturated polyester in the normal manner.

In the second method, propylene glycol and maleic anhydride are polyesterified and the polyester is then heated under reflux with DCPD at about 170 °C.

Latterly, attention has been concentrated on the acid catalysed addition of alcohols or acids to the double bond in the strained ring of DCPD (eqn. (10)). Three procedures have been described,[10,11] all rely on the fact that the acidity of a maleic anhydride/glycol 'half ester', or of maleic acid itself, is sufficient to catalyse the addition to DCPD, as shown in equations (11) to (13).

Half ester preparation

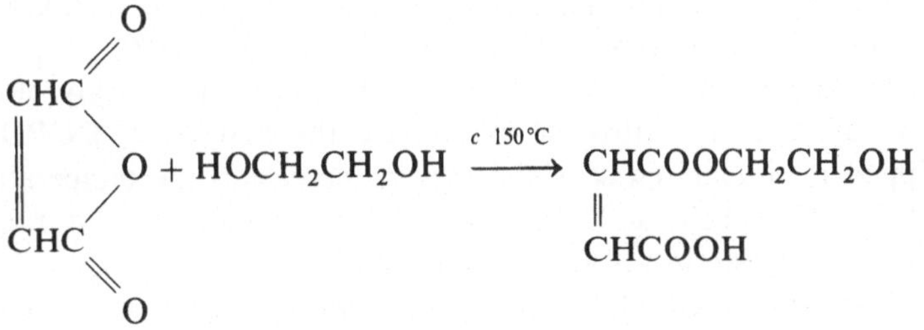

Reaction of half ester with DCPD

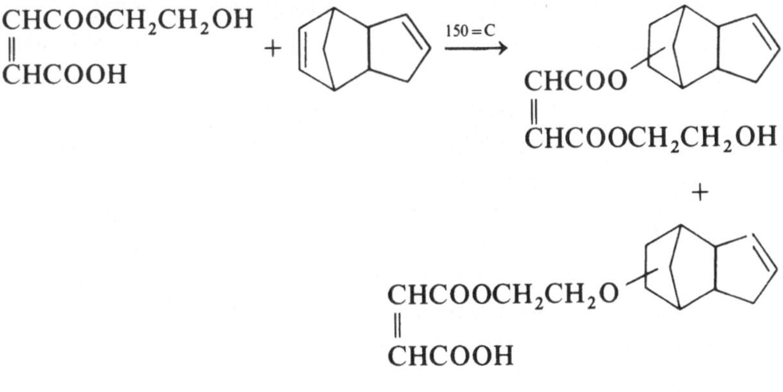

Hydrolysis method

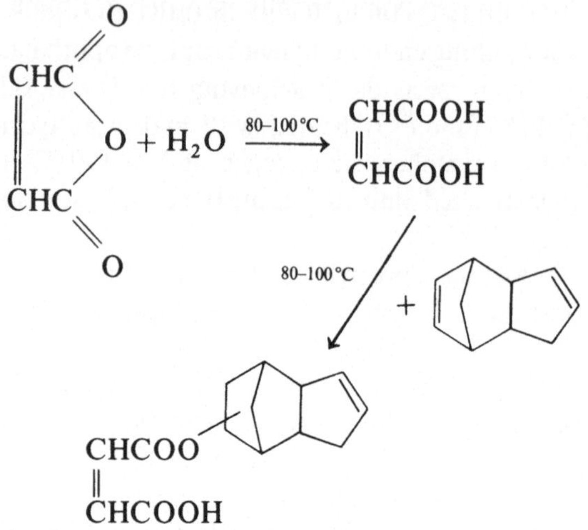

$$
\begin{array}{c}
\text{CHC} \\
\| \\
\text{CHC}
\end{array}
\Big\rangle\text{O} + \text{H}_2\text{O}
\xrightarrow{80\text{--}100\,°\text{C}}
\begin{array}{c}
\text{CHCOOH} \\
\| \\
\text{CHCOOH}
\end{array}
$$

80–100 °C

+

CHCOO⎯
‖
CHCOOH

(13)

In the so called 'beginning' method, DCPD, maleic anhydride and propylene glycol, in mole ratio of, for example, 0·25:1:1·1, are heated to c. 150 °C in an atmosphere of nitrogen and are held at this temperature under reflux for 30–60 min. The condenser is then set for distillation, and polyesterification is completed in the usual way at 200 °C.

In the 'half ester' method, maleic anhydride and propylene glycol are heated at 150 °C under nitrogen to prepare the half ester. DCPD is then dripped in, under reflux whilst slowly increasing the temperature to about 175 °C. When no DCPD refluxes, the temperature is raised to 200 °C and polyesterification is completed.

In the hydrolysis method[11] maleic anhydride is converted to maleic acid by heating with an equimolar quantity of water at 80–90 °C. DCPD is then added and, when reaction is complete, the glycol and, optionally, a modifying acid are charged prior to polyesterification at 200 °C.

Polyester resins prepared by the last three methods have lower molecular weights than general purpose orthophthalate resins and Diels–Alder addition types. This is due to the chain terminating nature of the DCPD addition reaction. At a given concentration in styrene they are, therefore, significantly less viscous than the other resins.

DCPD modified polyesters, whether Diels–Alder addition or DCPD addition types, give cured products having outstanding resistance to heat and ultraviolet light. They also have high heat distortion temperatures and enhanced chemical resistance relative to orthophthalate resins (see Ref. 10 for figures).

At the present time, with the high relative costs of maleic anhydride and phthalic anhydride, modification with DCPD looks an attractive proposition. The DCPD addition route is more economical than the Diels–Alder addition, since the latter requires a higher proportion of expensive maleic anhydride in the initial charge to produce a resin with reasonable reactivity. In the DCPD addition, the maleate unsaturation remains intact and there is scope for incorporation of modifying acids or phthalic anhydride.

3.5. GLYCOLS

3.5.1. Glycols in Standard Resins

The vast majority of unsaturated polyester resins are based on 1,2-propylene glycol, which currently† costs c. £370/tonne. Ethylene glycol (c. £500/tonne) based polyesters have reduced solubility in styrene, but this glycol is preferred in fire resistant formulations, particularly with a halogenated acid component. The latter usually improves solubility in styrene, and the polyesters have less tendency to burn than those containing propylene glycol.

Diethylene glycol (c. £450/tonne) is used where a cured resin with enhanced flexibility is required.

The glycols described in this section are all significantly more expensive than propylene glycol. However, the improved properties obtainable have enabled them to gain a foothold where high performance resins are necessary.

3.5.2. Neopentyl Glycol (NPG)

$$\begin{array}{c} CH_3 \\ | \\ HOCH_2-C-CH_2OH \\ | \\ CH_3 \end{array}$$

NPG costs c. £500/tonne. It has frequently been advocated for use when cured products with improved hydrolytic stability are required. In the USA where gel coated polyester is widely used to make baths, etc, the preferred gel coat resins for this application are now based mainly on isophthalic acid, maleic anhydride and neopentyl glycol. In the UK,

† All prices in this chapter relate to the last quarter of 1979.

baths and wash basins are more frequently made with a vacuum formed acrylic outer surface, reinforced behind with fibre glass, plus a general purpose polyester resin. There has, therefore, been less demand for these high performance gel coat resins here.

NPG gel coats are also making a significant impact on the marine market. They are claimed to reduce the incidence of gel coat blistering which can occur when boats have been in service for some time.[12] However, there is no single, simple solution to this problem as workshop practice, choice of back up resin and the type of glass reinforcement used are all contributory factors.[13,14]

The symmetrical structure of NPG can be used to produce resins with novel properties. Polyesters made from a glycol component (comprising at least 80 mole % NPG, the remainder preferably consisting of other symmetrical glycols such as ethylene glycol) and an acid component (made up of mainly fumaric acid and 12–25 mole % isophthalic acid) yield blends with styrene which are solid at normal ambient temperatures. This is believed to be due to the increased crystallinity of these polyesters.[15]

The novel polyester/styrene mixtures have melting points ranging from about 15 °C to 100 °C. Some have a buttery consistency at ambient temperature; others can be ground to yield free flowing powders. One application for these 'solid' polyesters is the preparation of a sheet moulding compound which does not require the addition of a maturation agent (see Section 3.8.1) to produce the necessary thickening. The only disadvantage would seem to be that the resin needs to be heated in order to liquefy it during the impregnation stage.

A further application for the above, highly crystalline systems involves blending them with normal liquid unsaturated polyester resins in order to introduce thixotropy. Compositions containing between 10 % and 50 % of the 'solid' polyester can be used.

3.5.3. 2,2,4-Trimethyl-1,3-Pentanediol (TMPD)

$$\text{HOCH}_2 - \underset{\underset{\text{CH}_3}{|}}{\overset{\overset{\text{CH}_3}{|}}{\text{C}}} - \text{CH} - \underset{}{\overset{\overset{\text{CH}_3}{|}}{\text{CH}}} - \text{CH}_3$$
$$\qquad\qquad\qquad\; \underset{\text{OH}}{|}$$

TMPD gives hydrolysis-resistant polyesters as a result of its sterically hindered hydroxyl groups. It currently costs c. £640/tonne in the UK, but enjoys a slight price advantage over NPG in the USA.

To prepare polyesters from TMPD, a stabiliser such as triethanolamine (c. 0·2% of the acid + TMPD charge) is usually added at the start of the reaction. This minimises cyclisation reactions.

Extensive data on TMPD based polyester formulations suitable for a wide range of end uses have been published by Eastman Chemical workers.[16,17]

Significant applications for TMPD polyesters in the USA include corrosion resistant storage tanks, gel coat resins for sanitary ware and the preparation of cultured marble.

3.5.4. 1:4 Cyclohexane Dimethanol (CHDM)

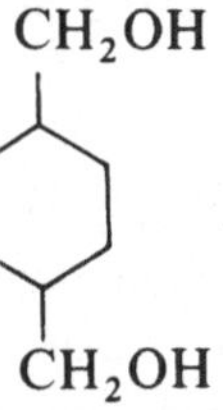

$$CH_2OH$$

$$CH_2OH$$

CHDM costs c. £1050/tonne. It reacts readily with dicarboxylic acids and anhydrides to give polyesters having excellent hydrolytic stability. The symmetrical structure of this glycol can be used to produce highly crystalline polyesters in a similar manner to that already described for NPG. For this application at least 55 mole % of the glycol component should be CHDM and at least 50 mole % of the acid component should be fumaric acid. When hot solutions of the products in styrene are cooled to ambient temperature, solid blends having similar applications to those based on NPG are obtained.[18]

In a further interesting application, polyesters made from, for example, CHDM, maleic anhydride and tetrahydrophthalic anhydride are claimed to give cured products with excellent chemical resistance, thermal stability and electrical properties.[19] The absence of aromatic structures in the polyester molecule is believed to be responsible for the high performance of these resins relative to general purpose grades based on phthalic anhydride. In this respect they are comparable with the dicyclopentadiene modified resins already discussed.

To date, there is little evidence that CHDM is being used to any significant extent in commercially available unsaturated polyester resins.

3.5.5. Dibromo Neopentyl Glycol (DBNPG)

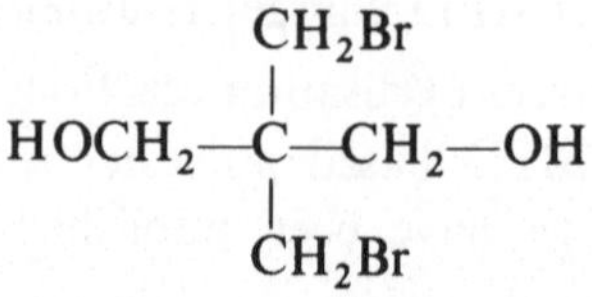

$$CH_2Br$$
$$\mid$$
$$HOCH_2\!\!-\!\!C\!\!-\!\!CH_2\!\!-\!\!OH$$
$$\mid$$
$$CH_2Br$$

DBNPG costs c. £1500/tonne. It is increasingly being used as a source of 'built in' bromine in unsaturated polyester formulations designed to have a reduced tendency to burn.[20] A major disadvantage is that some HBr is released during polyesterification. Glass lined or highly acid resistant stainless steel vessels are therefore needed to manufacture satisfactory products, although the possibility of vessel corrosion can be reduced by the use of, for example, endomethylene tetra hydro phthalic anhydride as part of the modifying acid charge. This acts as an HBr scavenger.[21]

With appropriate production vessels, light coloured products with high levels of combined bromine can be prepared rapidly at a maximum temperature of about 180 °C. The recommended procedure is to use DBNPG as the sole glycol. The products can then be blended with a general purpose grade if a lower bromine content is required.

In combination with suitable fire retardant additives, (see Section 3.7) cured products with much reduced tendency to burn can be obtained.

3.5.6. Bis (2-Hydroxyethyl Ether) of Tetrabromo Bisphenol A

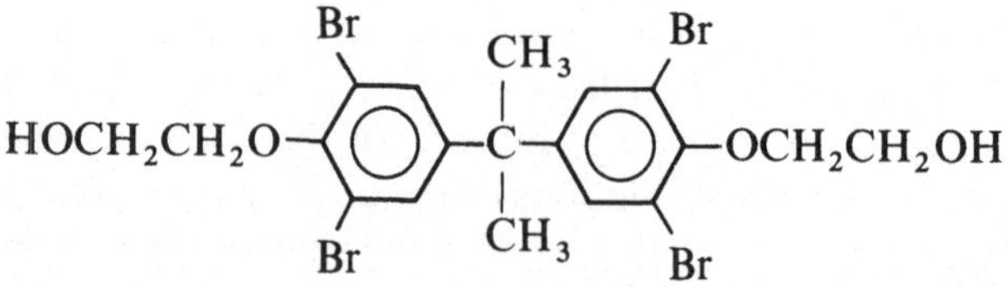

This ethylene oxide/brominated Bisphenol A reaction product is also being offered as a source of 'built in' bromine. It esterifies readily and, in contrast to DBNPG, does not release HBr at polyesterification temperatures. To date, nothing has been heard of its commercial use in the unsaturated polyester resin field. It costs c. £2000/tonne.

3.6. NEW MONOMERS

There is no sign of a serious competitor to styrene (currently c. £420/tonne) for the vast majority of applications. Methyl methacrylate is used to adjust refractive index downwards where this is required, e.g. in resins for the manufacture of transparent roofsheet.

Bromostyrene (c. £1300/tonne) is proposed as a source of 'built in' bromine in flame retardant formulations. Various glycol acrylates and methacrylates are also available.

3.7. FLAME RETARDANT FORMULATIONS

3.7.1. General Principles and Mechanisms of Fire Retardance
The most common way of making unsaturated polyesters burn less readily is to introduce organo halogen groupings either in the form of additives (e.g. chlorinated paraffin wax) or by chemical modification of the polyester (e.g. by the use of hexachloroendomethylene tetrahydrophthalic acid (HET acid) as part of the dicarboxylic acid charge).

The fire resistance of these halogen containing formulations is greatly improved by the addition of antimony trioxide and/or phosphorus compounds.

Various mechanisms have been proposed to explain the fire retarding properties of halogenated formulations. It is believed that HBr or HCl released on combustion interferes with the free radical reactions proceeding in the flame. The important reactions in the flame are:

$$\cdot OH + CO \rightarrow CO_2 + H \cdot \qquad (14)$$

$$H \cdot + O_2 \rightarrow \cdot OH + O \cdot \qquad (15)$$

The HBr and/or HCl reacts with the hydroxyl radicals, replacing them with the less reactive $Br \cdot$ or $Cl \cdot$ and $R \cdot$ where RH is an organic compound containing hydrogen in the flame, according to the following equations:

$$\cdot OH + HBr \rightarrow H_2O + Br \cdot \qquad (16)$$

$$Br \cdot + RH \rightarrow HBr + R \cdot \qquad (17)$$

The synergistic action of antimony trioxide in halogen containing formulations may be due to the endothermic formation of volatile antimony halides, at least one of which, $SbCl_3$, is believed to be an effective free radical trap. These can pass through a number of decompositions and recombinations, eventually giving the oxide again as a fine dust and releasing the halogen.

Zinc borate is being offered as a relatively cheap partial replacement for antimony trioxide. The latter currently costs c. £1750/tonne. It is thought that the borate functions by forming a protective glass matrix at the burning surface.

Phosphorus compounds are thought to act by accelerating the evolution of hydrogen halide from halogen containing formulations. A heat shield of polymeric *meta*-phosphoric acid is also said to be formed, which stops 'afterglow' caused by the highly exothermic solid phase oxidation of carbon.

$$2C + O_2 \rightarrow 2CO$$

The heat shield probably accounts for the higher levels of residual carbon char produced when phosphorus containing formulations are burned.

The use of alumina trihydrate and molybdenum compounds as fire retarding additives is discussed in 3.7.6 and 3.7.7.

Flame retardant polyester formulations have been reviewed[22,23] and extensive data on fire retardants collected.[24]

3.7.2. Formulation Guidelines

Flame retardant formulations are usually designed to pass a specified fire test. A common requirement in the UK is a Class 1 or Class 2 rating to the BS.476 part 7, 1971 Surface Spread of Flame test. A useful screening test on experimental formulations is to determine their oxygen index (ASTM D2863–76). The table below gives a very rough guide to the performance that can be expected in the BS.476 part 7 test for a given oxygen index.

Oxygen index %	Rating by BS.476 part 7, 1971
25 minimum	Class 2
c. 38 minimum	Class 1

N.B. A general purpose polyester specimen would have an oxygen index of about 19–20%.

An oxygen index of about 25% is readily obtainable from a combination of a general purpose resin and one or more fire retardant additives. A typical composition of this type would comprise 100 parts by weight of general purpose polyester, 10 parts of chlorinated paraffin wax, containing 70% combined chlorine, and 5–10 parts of antimony trioxide. To achieve the higher indices it is normal practice to use a halogenated resin plus additives, as a general purpose resin would require excessive levels of the latter which could reduce the strength of the cured resin and might also adversely affect weathering properties. A suitable halogenated resin might only require the addition of antimony trioxide to give the desired performance in the fire test.

The additives used depend on whether the particular application needs a clear resin or will tolerate an opaque one. If a clear resin is required antimony trioxide cannot be used as it functions as a white pigment. When an opaque resin is acceptable antimony trioxide will almost certainly be one of the additives used.

The cost of the formulation, as always, is a major factor. Toxicity of the additives is another. Tris (2,3-dibromo propyl) phosphate was frequently used in fire resistant polyester formulations until American tests branded it as a carcinogen. It is now prudent to insist that a negative Ames test for mutagenicity is obtained on a new halogen containing additive or raw material before expensive development work is started.

3.7.3. Polyesters with 'Built-In' Halogen

Where these are required, the preferred halogen containing building blocks are HET acid, tetrachlorophthalic anhydride and, less frequently, tetrabromo phthalic anhydride.

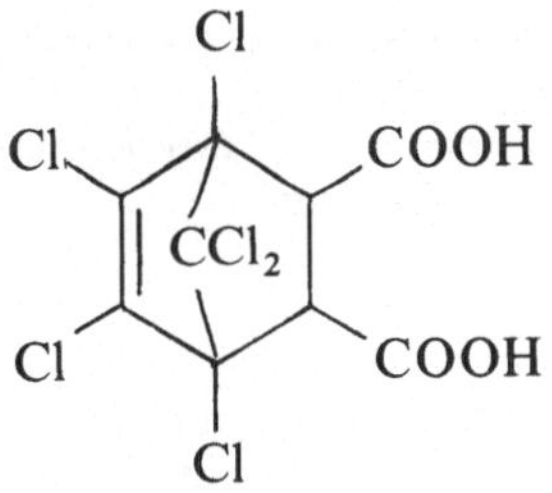

hexachloro endomethylene tetrahydro phthalic acid
(HET acid)

Usage of dibromo neopentyl glycol is increasing (see Section 3.5.5). The diol shown below prepared from decachlordiphenyl and ethanolamine is used in France.[25]

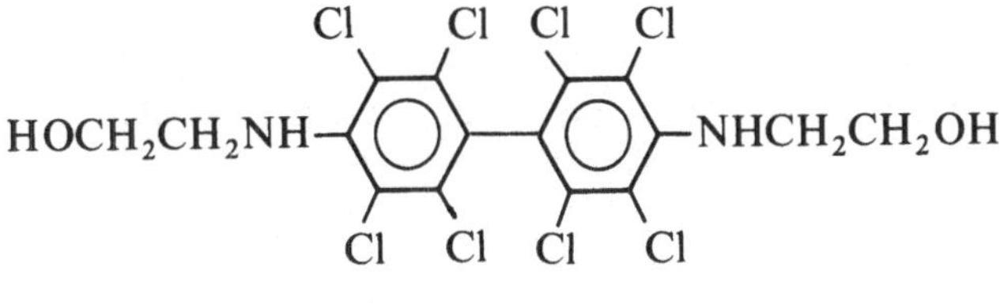

The corrosion difficulties associated with processing NPG polyesters have already been discussed. HET acid also tends to liberate HCl if the polyesterification temperature goes above about 180 °C. The remaining materials, with aromatic bound halogen, do not have this problem. Tetrabromo phthalic anhydride, however, as supplied, normally contains traces of sulphuric acid which must be neutralised. This is achieved by adding a neutraliser such as sodium acetate or diethanolamine to the polyesterification charge.[26]

3.7.4. Halogenated Additives

A wide range of organohalogen compounds are available.[24] As a general rule, bromine compounds are more effective than their chlorine equivalents, since the chlorine–carbon bond is stronger than the corresponding bromine linkage. Similarly, aliphatic halogen compounds are more effective than aromatic ones, since the aliphatic halogen is more labile. The thermal stability of the additive needs to be considered when formulating heat curable systems.

It is advantageous if the halogen compound is soluble in the polyester styrene solution. Insoluble materials have restricted application.

A selection of the halogenated compounds currently available is given

TABLE 3.1
HALOGENATED ADDITIVES

Additive	% Halogen	Physical form	Solubility in styrene	Approximate price £/tonne (late 1979)
Tetrabromo vinyl cyclohexane	74	m.p. 70–77 °C	soluble	2 500
Hexabromocyclo dodecane	73	m.p. 180 °C	insoluble	3 000
Pentabromo ethyl benzene	80	m.p. 136–138 °C	soluble	2 400
Hexabromobenzene	86	m.p. 330 °C	insoluble	—
Tetrabromoxylene	75	m.p. 250 °C	slightly soluble	1 500
Decabromodiphenyl oxide	83	m.p. 304 °C	insoluble	2 700
Octabromodiphenyl oxide	80	m.p. 75–125 °C	soluble	2 400
Pentabromodiphenyl oxide	70	viscous liquid	soluble	2 300
Pentabromotoluene	81	m.p. 280 °C	insoluble	2 900
Tetrabromo bisphenol A	59	m.p. 180 °C	insoluble	1 300
Tetrabromo phthalic anhydride[a]	67	m.p. 270 °C	soluble	2 000
Chlorinated paraffin wax	up to 70	viscous liquids	soluble	500

[a] Used here as an additive.

in Table 3.1. As can be seen, chlorinated paraffin wax is, by far, the cheapest source of halogen.

3.7.5. Additives Containing Phosphorus or Phosphorus and Halogen

Clear fire resistant polyester formulations normally contain both halogen and phosphorus compounds. Some or all of the halogen may be built into the polyester molecule as already described in Section 3.7.3. The phosphorus is almost always in an additive, which may also contain further halogen. There are a large number of phosphorus containing additives available.[24] A selection of these is given in Table 3.2.

TABLE 3.2
PHOSPHORUS BASED FIRE RETARDANT ADDITIVES (ALL LIQUIDS)

Additive	% Phosphorus	% Halogen
Tris (2-chloroethyl) phosphate	10·85	37·26
Tris (1,3 dichloropropyl) phosphate	7·2	49
Tris (monochlorpropyl) phosphate	9·4	32
Triethyl phosphate	17	—
Triphenyl phosphate	10	—

It should be noted that phosphorus compounds having a significant P–OH content are unsuitable for use in polyester systems promoted with cobalt salts as interaction occurs and the cobalt is deactivated.

3.7.6. Alumina Trihydrate

This filler has two main advantages over other fire retarding additives. It is cheap (currently from about £180 to £220/tonne, depending on particle size and quality) and it also reduces smoke emission.

In a fire situation, the trihydrate releases 34·5% of its weight as water vapour via a strongly endothermic reaction. The water dilutes any combustible gases present, while the endothermic reaction removes heat from the burning polymer.

Alumina trihydrate is used with both general purpose and halogenated polyesters and with other additives (e.g. antimony trioxide or trichloroethyl phosphate). Relatively high loadings of the filler (40–60% of the composition) are required to produce adequate fire resistance when it is the sole fire retarding component. Published data[27] indicates that a general purpose polyester formulation containing 60% by weight of alumina trihydrate would have an oxygen index of about 38%. This compares with an index of about 20% for a general purpose polyester without any additive.

Not surprisingly, polyesters with these high loadings of filler are more difficult to handle than unfilled resins, and laminators are less than enthusiastic about using them. It has been shown[28] that silane treated alumina trihydrate gives much lower viscosities when dispersed in polyester resins than those obtained with the uncoated filler at the same level of addition. Possibly use of the coated alumina will increase the acceptability of this type of formulation.

3.7.7. Molybdenum Compounds as Fire Retardants and Smoke Suppressants

Molybdenum compounds, in particular molybdenum trioxide and ammonium dimolybdate, have been found to be effective flame retardants for halogen containing polyester formulations. While less effective on its own, relative to antimony trioxide, molybdenum trioxide can be used to replace up to 50% by weight of the latter without loss of performance. Whereas antimony trioxide increases smoke generation, molybdenum compounds reduce it substantially (up to 50%). They are believed to function as char promoters in the solid phase and possibly catalyse the evolution of HCl and HBr from the halogenated components. Ash from burnt molybdenum containing formulations contains over 90% of the molybdenum added.[29,30]

Molybdenum trioxide costs approximately twice as much as antimony trioxide.

3.8.   MOULDING COMPOUNDS

3.8.1. General Description

The main development activity in this growing field of application of unsaturated polyesters has concentrated on improving two properties; shrinkage and maturation rate. The second property applies mainly to sheet moulding compounds (SMC), and before going any further it would be useful to explain what moulding compounds actually are.

There are two types: dough moulding compounds (DMC) and SMC. Both are designed for hot press moulding.

In its simplest form, DMC consists of resin, chopped glass, mineral filler, mould release agent, and catalyst. These are thoroughly mixed to a dough-like consistency in a Z-blade mixer, the glass being added last, and care being taken to avoid breaking down the fibres. The catalyst is a heat activated organic peroxide, with decomposition temperature chosen to suit the moulding temperature to be used. At ambient temperature the moulding compound has a shelf life of several weeks. A simple DMC formulation is shown in Table 3.3.

SMC is made by dispensing premixed resin, fillers, maturation agent, mould release agent and catalyst onto two moving sheets of polythene

TABLE 3.3
A TYPICAL DMC FORMULATION

Component	Parts by weight
Polyester resin	25
Filler (calcium carbonate)	55
Zinc stearate (lubricant)	1·5
t-Butyl perbenzoate	0·25
Chopped glass ($\frac{1}{4}$ in) (6·3 mm)	20

film, the lower of which also receives chopped glass rovings or glass mat. The two coated sheets are brought together, to blend the coatings and glass, by passing between two rollers. The compacted sheet is then wound into a roll and set aside to thicken. A simple SMC formulation is shown in Table 3.4.

The function of the maturation agent in SMC is to thicken up the mix so that the sheet compound becomes tack free and stiff enough to handle. The polythene can then be stripped off when the moulding operation is ready to start.

The preferred maturation agent is magnesium oxide, although a variety of suitable inorganic thickening agents have been described.[31] Other

TABLE 3.4
A TYPICAL SMC FORMULATION

Component	Parts by weight
Polyester resin	28
Precipitated calcium carbonate	42
Chopped glass roving ($\frac{1}{2}$–2 in) (12·5–50 mm)	30
Zinc stearate (lubricant)	1·4
Magnesium hydroxide (thickening agent)	0·7
t-Butyl perbenzoate	0·28

favoured agents are magnesium hydroxide and calcium hydroxide. The mechanism of thickening has received considerable attention,[32,33] it is believed to be salt formation with the acid end groups of the polyester.

$$2RCOOH + MgO \rightarrow RCOOMgOOCR + H_2O$$

Where R = polyester chain.

If the maturation rate is too fast complete impregnation of the glass during SMC production is made more difficult. Slow maturation causes storage problems. Normally SMC should become stiff enough to handle 24–48 h after mixing, when stored at normal ambient temperatures.

Maturation rate is affected by many factors. Resin molecular weight, acid value, moisture content, and free glycol content can all have a significant influence. Fillers, release agents, and other minor additives must also be carefully selected. Polyesters intended for use in SMC are, therefore, made to extremely tight acid value and molecular weight specifications. Care is taken to control levels of water and free glycol in the products, in order to produce a compound with a consistent maturation rate.

3.8.2. Shrinkage Control

Moulding compounds described so far in this section suffer from one major defect, i.e. shrinkage. All polyesters shrink during use, typically by about 8 % by volume, but this is not a serious problem in hand lay-up operations, etc. However, when making complex articles to fine tolerances with a good surface finish, a moulding material that shrinks is unacceptable and shrinkage control additives must be incorporated.

These additives are thermoplastics which are soluble in the styrene of the polyester resin and may be soluble or partially soluble in the unpolymerised polyester/styrene solution. There effectiveness as shrinkage controllers depends on their insolubility in the cured resin.

A wide range of such thermoplastics have been patented. These include polymers or copolymers from one or more C_1–C_4 alkyl esters of acrylic or methacrylic acid,[34] cellulose acetate butyrate and/or cellose acetate propionate,[34] polyvinyl acetate[35] polycaprolactones,[36] and saturated liquid polyesters, e.g. polypropylene adipate, plus sufficient of a thermoplastic polymer capable of being plasticised by the liquid polyester to prevent exudation of it during the moulding operation.[37]

In general, a highly reactive polyester (i.e. one made using a high proportion of maleic anhydride) is required in formulations including a shrinkage controlling thermoplastic. In some of the patents just described, the preferred polyester is the reaction product of maleic anhydride and propylene glycol.

The mechanism of shrinkage control has been investigated.[38,39] The following sequence of events is believed to occur. Once polymerisation at the moulding temperature has got under way, the thermoplastic, if it was in solution, begins to separate. In some systems, the thermoplastic may already be present as a well dispersed second phase in the initial moulding compound. In either case, the droplets of thermoplastic, which also contain some monomer and possibly polyester as well, expand thermally and thus compensate for the polymerisation shrinkage occurring in the continuous phase.

Microscopic voids are generally but not always found in the dispersed thermoplastic phase when cured, unfilled samples are examined. These are said to result from either migration of monomer from the dispersed droplets or from polymerisation in the droplets with subsequent shrinkage.

The differences in the performance of various thermoplastics as shrinkage control agents are believed to be related to differences in their

TABLE 3.5

NON-SHRINK DMC FORMULATION

Component	Parts by weight
Polyester resin[a]	35·0
Polypropylene adipate (Hexaplas PPA)[b]	7·0
Polyvinyl chloride (Breon 121)[c]	1·5
Benzoyl peroxide paste	1·0
Internal lubricant	2·0
Mineral filler	38·5
Chopped glass fibre ($\frac{1}{4}$ in)	15·0

[a] From maleic anhydride/isophthalic acid/phthalic anhydride/diethylene glycol/propylene glycol in mole ratio 5:1:1:0·5:7·3. Styrene content 40% by weight.

[b] Omission of the polypropylene adipate from the above formulation gave mouldings with a shrinkage of 0·0035 in/in. These showed considerable surface ripple and inferior gloss, thicker sections revealed extensive internal voids.

[c] The polyvinyl chloride is plasticised by the polypropylene adipate.

A moulding tested to British Standard 2782, 1958 showed a shrinkage of 0·001 in/in. The surface of the moulding was free from surface ripple and had a high gloss.

thermal coefficients of expansion, glass transition temperatures and polarities. In the case of thermoplastics, which are soluble in the unpolymerised polyester styrene solution, precipitation during the cure can also contribute an increase in volume.

A typical 'nil shrinkage' DMC formulation is shown in Table 3.5. This example is taken from Ref. 37.

Shrinkage control can be adversely affected when a maturation agent such as magnesium oxide is incorporated in some systems. This is said to occur with the thermoplastic resins described in Ref. 34. Low shrinkage systems of this type which can be chemically thickened are described in a later patent.[40] The preferred thermoplastic is a copolymer with pendant acid groups, e.g. from methyl methacrylate, ethyl acrylate and acrylic acid.

3.8.3. Toughened DMC and SMC

Moulding compounds which give moulded products with improved impact strength and resistance to internal cracking have been obtained by incorporating a butadiene acrylonitrile copolymer with terminal and pendant vinyl groups.[41] DMC containing 5–12·5 parts copolymer per 100 parts polyester and, SMC containing 5–10 parts per 100 parts polyester plus shrinkage control agent, were investigated. The copolymer forms discrete particles in the cured compound producing a structure very similar to that of high impact polystyrene.

3.9.   ESTERCRETE®

This polymer cement comprises a polyester resin, Portland cement and a free radical forming catalyst which is insoluble in polyester resin but readily soluble in water.[42] It is cured by the addition of water (c. 8% by weight).

The preferred catalyst is ammonium persulphate, and the cement is a water repellent stearic acid-coated grade which disperses readily in the resin. On addition of water to the polymer cement, the catalyst dissolves, and calcium hydroxide is released by reaction of water with the cement. In turn, this causes decomposition of the catalyst to produce free radicals which initiate addition polymerisation of styrene with the polyester resin. The overall effect is a gradual transition from the liquid cement to a solid with a low rate of heat evolution. Consequently, most of the shrinkage

® Estercrete is a registered trade name of the Cement Marketing Co. Ltd.

strains created are relieved by plastic flow. Adhesion to normal concrete is extremely good.

Minor modifying additives include finely ground dicarboxylic acids. These improve the shelf life by neutralising any free lime which develops in the cement as a result of water in the resin. Other additives can reduce the settlement of the cement on storage and help redispersion.[43,44,45] The basic formulation and typical properties are shown in Table 3.6.

TABLE 3.6
ESTERCRETE® FORMULATION AND PROPERTIES

Basic formulation	Parts by weight
Polyester resin	60
Portland cement	40
Catalyst	2
Stabilising acid	0·25
Anti-sedimenting agent	0·2
Water to be added for cure	9

Properties	Typical results
Specific gravity	1·45
Gel time at 20°C with the specified water addition	60–90 min
Set time at 20°C	3–5 h
Storage life at 20°C	3 months
Appearance	dark grey liquid

The polymer cement is used in combination with aggregates and sand. These formulations have compressive strengths after 24 h cure of the same order as those of Portland cement mixes after 28 days. Chemical resistance is superior to that of normal concrete and the polymer cement has been used successfully for factory floors, e.g. in breweries, where the traditional product had proved to be unsatisfactory. Other applications include airstrips, road surfacing—particularly on bridges and other areas where a fast hardening topping is required to reduce traffic dislocation, and mortars for pointing and surfacing.

3.10. POLYESTER FOAM

A novel foaming system patented by Farbenfabriken Bayer AG[46] is being used to make wall panels, prefabricated bathroom walls, etc.[47]

The foaming agent is a carbonic acid ester anhydride which is decomposed at normal ambient temperatures to the corresponding carboxylic acid ester and CO_2 in the presence of certain metal salts, e.g. cobalt naphthenate.

The carbonic acid ester anhydrides are prepared by reacting the appropriate carboxylic acid sodium salt with a chlorocarbonic acid ester. A preferred foaming agent is isophthalic acid bis (carbonic acid methyl ester anhydride). This is prepared by the reaction shown below:

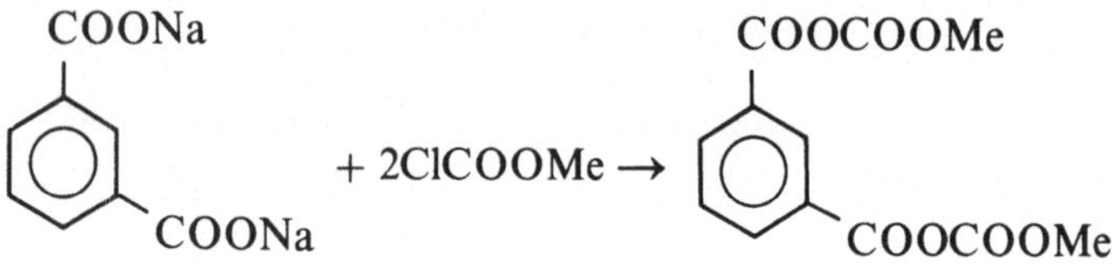

$$
\begin{array}{c}
\text{COONa} \\
\bighexagon \\
\quad\text{COONa}
\end{array}
\;+\; 2\text{ClCOOMe} \;\rightarrow\;
\begin{array}{c}
\text{COOCOOMe} \\
\bighexagon \\
\quad\text{COOCOOMe}
\end{array}
\qquad (18)
$$

When the dianhydride is treated with one of the metal salts listed in the patent,[46] decomposition takes place according to the following equation.

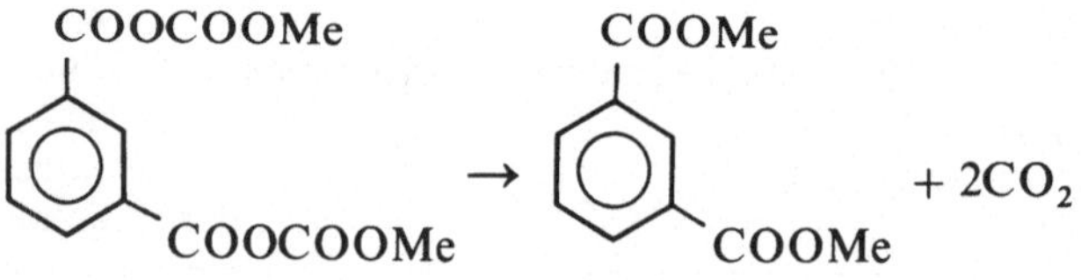

$$
\begin{array}{c}
\text{COOCOOMe} \\
\bighexagon \\
\quad\text{COOCOOMe}
\end{array}
\;\rightarrow\;
\begin{array}{c}
\text{COOMe} \\
\bighexagon \\
\quad\text{COOMe}
\end{array}
\;+\; 2\text{CO}_2
\qquad (19)
$$

If the carbonic acid ester anhydride is derived from an unsaturated acid such as acrylic acid, the acrylic ester produced by the foaming reaction will be copolymerised with the polyester. This could be an advantage, as no residual material will be left to contaminate the foam.

To prepare a foam, the polyester resin which already contains the metal salt is mixed with the foaming agent and a suitable peroxide catalyst. In the building applications already mentioned, the above mix is poured into a mould which has been partially filled with special lightweight aggregate such as expanded forms of glass or slate. When the resin has penetrated the filler, further aggregate is added to fill the mould, which is then closed before foaming starts. The same type of operation can be used to make sandwich panels using GRP, plasterboard, etc., as the skin material.

3.11. FORMULATIONS GIVING REDUCED STYRENE EMISSION

When polyester resins are applied to open moulds by hand lay-up or spray-up techniques some styrene evaporates. Where ventilation is poor, high atmospheric levels of styrene can build up. To date, there is no evidence to suggest that long term exposure to styrene vapour is a health hazard. Nevertheless, a threshold limit value (TLV) of 100 ppm for styrene vapour in the atmosphere is now in force.[48]

To ensure that workshop levels remain below the TLV good ventilation is essential. Most resin suppliers now market so called 'environmental' or 'low emission' grades which give rise to significantly less styrene vapour than unmodified equivalents. However, it should be emphasised that where ventilation is poor, an environmental resin is unlikely to solve a problem of high styrene levels. Efficient ventilation must be the first priority.[49] The environmental grades are a means of reducing levels of styrene still further.

Early environmental resins contained a small amount of paraffin wax, as an evaporation inhibiting additive. This tends to form a film at the resin surface during application and fully separates at gelation. The waxy layer can cause delamination problems, and additives preferred in later formulations include a combination of a high and a low melting point wax,[50] various polymers, e.g. polydibutyl fumarate, polybutyl acrylate,[51] and a combination of an evaporation inhibitor such as paraffin wax and an 'attachment promoter'.[52] This latter patent is particularly interesting. The attachment promoter can be an acyclic, hydrophobic ether or ester with at least two hydrocarbon groups, having at least one double bond in each, or an unsaturated isoprenoid compound, or an ether or ester of such a compound. Examples are linseed oil, dipentene and tri-methylol propane diallyl ether. The preferred wax content is 0·05–0·5%; the preferred attachment promoter addition is 0·1–2%. Examples give interlaminar strengths of compositions with and without the attachment promoter.

Other means of reducing styrene emission include lowering the styrene content of the resin and replacing some or all of the styrene with a less volatile monomer. Both these approaches have disadvantages. Reducing the styrene content would also involve lowering resin molecular weight to achieve a similar viscosity product which, in turn, could lead to inferior cured resin properties.

Styrene is considerably cheaper than any of the higher boiling point monomers which might be considered as replacements, e.g. vinyl toluene, methacrylates, etc. It is, therefore, unlikely that a polyester dissolved in one of these expensive monomers would be economically acceptable, except for very specialised applications. Moreover, the safety in use of alternative monomers, even at lower concentrations than styrene, would have to be established.

If permitted atmospheric styrene levels are drastically reduced in the future, the most likely outcome is a changeover to closed mould techniques, such as resin injection.

3.12. FUTURE TRENDS

Future development will have to keep pace with changes in the cost and availability of basic raw materials. At present polyesters are maintaining their competitive position in the market place. Nevertheless, close attention will have to be paid to resin costs to ensure that this position is not eroded.

Rising petrol prices and the efforts of the motor industry to produce cars with reduced fuel consumption should lead to increased use of polyesters and particularly SMC in car body construction. This will increase the demand for resins with more consistent and easily controlled maturation rates, and thus, may necessitate the development of new methods of maturation. Possibilities in this area include the highly crystalline systems already discussed,[15] or perhaps the use of an external gelling agent such as sodium stearate.[53]

High impact strength, rubber modified SMC and DMC have been briefly mentioned in this chapter (3.8.3). This should be a fruitful area for further development, with automotive applications, particularly, in mind.

The toughening of polyester formulations in general may become necessary as they could be faced with competition from glass reinforced thermoplastics. Possible ways of building improved impact properties into the polyester molecule include the use of butadiene polymers and copolymers with hydroxy or carboxyl functionality.

Legislation specifying resins with reduced tendency to burn for a variety of end products is likely and long overdue. This will almost certainly include a requirement for resin formulations which produce low levels of smoke in a fire.

As mentioned earlier, a drastic reduction in the TLV for styrene vapour in the atmosphere would probably lead to an increase in the use of closed moulds. Fortunately the knowledge and techniques for meeting this eventuality and many others are already available. These should provide a firm base from which further advances can be made.

REFERENCES

1. BOENIG, H. V. *Unsaturated Polyesters: Structure and Properties*, 1964, Elsevier Publishing Company, Amsterdam.
2. Amoco Chemicals Corporation. *How ingredients influence unsaturated polyester properties*, Bulletin IP-70, 1979.
3. British Patent No. 1000534 issued to Chemische Werke Hüls AG, 1965.
4. US Patent No. 3254060 issued to Allied Chemical Corporation, 1966.
5. US Patent No. 3355434 issued to Jefferson Chemical Company, 1967.
6. British Patent No. 1101247 issued to Scott Bader & Company Ltd, 1968.
7. British Patent No. 1375038 issued to Produits Chimiques Pechiney Saint Gobain, 1974.
8. ANON. *Hydrocarbon Process.*, December 1977, pp. 115–16.
9. ALEXANDER, J. *British Plastics Federation Reinforced Plastics Congress*, **39**, 1976, Paper A.
10. SMITH, P. I., MCGARY, C. W. and COMSTOCK, R. I. *22nd SPI Reinforced Plastics/Composites Conf.*, 1967, Paper 1-C.
11. NELSON, D. L. *34th SPI Reinforced Plastics/Composites Conf.*, 1979, Paper 1-G.
12. DAVIS, J. H. and HILLMAN, S. L. *26th SPI Reinforced Plastics/Composites Conf.*, 1971, Paper 12-C.
13. EDWARDS, H. R. *34th SPI Reinforced Plastics/Composites Conf.*, 1979, Paper 4-D.
14. British Plastics Federation, Publication No: 220/1, 1978.
15. British Patent No. 1319243 issued to Scott Bader Company Ltd, 1973.
16. GOTT, S. L., SUGGS, J. L. and BLOUNT, W. W. *28th SPI Reinforced Plastics/Composites Conf.*, 1973, Paper 19-C.
17. Technical data available from Eastman Chemical.
18. British Patent No. 1318517 issued to Scott Bader Company Ltd, 1973.
19. British Patent No. 1190116 issued to Koppers Company, Inc., 1970.
20. LARSEN, E. R. and WEAVER, W. C. *28th SPI Reinforced Plastics/Composites Conf.*, 1973, Paper 2-A.
21. German Patent No. 2819446 issued to Freeman Chemical Corporation, 1979.
22. NAMETZ, R. C. *Ind. Eng. Chem.*, 1967, **59**, 99.
23. BELL, K. M. and CAESAR, H. J. *British Plastics Federation Reinforced Plastics Congress*, 7, 1970, Paper 7.
24. *Flame Retardancy of Polymeric Materials* eds. Kuryla, W. C. and Papa, A. J., 1973, Marcel Dekker, New York.
25. French Patent No. 1336751 issued to Pechiney Ugine Kuhlmann, 1963.
26. GOINS, O. K., ATWELL, R. W., NAMETZ, R. C. and CHANDIK, B. *29th SPI Reinforced Plastics/Composites Conf.*, 1974, Paper 23-B.

27. BONSIGNORE, P. V. and MANHART, J. H. *29th SPI Reinforced Plastics/Composites Conf.*, 1974, Paper 23-C.
28. ATKINS, K. E., GENTRY, R. R., GANDY, R. C., BERGER, S. E. and SCHWARTZ, E. G. *32nd SPI Reinforced Plastics/Composites Conf.*, 1977, Paper 4-D.
29. SKINNER, G. A. *Fire and Materials*, 1976, **1**, pp. 154–9.
30. MOORE, F. W. and CHURCH, D. A. *Paper presented at the International Symposium on flammability and flame retardants, Toronto, Ontario*, 1976.
31. FEKETE, F. *27th SPI Reinforced Plastics/Composites Conf.*, 1972, Paper 12-D.
32. WARNER, N. K. *28th SPI Reinforced Plastics/Composites Conf.*, 1973, Paper 19-E.
33. BURNS, R., GHANDI, K. S., HANKIN, A. G. and LYNSKEY, B. M. *Plastics and Polymers*, December 1975, **43**, pp. 228–35.
34. British Patent No. 1201087 issued to Rohm & Haas Company, 1970.
35. German Patent No. 2104575 issued to Union Carbide Corporation, 1971.
36. US Patent No. 3549586 issued to Union Carbide Corporation, 1970.
37. British Patent No. 1098132 issued to The Distillers Company Ltd, 1968.
38. KROEKEL, C. H. and BARTKUS, E. J. *23rd SPI Reinforced Plastics/Composites Conf.*, 1968, Paper 18-E.
39. ATKINS, K. E., KOLESKE, J. V., SMITH, P. L., WALTER, E. R. and MATTHEWS, V. E. *31st SPI Reinforced Plastics/Composites Conf.*, 1976, Paper 2-E.
40. British Patent No. 1276198 issued to Rohm & Haas Company, 1972.
41. MCGARRY, F. J., ROWE, E. H. and RIEW, C. K. *32nd SPI Reinforced Plastics/Composites Conf.*, 1977, Paper 16-C.
42. British Patent No. 1065053 issued to Cement Marketing Company Ltd, 1967.
43. British Patent No. 1091325 issued to The Distillers Company Ltd, 1967.
44. British Patent No. 1157292 issued to British Resin, 1969.
45. British Patent No. 1292333 issued to BP Chemicals Ltd, 1972.
46. British Patent No. 1160476 issued to Farbenfabriken Bayer AG, 1969.
47. ANON. *Mod. Plastics International*, December 1973, pp. 20–2.
48. Guidance Note EH 15/17—TLV, Health & Safety Executive, HM Factory Inspectorate.
49. EVANS, P. D. *Reinforced Plastics*, **12**, 1976, 364.
50. Europatent No. 941 issued to BASF AG, 1979.
51. Belgian Patent No. 859773 issued to BIP Ltd, 1978.
52. Norwegian Patent No. 783398 issued to AB Syntes, 1979.
53. British Patent No. 1440345 issued to BP Chemicals Ltd, 1976.

Chapter 4

PHENOL–ARALKYL AND RELATED POLYMERS

GLYN I. HARRIS

Advanced Resins Ltd, Cardiff, UK

SUMMARY

The synthesis of polymers by means of the Friedel–Crafts condensation reaction dates from 1881. More recently, a wide variety of aromatic compounds have been polymerised by means of catalysts such as aluminium chloride and stannic chloride. The condensation reactions involve the elimination of small molecules, usually hydrogen halide or methanol.

The polymerisation reactions can be made to produce crosslinked networks. The curing process, however, is liable to produce voids which impair the mechanical properties of the resultant products. The more promising materials are derived from phenol and αα'-dimethoxy-para-xylene; in this case, the phenolic component facilitates crosslinking by either hexamethylene tetramine, or a diepoxide.

These resins have now been developed commercially. They are used in mouldings and laminates because of their good chemical resistance, electrical properties and high-temperature performance. They also have some potential as coatings. Blending with conventional phenol–formaldehyde resins improves the properties of the latter.

4.1. INTRODUCTION

The possibility of making polymers by a Friedel–Crafts condensation dates back to 1881[1] when polymeric materials were reported to be formed by the action of aluminium chloride on benzyl chloride. These polybenzyl polymers were of periodic academic interest[2-5] up until the 1950s when

87

they became the subject of more intense examination as part of the search for polymers to meet increasingly demanding industrial and military applications. In this chapter the progressive polymer development from polybenzyl through the so called Friedel–Crafts resins to the now commercially established phenol–aralkyl resins is reviewed.

4.2. POLYBENZYL POLYMERS

The first serious examination of the polymerisation of benzyl chloride with small quantities of aluminium, ferric and stannic chlorides was reported by Jacobson[2] in 1932. Aluminium trichloride yielded mainly an insoluble hydrocarbon and a small amount of soluble resin having the same empirical formula, namely $(C_7H_6)_n$.

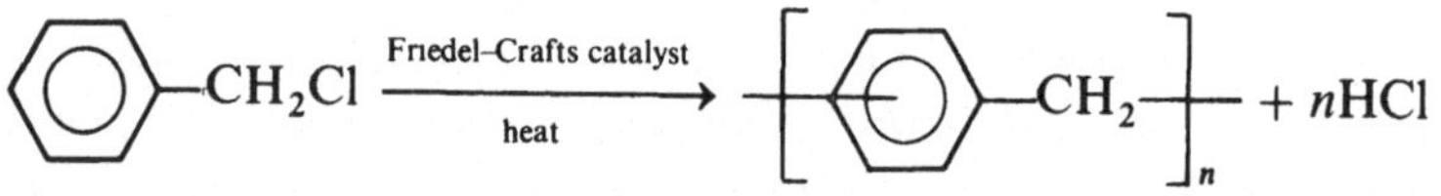

FIG. 4.1.   Preparation of polybenzyl.

The proportions of these were reversed when ferric chloride was employed, while stannic chloride yielded entirely the soluble polymers. The latter had molecular weights in the range 1260 to 2250 and were found to be relatively resistant to attack by oxidising agents. The properties of the soluble polymers were accounted for by assuming that they are hydrocarbon chains containing the group $—C_6H_4—CH_2—$ as the structural unit. The investigations of Shriner and Berger[5] led them to the view that polybenzyl prepared from the polymerisation of both benzyl halides and benzyl alcohol were predominantly linear *para*-substituted polymers. Detailed studies by Haas and co-workers[6] using infrared, X-ray and chemical methods were interpreted in favour of a more complex structure. They proposed that these polymers contain a nucleus of almost completely substituted phenyl rings and have a periphery of pendant benzyl groups. On the basis of the results of oxidative and thermal degradation, a highly branched, globular structure was proposed by Parker.[7] Similar findings were also reported by Ellis and White,[8] while Ellis and co-workers[9] attributed the fluorescence of the polybenzyl formed by treating benzylchloride in the presence of air at room temperature to benzyl substituted

9-phenyl or 9:10-diphenylanthracene. The only recently reported[10] preparation of a linear polybenzyl has been that of treating polybenzyl chloride with aluminium chloride at low temperatures ($-100\,°C$).

The oxidative degradation mechanism for polybenzyl was established by Conley,[11] but more important, Parker[12] at the Royal Aircraft Establishment (RAE) established that a polybenzyl, polymerised with the aid of aluminium chloride, was more thermally stable than the best phenolic resin. By 1964 investigators[13] were, however, resigned to the view that these polymers in an unmodified state were of little practical use, the thermoplastic form being extremely weak and brittle, while the insoluble infusible forms are impossible to process.

4.3. POLYDIPHENYL ETHER RESINS

One of the alternative classes of monomers to benzyl halides, investigated particularly in the USA, was the bis-chloromethylated diphenyl ethers. Doebens[14] and later Geyer and co-workers[15] prepared polyaromatic ethers by heating monomers such as 4,4'-di-(chloromethyl) diphenylether with selected metallic halides. The condensation reaction proceeds with formation of methylene bridges and the concurrent elimination of hydrogen chloride as illustrated in Fig. 4.2. The crosslinking as well as the further polymerisation can occur through continued reaction of the residual halomethyl groups with other diphenyl ether moieties. Since the reaction to form crosslinking methylene bridges is the same as that involved in polymer chain extension, the final condensation polymer is, in practice, nearly always a highly crosslinked resinous polymer insoluble in most

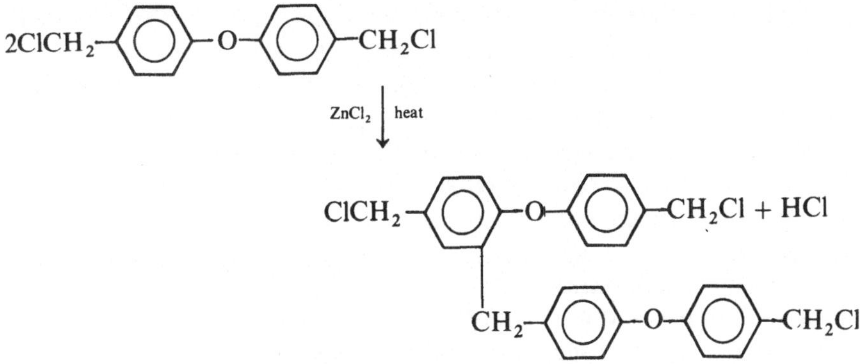

FIG. 4.2.   Condensation of 4,4'-di-(chloromethyl)diphenylether.

polar and non-polar solvents. Only when there is a deficiency of halomethyl groups or reactive sites, are soluble polymeric materials obtained. The practical value of these polymers is, however, limited by the unacceptability of the hydrogen chloride released during the final cure of these polymers. This problem was resolved by Sprengling and co-workers[16] at Westinghouse Electric Corporation by using methoxymethyl- rather than chloromethyl-substituted diphenylether. They prepared monomeric compounds of the formula:

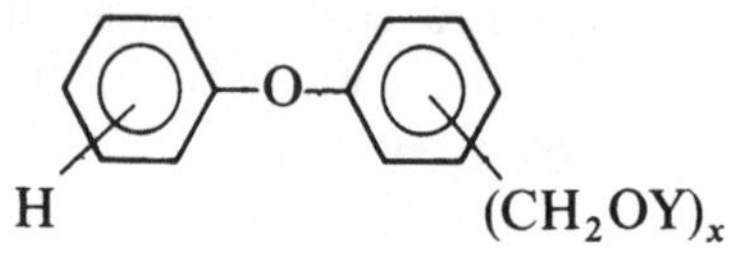

where Y is a lower alkyl substituent, generally methyl, and x has an average value of about 0·8 to about 3·0.

Polymers derived from these monomers were reported to have the general formula shown in Fig. 4.3, where n is an integer in the range one to nine. The reaction is catalysed by Friedel–Crafts catalysts such as aluminium trichloride, zinc chloride, boron trifluoride, etc.; solids

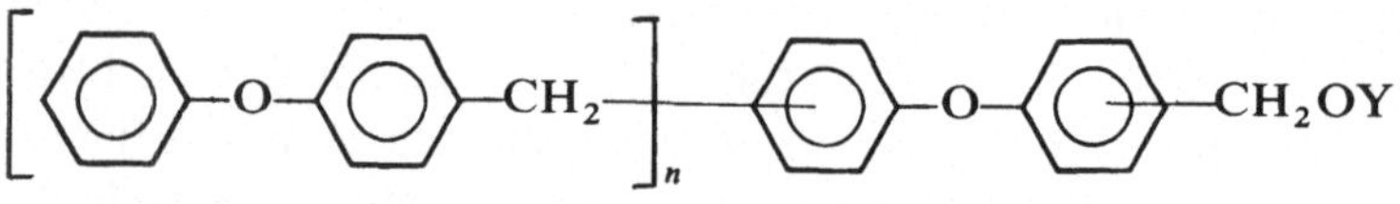

FIG. 4.3.   Condensation of methoxymethyldiphenylether.

such as silica, diatomaceous earths, bentonites, some metals in the form of their organic soluble chelates, notably ferric acetyl acetonate and soluble acids such as p-toluene sulphonic acid. The change from using chloromethyl- to methoxymethyl-substituted diphenyl ethers is also reported to decrease chain branching and consequently reduce the tendency for undesirable gel formation during an early stage in the polymer conversion. These observations were exploited by the Westinghouse Corporation for the manufacture of the first commercially promoted Friedel–Crafts resin. Chloromethyldiphenyl ether and a small amount of diphenyl ether are condensed to give a low molecular weight soluble pre-polymer which is chlorine-free. This is, subsequently, partially condensed with a methoxymethyldiphenyl ether to yield a soluble and fusible resinous composition which readily cures in the presence of suspended silica to give a hard, tough resin and methanol as the sole by-product.

This new resin[17-19] was launched by the Westinghouse Electric

Corporation under the tradename 'Doryl' in about 1961 in the form of varnishes and prepregs. The new materials were claimed to be capable of operating under Class 'H' (180 °C) electrical insulating conditions and to have several other outstanding characteristics, namely:

(1)   bond strength maintained over 3000 h at 250 °C;
(2)   electric strength retention for 10 years at 250 °C in form of insulation tube;
(3)   imperviousness to chemical and solvent attack;
(4)   after initial cure, the material maintains a hornlike structure even after extreme and abrupt temperature changes, and
(5)   considerably lower cost than previous Class 'H' materials, long shelf life and ease of application.

Despite their many attractive properties, the 'Doryl' resins have, in the years since their launching, enjoyed only limited commercial success. This has been, at least partially, due to the curing catalyst being heterogeneous and, so, requiring it to be stirred into the system immediately before use.

4.4.   FRIEDEL–CRAFTS RESINS BASED ON αα'-DICHLORO-p-XYLENE

In the early 1960s efforts were made in the UK to improve the processability and physical properties of polybenzyl polymers by crosslinking them with αα'-dichloro-p-xylene in the presence of a mild Friedel–Crafts catalyst such as stannic chloride. This approach proved relatively successful and Phillips[20] reported the preparation of asbestos felt reinforced composites with these resins. The laminates were found to be very similar in appearance to conventional Durestos® boards with phenolic resins. They differed in two respects.

(1)   *Resistance to concentrated alkali* Even after twelve weeks immersion in 40 % caustic soda the specimens were unchanged, whereas the phenolic laminates were swollen and weak.
(2)   *Strength retention after heat ageing* There was no substantial loss in room temperature flexural strength after exposure to 240 °C for 1000 h, whereas the Durestos grade had virtually no strength after 250 h exposure.

® Durestos is a registered tradename of Turner Bros. Asbestos Co.

The implications of the above investigation proved far reaching both from the synthetic viewpoint and from the performance appraisal and expectations for these resins. Since phenyl, phenylene and phenoxy groups are present in a variety of compounds it encouraged the synthesis of a wide range of Friedel–Crafts resins. This is exemplified in Fig. 4.4 by the condensation of diphenyl with $\alpha\alpha'$-dichloro-p-xylene in the presence of trace amounts of stannic chloride to give a pre-polymer and hydrogen chloride. Crosslinking of the pre-polymer to a hard resinous product was readily achieved by mixing with further quantities of $\alpha\alpha'$-dichloro-p-xylene and heating.

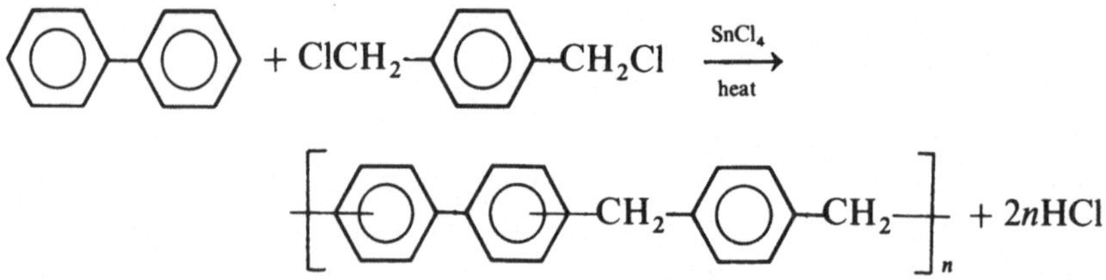

FIG. 4.4. Condensation of diphenyl and $\alpha\alpha'$-dichloro-p-xylene.

The reactivity of a wide range of organic compounds was examined by Phillips[20] who found they fell into three categories summarised in Table 4.1. The first class included dihydric phenols which give a very strongly exothermic reaction with $\alpha\alpha'$-dichloro-p-xylene on heating together in the absence of a solvent. For controlled reactions the condensation of the co-reactants in the first two categories was found to be best undertaken in a 1,2-dichloroethane solution.

The reaction kinetics and structure of the condensation products formed from the reaction of $\alpha\alpha'$-dichloro-p-xylene with benzene[21,22] and diphenylmethane[23,24] in 1,2-dichloroethane were examined by Grassie and Meldrum. They found in the first reaction system that the rate constant for the reaction of the second chloromethyl group in $\alpha\alpha'$-dichloro-p-xylene was twenty times as great as the first rate constant. This was offered as an explanation to account for the observation that the high molecular weight polymers, isolated by gel permeation chromatography, have no detectable chloromethyl groups even though they must undoubtedly exist in low concentrations as intermediates between successive hydrocarbons. It was established in the condensation of $\alpha\alpha'$-dichloro-p-xylene (P_1) with benzene (P_0) that when the concentration of first condensation products, namely p-benzylchloromethylbenzene (P_2) and dibenzylbenzene (P_3), become large then more complex products begin to appear. They are formed

TABLE 4.1

REACTIVITIES OF VARIOUS AROMATIC COMPOUNDS TO αα′DICHLORO-*p*-XYLENE

Qualitative guide to chemical reactivity	*Co-reactant*
1. Reacts without catalyst	Resorcinol
	Catechol
	Diphenyl ether
	Diphenylene oxide
2. Reacts with aid of stannic chloride catalyst	Naphthalene
	Fluorene
	Anthracene
	Mesitylene
	Phenanthrene
	o, *m* and *p* Terphenyl
	Triphenylene
	Diphenylmethane
	Triphenylmethane
	Diphenylsulphone
3. Unreactive with stannic chloride catalyst present	Monochlorobenzene
	o-Dichlorobenzene
	Nitrobenzene
	Benzophenone

by two routes. The first occurs when the concentrations of P_1 and P_0 are still quite high, and products will be formed predominantly by the reaction of P_1 with an earlier product, followed by reaction of benzene (P_0) with the free chloromethyl group as in the sequence shown in Fig. 4.5.

The second route becomes important when an appreciable concentration of products accumulate in the reaction mixture. The pendant chloromethyl

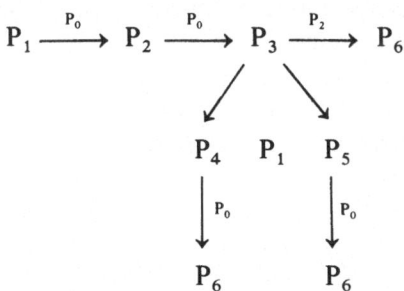

FIG. 4.5. Condensation polymers from αα′-dichloro-*p*-xylene (P_1) and benzene (P_0).

groups are then considered to react increasingly with aromatic nuclei to result, for example, in the formation of seven nuclei products (P_7) by reaction of P_4 and P_5 with P_3 or of P_2 with P_6. Thus the reaction rapidly becomes highly complex with a proliferation of isomers at each level of molecular weight.

The parallel studies[23,24] on the condensation of αα'-dichloro-p-xylene with diphenylmethane established, as might be expected, that there were many common features. In each case there are two kinds of products. Hydrocarbons are again obtained in much larger concentrations than chloromethylated products which can likewise be attributed to the second chloromethyl group in the αα'-dichloro-p-xylene molecule, being very much more reactive than the first. The frequency of occurrence of chloromethyl groups in product molecules was, however, observed to increase from about one in every 24 aromatic nuclei to about one in six nuclei when the initial ratio of αα'-dichloro-p-xylene to diphenylmethane is changed from 1:2 to 4:1. From this it was concluded that the extent of reaction required for gelation would be quite sensitive to the relative initial proportion of reactants and this has been confirmed in practice.

4.5. ORGANOMETALLIC RESINS

The only serious attempt to promote the use of the Friedel–Crafts resins based on αα'-dichloro-p-xylene was undertaken by the RAE for military applications. Phillips[13,20] successfully extended his condensation studies with αα'-dichloro-p-xylene to include some phenyl-substituted organometallic compounds. Octaphenylcyclotetrasiloxane and αα'-dichloro-p-xylene, dissolved in a mixture of o-chlorobenzene and 1,2-dichloroethane, were found to react immediately the system was heated to reflux and catalysed with stannic chloride. Likewise triphenylphosphate and αα'-dichloro-p-xylene were found to react slowly in an o-dichlorobenzene solution.

The impregnation of asbestos felts with siloxane-based resin and their subsequent processing into laminates showed two disadvantages. The impregnated felts were stiff and rather awkward to handle, moreover, during the post cure stage there was persistent blistering so that many laminates were spoiled. By contrast, asbestos felts impregnated with the phosphate resin were soft and rather sticky to the touch, but it was observed that the boards made from them never blistered. Accordingly a number of siloxane/phosphate blends were examined and, with a ratio of 10:3 siloxane/phosphate, a firm and yet flexible impregnated felt was reported to be obtained.[20,25] This cured reasonably quickly to give laminates which did not blister during postcure. The resultant composites when subject to extensive heat-ageing are characterised by:

(1) higher strength, at room temperature and at temperatures up to 400 °C, than commonly found with conventional asbestos reinforced resin composites,

(2) retention of an inorganic residue which prevents gross delamination on prolonged exposure at 300 °C or more, and

(3) no catastrophic failure even on exposure to 350 ° and 400 °C.

Infrared studies[25] show that the methylene bridges in the resins are oxidised to carbonyl groups during ageing at high temperatures and the thermal stability of the siloxane/phosphate laminate is below that of a conventional asbestos felt reinforced polysiloxane resin composite. Moore[26] extended evaluation studies to chemical resistance testing. He found that silica cloth reinforced boards maintain, at room temperature and 90 °C, an overall strength retention of 77·5 % after 5000 h exposure.

The siloxane/phosphate composites were proposed for use in the field of guided weapons because of their ability to withstand 5 h exposure up to 400 °C. The commercial use of these laminates has, however, been inhibited due to handling difficulties and, in particular, to 'sweating' which is caused by the slow escape of some of the trapped hydrogen chloride produced during the condensation. This, coupled with the general processing problems due to the release of corrosive hydrogen chloride gas during the curing reaction, has prevented any serious commercial exploitation of these and related resins based on αα′-dichloro-p-xylene.

4.6.   FRIEDEL–CRAFTS RESINS BASED ON αα′-DIMETHOXY-p-XYLENE

The processing problems associated with the manufacture and cure of Friedel–Crafts resins based on aralkyl chlorides encouraged a search for other classes of monomers. The aralkyl ethers were found, independently in 1964 by Harris[27] at Midland Silicones Ltd and Phillips[28] at the RAE, to be attractive alternative resin intermediates. The condensation of αα′-dialkoxy-p-xylenes with aromatic compounds has the twin advantage over aralkyl chlorides of being a less exothermic reaction and yielding a relatively innocuous alcohol as a by-product. This type of

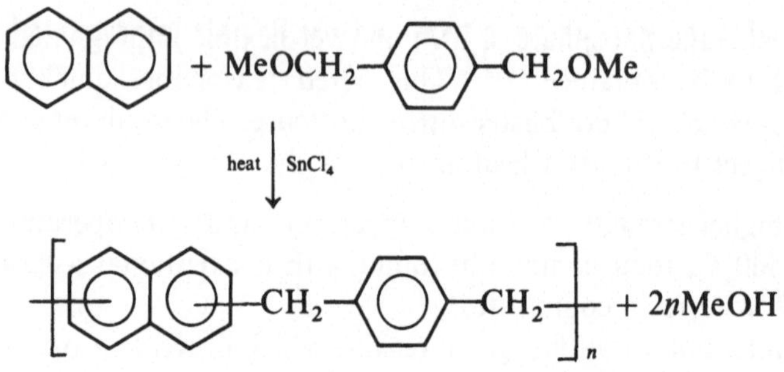

FIG. 4.6. Condensation of naphthalene and $\alpha\alpha'$-dimethoxy-p-xylene.

condensation is illustrated in Fig. 4.6 by the condensation of $\alpha\alpha'$-dimethoxy-p-xylene and naphthalene to yield a pre-polymer and methanol as the by-product.

Harris[29] investigated the reactivity of a wide range of aromatic, heterocyclic and organometallic compounds with $\alpha\alpha'$-dimethoxy-p-xylene. The results of this study are summarised in Table 4.2.

The technique employed was to heat, and subject to slow agitation, a mixture of $\alpha\alpha'$-dimethoxy-p-xylene and the co-reactant together with 0·001 moles of stannic chloride/mole of the aralkyl ether. The reactivity and overall course of the reaction was followed in each case by the elimination of methanol. The $\alpha\alpha'$-dimethoxy-p-xylene reacted with all the aromatic hydrocarbons examined except anthacine and with all the

TABLE 4.2
REACTIVITY TO $\alpha\alpha'$DIMETHOXY-p-XYLENE

Reactive	Unreactive
Anisole	Anthracene
Benzene	Benziminazole
N-benzylaniline	Chlorobenzene
Dibenzylether	Diphenylurea
Diphenyl	N-methylaniline
Diphenylamine	Nitrobenzene
Diphenylether	Tetraphenylsilane
p-Diphenylbenzene	Tetramethyltetraphenyltrisiloxane
Naphthalene	Poly(methylphenylsiloxane)
Octaphenylcyclotetrasiloxane	
Terphenyl	
1,3,5 Triphenylbenzene	
Triphenyl phosphate	

other aromatic compounds except those containing electron withdrawing substituents. By contrast, organosilicon compounds containing phenyl groups were found to be generally unreactive. This is not surprising in view of the cleavage of silicon–aryl bonds previously reported[30,31] during attempts to carry out Friedel–Crafts reactions on methylphenyl-silanes. The exception is octaphenyltetrasiloxane in which the silicon–aryl bonds are probably protected by steric hindrance.

The most promising resins in this group have been found to be those based on the condensation of αα′-dimethoxy-p-xylene and diphenyl ether. This is due to the inherent heat stability of the diphenyl ether moiety and the ease of the condensation reaction between these two intermediates in the presence of a relatively weak Friedel–Crafts catalyst such as stannic chloride. The reaction proceeds smoothly, with the loss of methanol on heating the system to about 165 °C. It is advanced to a suitable application viscosity and the final cure is best achieved by heating after the addition of a stronger Friedel–Crafts catalyst such as ferric chloride.[32] Harris and Edwards[33] studied the effect of functionality on the physical properties of these resins by preparing a series with the αα′-dimethoxy-p-xylene content ranging from 50 to 70 mole %. The changes in flexural strength of glasscloth (Marglass® 7T/methacrylato-silane finish) reinforced laminates based on these resins on heat ageing for up to 1000 h at 250 °C are plotted in Fig. 4.7. All the laminates are characterised by excellent initial flexural strength at room temperature but low strength retention at 250 °C. The increase in functionality is reflected in a faster build-up in high-temperature strength and a more rapid decline in the room temperature strength on heat ageing at 250 °C.

The overall assessment of these resins is that they offer good thermal stability and would be suitable for applications requiring prolonged exposure to temperatures up to at least 200 °C. The slow build-up in high-temperature mechanical strength and the need to use a strong Friedel–Crafts catalyst to achieve a rapid and controlled cure has, unfortunately, prevented their widespread commercial acceptance in reinforced composites.

Other investigations on the above system have been reported by Paxton[34] of Associated Electrical Industries Ltd. He developed a solvent-less varnish which was promoted commercially under the tradename Caldura®. It was recommended for the impregnation of the windings of

® Marglass is the registered trademark of Marglass Ltd.
® Caldura is the registered trademark of Associated Electrical Industries Ltd.

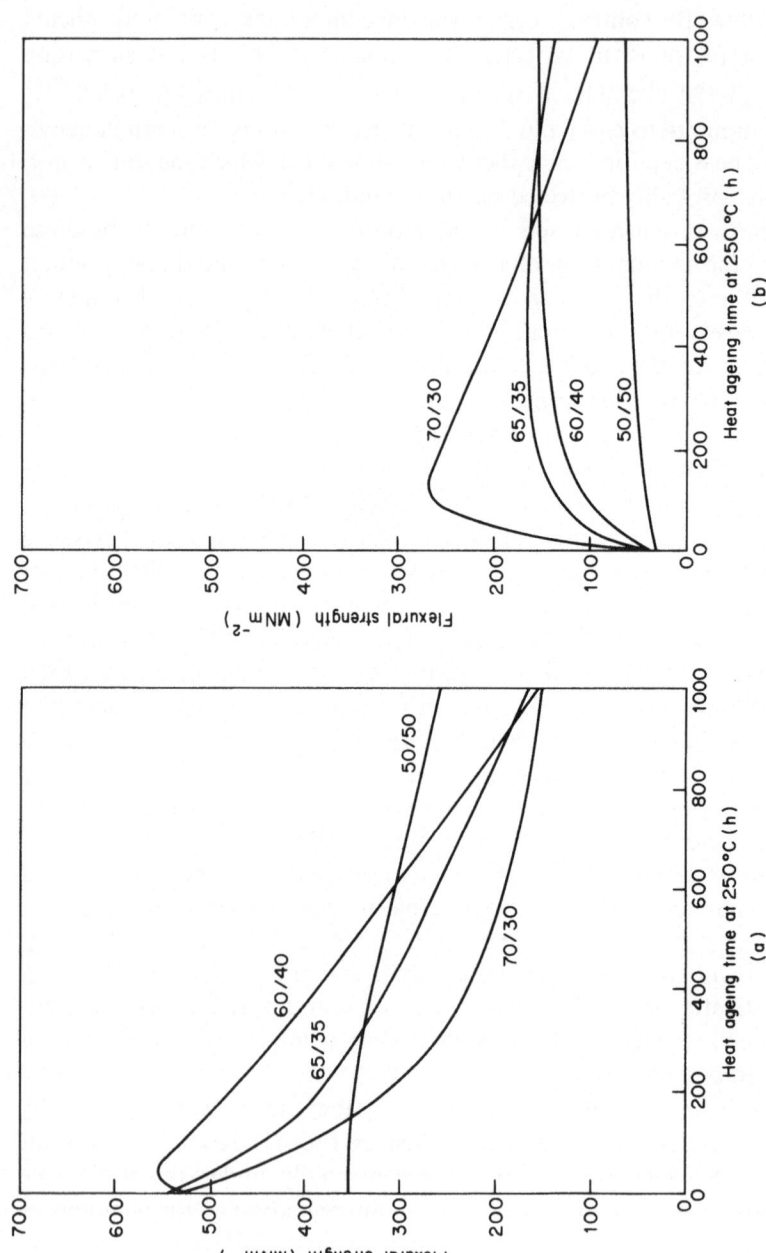

FIG. 4.7. Flexural strengths of glasscloth reinforced αα'-dimethoxy-p-xylene/diphenylether resin laminates on exposure at 250°C for up to 1000h, (a) tested at room temperature, (b) tested at 250°C.

electrical machines designed to operate at temperatures up to 240 °C, and is also claimed to be compatible with polyimide wire enamels and to give good hot bond strengths. The cured resin is also reported to be resistant to aircraft lubricants and hydraulic fluids.

A detailed evaluation of carbon fibre reinforced composites based on resins prepared from the condensation of αα'-dimethoxy-p-xylene with toluene,[35] xylene,[36] terphenyl[35] and diphenyl ether[37] was undertaken by Parker. As was to be expected with matrix resins, which give off volatiles during cure, the composites tended to have high void contents,[38] averaging 4% and ranging from 1–20%. All the mechanical properties were adversely affected by the high void contents. In longitudinal tension, 50–60% of the fibre strength, and 70% of the fibre modulus, was realised in the composite. The transverse tensile strength was low, while the transverse modulus was almost the same as for the resin itself. The interlaminar shear strength was also noticeably affected by void content. Extrapolation of the results to zero void contents showed that potentially the shear strength could approach the values of typical epoxy resin composites. Very little difference was observed between the diphenyl ether and the terphenyl resin based composites except that the former were faster curing. Composites of resins cured with ferric chloride were found to have excellent retention of strength at elevated temperatures after postcure. When boron trifluoride was used as the catalyst, the composites gave better interlaminar shear strength than with ferric chloride, but were poorer in strength retention at high temperature. From Arrhenius plots it was calculated that 75% strength retention lives, for these composites, were one year at 200 °C and 10 years at 150 °C.

The possibility of making unsaturated condensation polymers by the Friedel–Crafts reaction has been reported by Huck and Pritchard.[39] They prepared a low molecular weight polymer containing substantial olefinic unsaturation by heating αα'-dimethoxy-p-xylene and stilbene in the presence of a Lewis acid catalyst. When this condensation polymer was mixed with maleic anhydride in the ratio of one mole of the latter to each mole of retained unsaturation and heated with benzoyl peroxide, a hard intractable resin was obtained. Unfortunately, the high cost of stilbene would appear to preclude a commercial future for this polymer even if the physical properties prove attractive.

4.7. PHENOL–ARALKYL RESINS

The polyaromatic resins synthesised from αα'-dimethoxy-p-xylene, described in the last section, proved far more acceptable for processing but, nevertheless, have at least two serious deficiencies. These are, the relatively low mechanical strength of the reinforced composites at elevated temperatures, and the slow curing rates which preclude their widespread use in the form of moulding compounds. To overcome these shortcomings, Harris[40 41] prepared a further series of resins based, essentially, on the condensation of aralkyl ethers and phenols, so presenting the opportunity of using hexamine (hexamethylene tetramine) or a di- or poly-epoxide to be used as the curing agent. A simple example is illustrated in Fig. 4.8,

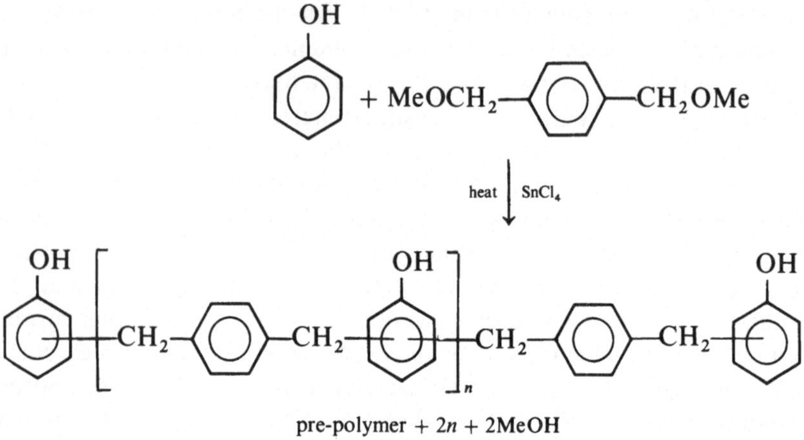

pre-polymer + 2n + 2MeOH

FIG. 4.8. Condensation of phenol and αα'-dimethoxy-p-xylene.

namely the condensation of phenol and αα'-dimethoxy-p-xylene to give a pre-polymer, which on heating with hexamine quickly cures with the release of ammonia as a gaseous by-product. This novel class of thermosets —the phenol–aralkyl resins—was launched commercially in 1970 by Albright & Wilson Ltd under the tradename Xylok. A detailed review of the preparation, structure and properties of this important addition to the class of high performance thermosetting resins represents the subject matter of the remainder of this chapter.

4.8. PREPARATION AND HEXAMINE CURE OF PHENOL–ARALKYL RESINS

The condensation reaction between αα'-dimethoxy-p-xylene and a phenol generally occurs readily in the presence of very low levels of Friedel–Crafts catalyst at temperatures in the range 120°–165 °C. Stannic chloride is a very suitable reaction catalyst at an addition of 0·001 moles/mole of the aralkyl ether. Other catalysts which may be used include ferric, zinc and cupric chloride as well as sulphuric acid and even some acidic clays. In a typical condensation, such as that between phenol and $\alpha\alpha'$-dimethoxy-p-xylene, exemplified in Fig. 4.8, the ratios of reactants are arranged such that the condensation can be taken to completion. The resultant pre-polymers have number average molecular weights, as determined by vapour pressure osmometry in 2-ethoxyethanol, generally in the range 450 to 750. Molecular weight distribution investigations on the same pre-polymers by gel-permeation chromatography indicate a significant spread. Typical values are given in the table below.

%	Molecular weight range
42	>1 500
35	400–1 500
23	<400

Some of the low molecular weight fraction (<400) is due to residual free phenol which is ideally present at a level of 2·5 to 5·0 %. Lower levels of free phenol, as with phenolic novolacs, are undesirable since the rate of gelation with hexamine becomes unacceptably long.

The softening points of the pre-polymers formed from the condensation of phenol and $\alpha\alpha'$-dimethoxy-p-xylene generally fall in the range 70–100 °C and can be influenced by the catalyst used. This suggests that the catalyst, as in phenolic novolacs[42,43] can probably affect the structure by influencing the distribution of *ortho–ortho*, *ortho–para* and *para–para* linkages (Fig. 4.9).

Nuclear magnetic resonance studies[44] directed to measuring the chemical shifts of the methylene protons for the three linkages have only been partially successful. It has proved impossible to separate the *ortho–ortho* and *ortho–para* peaks using pyridine as the solvent. Even the use of additives such as europium compounds and deuterated methanol have failed to bring about splitting. Quantitative measurements of the peaks assigned to the combined *ortho–ortho* and *ortho–para* linkages and to the *para–para* linkage have given values of 71 % and 29 %, respectively.

The results of an extension of this NMR investigation to substitution

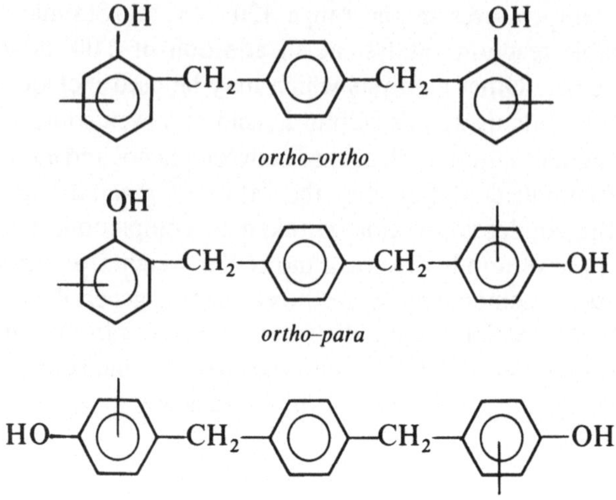

FIG. 4.9. Three possible linkages in phenol–aralkyl pre-polymers.

in a phenol–aralkyl pre-polymer as well as a conventional phenolic and a high '*ortho–ortho*' phenolic are summarised in Table 4.3.

These results indicate that about 65 % of the substitution in the phenol–aralkyl pre-polymer is in the *ortho* position. This is lower than in a high *ortho–ortho* phenolic, but still leaves a significant proportion of *para* positions free to react with hexamine during the curing stage.

The hexamine cure of phenol–aralkyl resins proceeds through di-methylene amino bridges which slowly decompose at elevated temperatures to give the more heat stable methylene and azomethine linkages.[45] The ultimate products are hard brown intractable solids.

TABLE 4.3
SUBSTITUTION IN PHENOLIC NUCLEI

Resin system	Condensation catalyst	Position of substitution on phenolic nuclei	
		% ortho	% para
Phenol/αα'-dimethoxy-p-xylene	Stannic chloride	65	35
Phenol/formaldehyde	Hydrochloric acid	44	56
Phenol/formaldehyde (high *ortho–ortho*)	Zinc acetate	89	11

4.9. NATURE AND SCOPE OF THE REACTION

The condensation reaction by which the phenol–aralkyl resins are formed falls into the broad terms of a Friedel–Crafts reaction and the most likely reaction mechanism is that outlined below.

$$RCH_2OX + SnCl_4 \rightleftharpoons RCH_2^+ + SnCl_4^- \tag{1}$$

$$ArH + RCH_2^+ \rightarrow ArCH_2R + H^+ \tag{2}$$

$$H^+ + XSnCl_4^- \rightarrow HX + SnCl_4 \tag{3}$$

where X = alkoxy

R = divalent aromatic hydrocarbon or hydrocarbonoxy radical.

The aralkyl ether is ionised under the influence of the catalyst, followed by electrophilic substitution of the aromatic nucleus of phenol or any other reactive aromatic compound by the carbonium ion. Evidence in support of this mechanism was provided by an examination of the influence of substituents in the aromatic co-reactant on the ease of reaction. Methyl or phenolic hydroxy substituents, for example, which are electron donating overall, increase the rate of reaction. A dihydroxy-phenol reacts more rapidly than phenol itself, which, in turn, reacts more readily than an aromatic hydrocarbon, such as a diphenyl. By contrast, halogeno, nitro and carboxy groups, which are strongly electron withdrawing either retard or completely inhibit the reaction. Even when sulphuric acid is used as the catalyst there is little reaction between benzoic acid and $\alpha\alpha'$-dimethoxy-p-xylene. Chlorobenzene is so unreactive that it can be used as the solvent if it is necessary to carry out the Friedel–Crafts condensation in solution.

Some indication of the wide range of phenol–aralkyl resins which can be prepared is given by the examples listed in Table 4.4. They can be considered to fall into three classes on the basis of the intermediates used. In the first class are the products of condensation of an aralkyl ether with a single co-reactant, namely mono- or di-hydric phenols. In addition to alkyl-substituted phenols it has been found, rather surprisingly, that the condensation reaction even occurs, if only at a slow rate with carboxy-, chloro- and nitro-substituted phenols as well as with salicyclic acid.

The second class of phenol–aralkyl resins is based on the condensation of an aralkyl ether with two phenols. By using phenols of different functionalities it is possible to formulate resins with predictable physical properties and chemical reactivities. When there is a considerable difference in the

TABLE 4.4
EXAMPLES OF PHENOL–ARALKYL RESINS

		Intermediates
Class	Aralkyl ether	Co-reactant
1.	αα'Dimethoxy-p-xylene	Phenol
	αα'Dimethoxy-p-xylene	Phenol
	αα'Dimethoxy-m-xylene	Phenol
	αα'Dimethoxy-p-xylene	Diphenylolpropane
	αα'Dimethoxy-p-xylene	p-t-Butylphenol
	αα'Dimethoxy-p-xylene	Resorcinol
	αα'Dimethoxy-p-xylene	2-Naphthol
	αα'Dimethoxy-p-xylene	Salicylic acid
	αα'Dimethoxy-p-xylene	2-Chloro-phenol
	αα'Dimethoxy-p-xylene	4-Chloro-phenol
	αα'Dimethoxy-p-xylene	2,4-Dichlorophenol
	αα'Dimethoxy-p-xylene	4-Nitrophenol
2.	αα'Dimethoxy-p-xylene	Phenol and p-cresol
	αα'Dimethoxy-p-xylene	Phenol and 2-hydroxydiphenyl
	αα'Dimethoxy-p-xylene	Phenol and 2-naphthol
	αα'Dimethoxy-p-xylene	Phenol and diphenylolpropane
	αα'Dimethoxy-p-xylene	Phenol and 44' dihydroxydiphenyl
	αα'Dimethoxy-p-xylene	Phenol and salicyclic acid
	αα'Dimethoxy-p-xylene	p-Cresol and resorcinol
	αα'Dimethoxy-p-xylene	Diphenylolpropane and 2-hydroxydiphenyl
3.	αα'Dimethoxy-p-xylene	Phenol and diphenyl
	αα'Dimethoxy-p-xylene	Phenol and diphenyl ether
	αα'Dimethoxy-p-xylene	Phenol and diphenylamine
	αα'Dimethoxy-p-xylene	Phenol and carbazole
	αα'Dimethoxy-p-xylene	Phenol and diphenylsulphone
	αα'Dimethoxy-p-xylene	2-Hydroxydiphenyl and diphenyl ether
	αα'Dimethoxy-p-xylene	p-Cresol and triphenyl phosphate
	αα'Dimethoxy-p-xylene	p-Cresol and octaphenylcyclotetrasiloxane

ease of electrophilic substitution of the phenols, the more reactive species rapidly become poly-substituted and so lead to premature gelation of the system. In such cases, an alternative synthetic procedure can be used. The aralkyl ether and the less reactive phenol are condensed to a pre-determined stage before the second phenol is added and the reaction taken to completion. Resins based on resorcinol and phenol itself have been prepared by this method.

Similar considerations apply to the third class of phenol–aralkyl resins listed in Table 4.4. In this series, the co-reactants are phenols with other classes of aromatic compounds, or alternatively heterocyclic or organo-metallic compounds. The non-phenolic co-reactants are introduced with the object of improving specific properties. Hydrocarbons, such as diphenyl, lower the permittivity and water absorption, whilst compounds such as diphenyl sulphone, diphenylamine or carbazole may be used to introduce sulphone or amino groups and so improve the prospects of bonding to metal substrates. Organometallic intermediates such as triphenylphosphate or octaphenylcyclotetrasiloxane may be introduced into these polymers to improve flexibility and heat stability or to give a polymer which, on complete thermal decomposition, still leaves an inorganic residue.

4.10. COMPARISON OF STRUCTURE AND PROPERTIES

The structure of the simple phenol–aralkyl pre-polymer shown in Fig. 4.8 has a close relationship to that of the phenolic novolacs. The essential difference is that the phenolic nuclei are linked by xylylene rather than methylene bridges (Fig. 4.10). This structural difference is reflected in a number of important physical properties, namely:

(1) chemical reactivity,
(2) thermal stability,
(3) permittivity and loss tangent,
(4) water absorption and
(5) chemical resistance.

The chemical reactivity of phenol–aralkyl resins in terms of their curing rate with hexamine is slower than for phenolic novolacs. This is to be

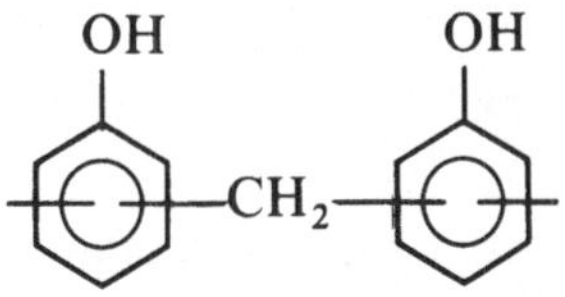

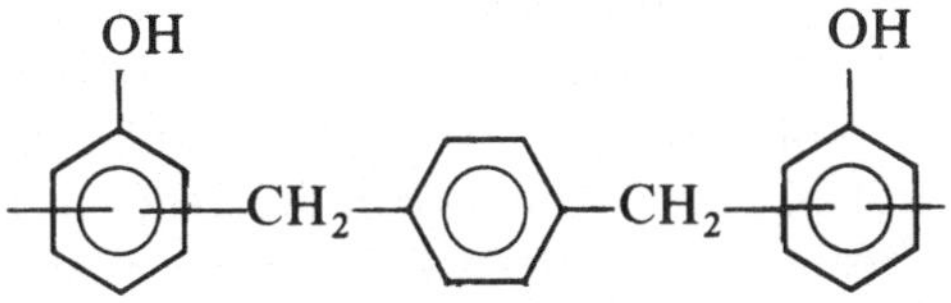

FIG. 4.10. Comparison of resin structures, (a) phenolic resin, (b) phenol–aralkyl resin.

expected since the number of reactive sites/unit length of chain is lower. Fortunately, in laminate manufacture this is not a problem, while the cure times of phenol–aralkyl moulding compounds can be advanced to at least those of phenolics by operating the moulding equipment at a temperature about 20 °C higher.

The phenol–aralkyl resins are far more heat stable than phenolic resins, and this can be explained in the context of the generally accepted oxidative degradation mechanism for the latter proposed by Conley and Bieron[46] given in Fig. 4.11. This involves two stages, namely a primary oxidation

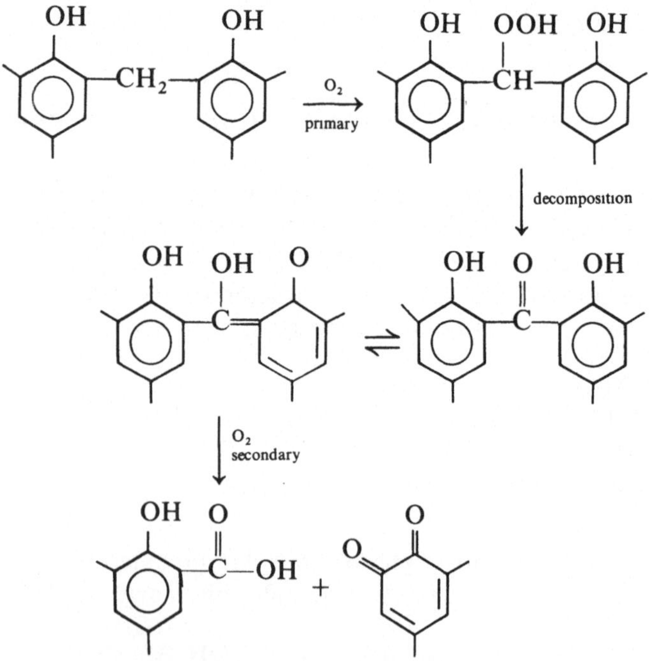

FIG. 4.11.   Oxidative degradation of phenol–formaldehyde condensates.

of the substituted dihydroxydiphenylmethane unit to substituted di-hydroxybenzophenone, followed by a secondary oxidation leading to chain scission. Infrared studies by Harris[47] indicate that oxidation of methylene to carbonyl groups proceeds at a similar rate for both phenolic and phenol–aralkyl resins. The overall difference in heat stability of the two classes of resins is, undoubtedly, due to the relative thermal oxidative stability of the two ketonic species shown in Fig. 4.12. Proton resonance can occur in the first species between two hydroxy groups and one carbonyl,

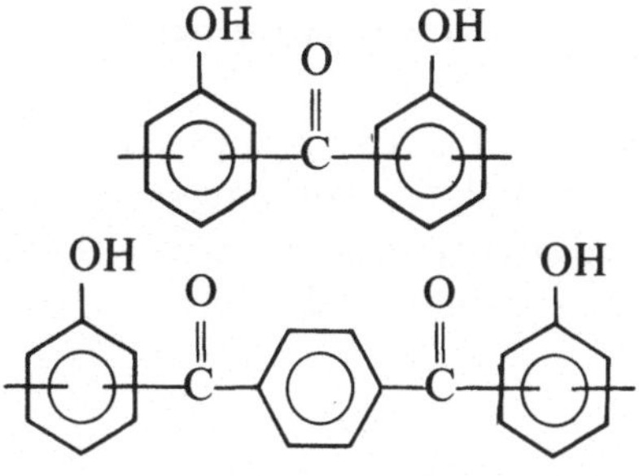

FIG. 4.12.   Primary oxidation products.

while in the ketonic species, derived from the phenol–aralkyl resin, there is only the possibility of resonance between one hydroxy and one carbonyl group. In addition, the presence of a non-phenolic aromatic ring probably acts as an energy sink. The overall effect is that on exposure to high temperatures the phenolic resin develops bonds with a high double bond character and is consequently more susceptible to secondary oxidation than the simple phenol–aralkyl resin.

A third difference between the two classes of resins is in the respect of permittivity and loss tangent. The lower number of phenolic hydroxy groups in the phenol–aralkyl resin is reflected in its less polar character and lower dielectric properties. Typical values for glasscloth reinforced laminates are shown in Table 4.5. The permittivity of the phenol–aralkyl laminate is nearly one unit less than that of the phenolic, and the lower dielectric loss of the phenol–aralkyl resin makes it a better insulating material than the phenolic resin.

The lower polar character of the phenol–aralkyl resin is also reflected in a water absorption which is about one-third that of a phenolic. This,

TABLE 4.5
COMPARISON OF PERMITTIVITY AND LOSS TANGENT

Property	Phenol–aralkyl resin	Phenolic resin
Permittivity at 1 MHz		
dry	4·77	5·73
wet	4·82	5·81
Loss tangent at 1 MHz		
dry	0·011	0·012
wet	0·013	0·038

in turn, leads to better electrical properties and dimensional stability in a damp environment.

The reduced number of hydroxy groups also influences the chemical resistance. Thus, for example, a glasscloth reinforced phenolic laminate is completely destroyed after 100 h immersion in 10 % caustic soda solution at 90 °C. By contrast, a phenol–aralkyl laminate shows a 75 % strength retention.

4.11. UPGRADING PHENOLIC NOVOLAC AND RESOL RESINS

While the difference in chemical structure results in some significant changes in physical properties, the overall close structural relationship between phenolic novolac and phenol–aralkyl resins results in them being completely compatible. The addition of phenol–aralkyl resins to upgrade the performance characteristics of phenolic novolacs is currently being utilised in a number of product areas including moulding compounds, lamp capping cements and friction linings. In each instance, an upgrading in the thermal stability is the most sought after improvement. Some

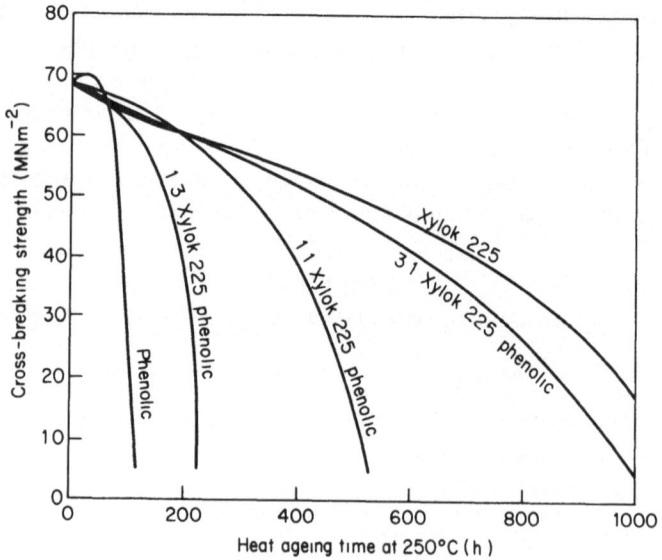

FIG. 4.13. Changes in the cross-breaking strengths with heat ageing of asbestos-filled mouldings, based on phenolic Xylok 225 and blended resins.

indication of the extent to which this can be achieved is provided by the plots in Fig. 4.13.

The mouldings based on 1:3, Xylok 225:phenolic have a useful life at 250°C, about twice that of phenolic mouldings, while mouldings based on 3:1, Xylok 225:phenolic show a strength retention on exposure at 250°C which is slightly inferior to the Xylok 225 mouldings. In addition to thermal stability, the solvent resistance, water absorption, dimensional stability and dielectric properties of the mouldings are improved by the introduction of the Xylok 225.

The upgrading of phenolic resins with phenol–aralkyl resins is not limited to phenolic novolacs. Phenolic resols can in many instances also be upgraded, but the results are far less predictable. It has been found by Harris and Golledge[48] that the addition of a phenol–aralkyl pre-polymer can, depending on the particular resol, upgrade properties such as:

(1)   thermal stability,
(2)   mechanical strength and
(3)   wet insulation resistance.

The indications to date are that the upgrading is only achieved with

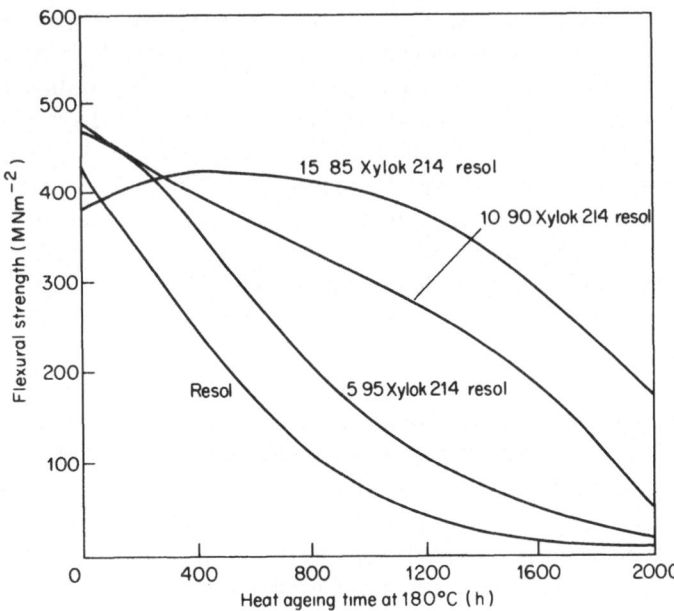

FIG. 4.14.   Flexural strengths of glasscloth reinforced phenol–aralkyl modified phenolic resol laminates.

resols having high methylol contents. An indication of the improvement in thermal stability of glasscloth reinforced resol laminates which can be achieved by the addition of 5, 10 and 15% of phenol–aralkyl prepolymer to such a resol is given in Fig. 4.14.

4.12. HIGH-TEMPERATURE STABILITY AND STRENGTH

In most applications it is the combination of properties offered by a reinforced composite rather than individual ones which influences selection. For high-temperature structural applications it is the combination of thermal stability and high-temperature mechanical strength and strength retention which is generally of paramount importance. The glasscloth reinforced phenol–aralkyl resins fall into this category on prolonged exposure at temperatures up to 250 °C. The results of a comparative study with other classes of post cured glasscloth laminates are shown in Fig. 4.15. The laminates based on a phenolic and a special high-temperature phenolic have a high initial flexural strength and show poor thermal stability. The methyl nadic anhydride-cured epoxy and epoxy-novolac, the silicone and the thermosetting acrylic have relatively poor mechanical strength at 250 °C but quite good strength retention. The laminate based on an American produced polyimide alone has given an overall performance approaching that of the phenol–aralkyl board, namely Xylok 210. That is, high initial flexural strength and a moderately good strength retention, i.e. about 60%. By comparison, the Xylok 210 laminate gives an initial flexural strength at 250 °C of the same order, and over 80% retention after 1000 h, dropping to just over 50% after 2000 h. Even at higher temperatures, the glasscloth reinforced phenol–aralkyl laminates give quite high strength retentions. They fall to 50% of their initial value after 750–1000 h at 275 °C and after about 300 h at 300 °C.

An Arrhenius plot reported by Buchi and Kultzow[49] on glasscloth reinforced phenol–aralkyl (Xylok 210) and a bis-maleimide system (Kerimid® 601) is given in Fig. 4.16. The results show that the phenol–aralkyl composite is the marginally more thermally stable, with a 20 000 h thermal classification temperature for 50% strength retention at about 183 °C. It, consequently, satisfies the Class 'H' insulation requirements of the International Electrotechnical Commission.

® Kerimid is a registered trademark of Societe Des Usines Chimiques Rhone–Poulenc.

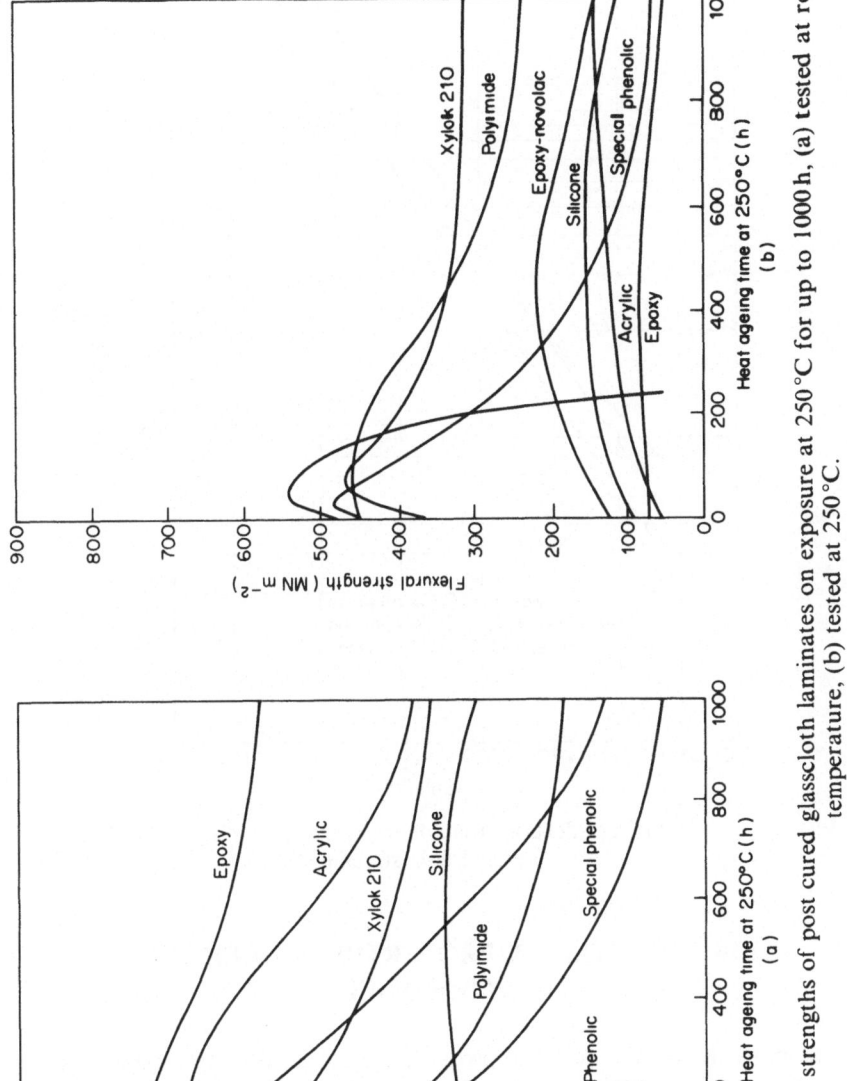

FIG. 4.15.   Flexural strengths of post cured glasscloth laminates on exposure at 250°C for up to 1000h, (a) tested at room temperature, (b) tested at 250°C.

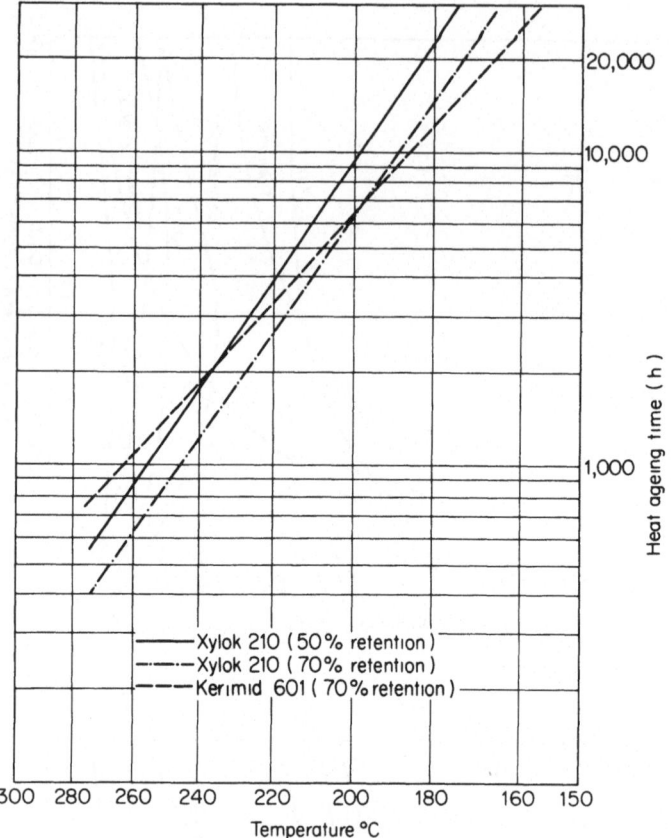

FIG. 4.16.   Retention of flexural strength of glasscloth reinforced Xylok 210 and Kerimid 601 laminates.

4.13.   ELECTRICAL PROPERTIES AND CHEMICAL RESISTANCE

The more important electrical properties of glasscloth reinforced phenol–aralkyl composites and a series of other post cured boards together with some data on chemical resistance are given in Table 4.6. The best combination of electrical properties are given by the phenol–aralkyl, silicone and epoxy composites, but only the phenol–aralkyl and silicone composites give virtually constant permittivity and loss tangent values over the temperature range 20° to 200 °C. The overall chemical, solvent and oil resistance of the phenol–aralkyl board is, however, very much

TABLE 4.6
COMPARISON OF ELECTRICAL PROPERTIES AND CHEMICAL RESISTANCE[a]

Property	Phenolic	Special phenolic	Epoxy	Glasscloth laminates		Silicone	Polyimide	Xylok 210
				Epoxy-novolac	Acrylic			
Electrical								
Electric strength at 20°C (MV m⁻¹)[b]	20	30	26	28	4	26	22	28–33
Electric-strength life at 250°C (h)	144	216	700	430	0	750	144	1000–1400
Insulation resistance (MΩ)	2.5×10^4	3.4×10^4	1.7×10^5	1.3×10^5	4.8×10^5	2.0×10^5	7×10^3	4.0×10^5
Comparative tracking index	195	170	180	240	115	370	180	185
Permittivity at 1 MHz								
dry	5·73	5·77	5·05	5·34	4·42	3·80	5·77	4·77
wet	5·81	5·93	5·08	5·37	4·46	3·88	5·87	4·82
Loss tangent at 1 MHz								
dry	0·0122	0·0261	0·0174	0·0165	0·023	0·003	0·0218	0·0107
wet	0·0380	0·0265	0·0208	0·0182	0·024	0·008	0·0243	0·0130
Chemical resistance								
Change in weight, %, after exposure for 168 h in								
water at 100°C	1·6	—	2·0	0·6	—	0·5	1·2	0·6
10% NaOH at 90°C	destroyed	—	2·8	24·0	—	−12·5	destroyed	2·5
10% HCl at 90°C	−8·2	—	4·9	−5·0	—	−14·3	−12·5	−5·7
30% antifreeze at 90°C	9·4	—	2·0	0·9	—	1·4	0·9	0·9
engine oil at 150°C	1·4	—	0·2	0·2	—	2·3	1·3	0·1
Skydrol 500B at 100°C	1·9	—	11·6	0·0	—	Delam	0·5	0·0
transformer oil at 100°C	0·2	—	−0·5	−0·6	—	1·9	0·0	−0·2
toluene at 110°C	0·9	—	6·7	0·6	—	Delam	−0·2	−0·2
trichloroethylene at 85°C	3·3	—	13·8	1·3	—	Delam	0·2	0·0

[a] Reproduced from Harris, G. I., Edwards, A. G. and Huckstepp, B. G., Friedel–Crafts resins composites for hostile environments, *Plastics and Polymers*, Dec. 1974; by permission of the publishers.
[b] The electric strength was measured on laminates 1·5 mm thick in air, between 38 mm and 76 mm-diameter brass electrodes under a voltage rise of 1 kV s⁻¹. Proof testing was carried out at 11·8 MV m⁻¹ for 60 s when the electric-strength life was being determined. The insulation resistance was measured according to BS 2782 and the comparative tracking index according to BS 3781. The dielectric measurements were made by Lynch's method (ERA 5183).

superior to that of the silicone. Components machined from glasscloth reinforced boards have found applications in a number of chemical plant applications including valve face plates of pumps handling highly corrosive mixtures of chlorosilanes and benzene, end stops of flowmeters and reducing pipe joints handling concentrated hydrochloric acid.

4.14. RADIATION AND FLAMMABILITY RESISTANCE

The radiation resistance of highly aromatic polymers is known to be good,[50,51] so the excellent results obtained at the Rutherford High Energy Laboratory on glasscloth reinforced phenol–aralkyl laminates are, perhaps, not surprising. The changes in flexural strength of these laminates on exposure to 10 000 Mrad of gamma radiation at a dose rate of 2 Mrad h^{-1} are summarised in Table 4.7. The flexural strength shows a fall of 8 %, while the flexural modulus increases by nearly 5 %.

TABLE 4.7

RADIATION RESISTANCE OF GLASSCLOTH REINFORCED PHENOL–ARALKYL LAMINATE

Property	Values after irradiation of				
	0 rad	1 628 Mrad	2 500 Mrad	4 000 Mrad	10 000 Mrad
Flexural strength (MN m^{-2})	605	635	670	675	555
Flexural modulus (GN m^{-2})	31·1	31·2	32·2	31·9	32·6

The natural self-extinguishing character of the phenol–aralkyl composites is quantitatively confirmed by a value of VE-0 with the UL-94 test and an oxygen index of 62 % when examined against ASTM D 2863. Both the phenol–aralkyl laminates and mouldings are also characterised by outstanding low smoke generating characteristics. Results of smoke generation testing, performed in accordance with National Bureau of Standards procedures, indicate times to reach total obscuration in excess of 90 s for the phenol–aralkyl/glasscloth composites. These materials pass current FAA requirements for smoke generation.

4.15. EPOXIDE CURED PHENOL–ARALKYL RESINS

Phenol–aralkyl resins cured with hexamine are suitable for the manufacturing of thin, but not thick section glasscloth reinforced composites, because of the processing difficulties arising from the release of ammonia as a by-product of the cure. This shortcoming led to a search for an alternative mechanism for curing phenol–aralkyl resins. Harris and Edwards[52] found that on heating phenol–aralkyl pre-polymers with selected cycloaliphatic diepoxides and imidazole accelerators, an effective cure can be achieved with the release of only negligible quantities of by-products. The crosslinking occurs between the phenolic hydroxy groups of the pre-polymer and the epoxide groups of the diepoxide, as shown in Fig. 4.17.

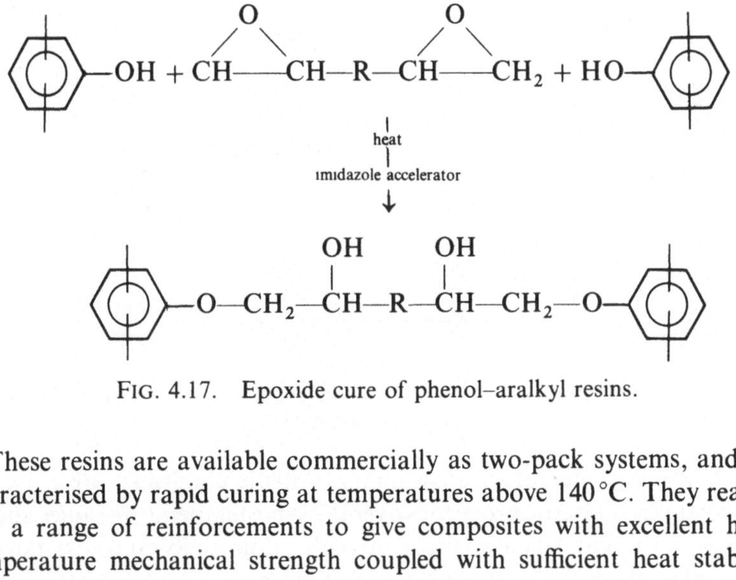

FIG. 4.17. Epoxide cure of phenol–aralkyl resins.

These resins are available commercially as two-pack systems, and are characterised by rapid curing at temperatures above 140 °C. They readily wet a range of reinforcements to give composites with excellent high-temperature mechanical strength coupled with sufficient heat stability to satisfy the requirements of Class 'H' (180 °C) electrical insulation.

The superior high-temperature flexural strength retention of an un-postcured glasscloth reinforced epoxide-cured phenol–aralkyl resin laminate (Xylok 237) relative to postcured epoxide, epoxide novolac and silicone composites is illustrated in Fig. 4.18.

Both the epoxide and epoxide novolac resins used for this comparative study were cured with methyl nadic anhydride, in order to develop maximum high-temperature mechanical strength. The superior strength retention of the glasscloth reinforced phenol–aralkyl laminate up to 180 °C

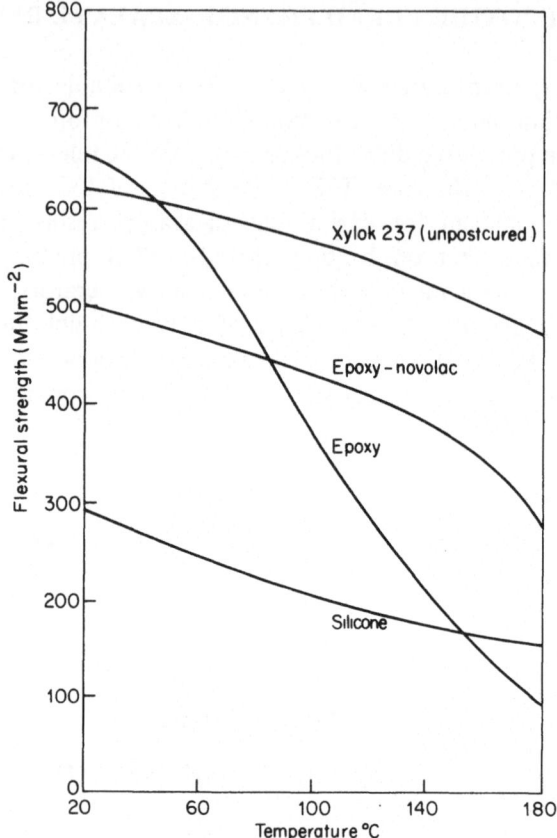

FIG. 4.18.   Changes in flexural strength of various glasscloth reinforced laminates,
with temperature.

reflects the high degree of crosslinking. This is confirmed, also, by the
relative stable dielectric properties over the same temperature range, shown
in Figs. 4.19. The small changes in both permittivity and loss tangent
of the phenol–aralkyl resin, when measured over the temperature range
20–180 °C, contrasts with the large changes occurring above 140 °C for
the epoxide–novolac composite.

A measure of the wetting characteristics of an epoxide-curing phenol-
aralkyl resin (Xylok 237) is indicated by the ease with which it coats
and impregnates Nomex® paper, which is notoriously difficult. Furthermore, it bonds the plies to give composites with the physical properties
summarised in Table 4.8.

® Nomex is the registered trademark of DuPont de Nemours.

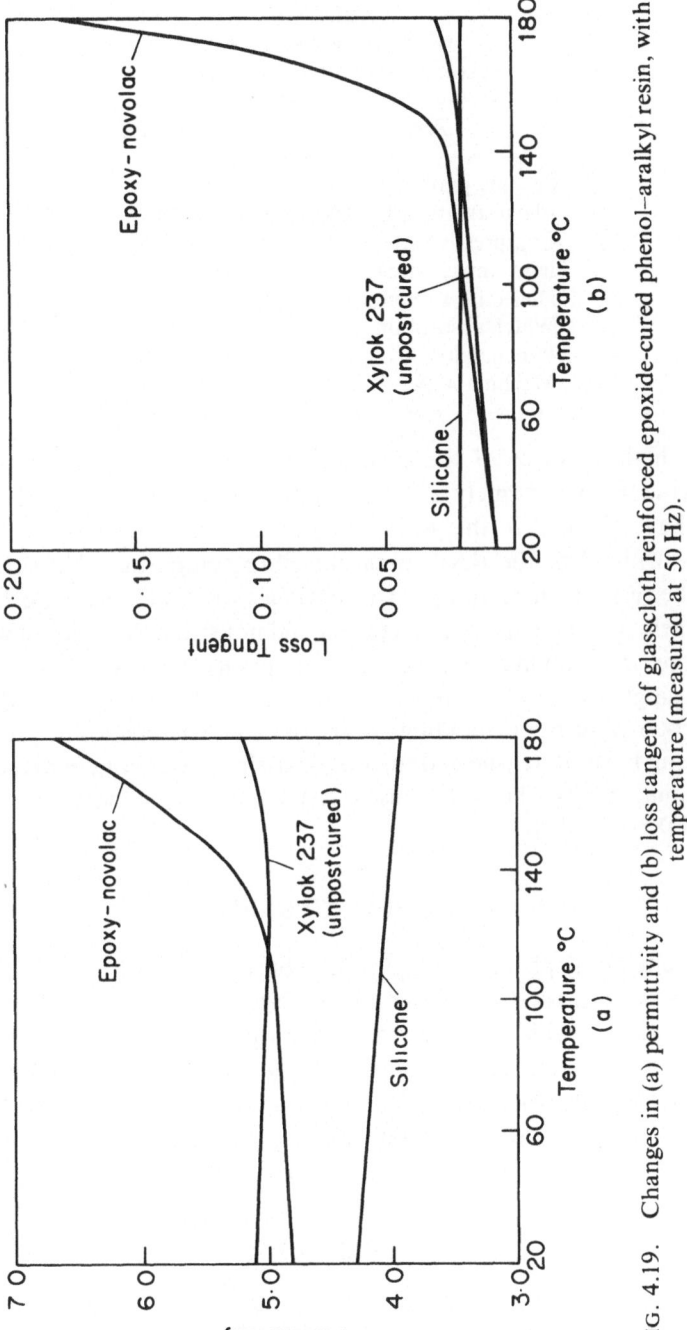

FIG. 4.19. Changes in (a) permittivity and (b) loss tangent of glasscloth reinforced epoxide-cured phenol–aralkyl resin, with temperature (measured at 50 Hz).

TABLE 4.8

PHYSICAL PROPERTIES OF
XYLOK 237/NOMEX 411 LAMINATES

Property	Value
Specific gravity	1·26
Flexural strength (MN m^{-2})	130·98
Compressive strength (MN m^{-2})	288·85
Izod impact strength (J/12·7 mm)	18·53
Dielectric strength (MV m^{-1})	60
Water absorption (%)	0·25
Punchability	Good
Machinability	Excellent

The high T_g value of the glasscloth reinforced epoxy-cured phenol–aralkyl laminates, namely 220 °C, coupled with good coating characteristics have resulted in this resin being extensively evaluated by Phillips and Murphy[53] at the RAE in carbon fibre composites. They reported that the resin in these composites continues to be characterised by very rapid gelation and cure, and develops good hot strength retention without postcure. This ability to produce high quality laminates after only a short, single stage cure is unusual in the carbon fibre field. The adhesion to carbon fibre is outstanding, while the room temperature properties including flexural, tensile and interlaminar shear strength are higher than is attainable with the widely used bisphenol A precondensate system, Shell DX10/BF$_3$400.

4.16.   CONCLUSION

To conclude, the phenol–aralkyl resins constitute an attractive new class of high performance thermosetting resins which can be used alone or in combination with phenolic resins. The opportunities offered, particularly by the second use, will undoubtedly grow as demands for easily processable high performance composites continue to increase. In that respect the phenol–aralkyl resins offer to phenolic resins—the old 'war horse' of the plastics industry—a new lease of life.

REFERENCES

1. FRIEDEL, C. and CRAFTS, J. M. *Bull. Soc. Chim. Fr.*, 1881, **25**, 52.
2. JACOBSON, R. A. *J. Amer. Chem. Soc.*, 1932, **54**, 1513.

3. BEZZI, S. *Gaz. Chim. tal.*, 1936, **66**, 491.
4. FUSON, R. C. and MCKEEVER, C. H. *Org. React.*, 1942, **1**, 63.
5. SHRINER, R. L. and BERGER, L. *J. Org. Chem.*, 1941, **6**, 305.
6. HAAS, H. C., LIVINGSTON, D. I. and SAUNDERS, M. *J. Polym. Sci.*, 1955, **15**, 503.
7. PARKER, D. B. V. *European Polymer J.*, 1969, **5**, 93.
8. ELLIS, B. and WHITE, P. G. *J. Polym. Sci.*, 1973, **11**, 801.
9. ELLIS, B., WHITE, P. G. and YOUNG, R. N. *European Polymer J.*, 1969, **5**, 307.
10. KENNEDY, J. P. and ISAACSON, R. B. *J. Macromol. Chem.*, 1966, **1**, 541.
11. CONLEY, R. T. *J. Appl. Polymer Sci.*, 1965, **9**, 1107.
12. PARKER, D. B. V. *RAE Tech. Note Chem.*, 1956, 1284.
13. PHILLIPS, L. N. *Trans. Plastics Inst.*, 1964, **32**, 298.
14. DOEBENS, J. D., US patent No. 2,911,380, 1959.
15. GEYER, G. R., HATCH, M. J. and SMITH, H. B., US Patent No. 3,316,186, 1967.
16. SPRENGLING, G. R., US Patent No. 3,405,091, 1968.
17. ANON. *Modern Plastics*, Oct. 1962, 110.
18. ANON. *Chem. Eng.*, 1962, **69**, 73.
19. ANON. *Dow Diamond*, 1963, **26**, 5.
20. PHILLIPS, L. N. *RAE Tech. Report*, 1963, CPM 3.
21. GRASSIE, N. and MELDRUM, I. G. *European Polymer J.*, 1969, **5**, 195.
22. GRASSIE, N. and MELDRUM, I. G. *European Polymer J.*, 1970, **6**, 499.
23. GRASSIE, N. and MELDRUM, I. G. *European Polymer J.*, 1970, **6**, 513.
24. GRASSIE, N. and MELDRUM, J. G. *European Polymer J.*, 1971, **7**, 17.
25. NIXON, B. *RAE Tech. Note*, 1964, CPM 71.
26. MOORE, B. J. C. *RAE Tech. Report*, 1965, 65161.
27. HARRIS, G. I. and MARSHALL, H. S. B., UK Patent No. 1,099,123, 1968.
28. PHILLIPS, L. N., UK Patent No. 1,094,181, 1967.
29. HARRIS, G. I. (unpublished work).
30. EVISON, W. E. and KIPPING, F. S. *J. Chem. Soc.*, 1931, 2774.
31. YAKUBOVICH, A. YA. and MOTSAREV, G. V. *J. Gen. Chem., Moscow*, 1953, **23**, 1414.
32. HARRIS, G. I., UK Patent No. 1,127,122, 1968.
33. HARRIS, G. I. and EDWARDS, A. G. (unpublished work).
34. PAXTON, J. C. *Procurement Executive, Ministry of Defence, D. Mat. Report*, 1973, 190.
35. PARKER, B. M. *RAE Tech. Report*, 1972, 72029.
36. PARKER, B. M. *RAE Tech. Report*, 1973, 72220.
37. PARKER, B. M. *RAE Tech. Memorandum Mat.*, 1975, 217.
38. PARKER, B. M. *RAE Tech. Report*, 1976, 76051.
39. HUCK, P. J. and PRITCHARD, G. *J. Polym. Sci.*, 1973, **11**, 3293.
40. HARRIS, G. I. and COXON, F., UK Patent No. 1,150,203, 1969.
41. HARRIS, G. I. *Brit. Poly. J.*, 1970, **2**, 270.
42. BENDER, H. L., FARNHAM, A. G. and GUYER, J. W., US Patent No. 2,464,207, 1949.
43. FRASER, D. A., HALL, R. W. and RAUM, J. L. J. *J. Appl. Chem., London*, 1957, **7**, 676.
44. HARRIS, G. I. and EDWARDS, A. G. (unpublished work).
45. ZINKE, A. *J. Appl. Chem., London*, 1951, **1**, 257.
46. CONLEY, R. T. and BIERON, J. F. *J. Appl. Polym. Sci.*, 1963, **7**, 103.

47. HARRIS, G. I. (unpublished work).
48. HARRIS, G. I. and GOLLEDGE, J., UK Patent Application No. 15953/77, 1977.
49. BUCHI, G. and KULTZOW, R. Paper presented at *SPE Antech. Conference on High Performance Plastics, Cleveland*, October, 1976.
50. ALEXANDER, P. and CHARLESBY, A. *Proc. R. Soc., A*, 1955, **230**, 136.
51. BAUMAN, R. G. and GLANTZ, J. A. *J. Polym. Sci.*, 1957, **26**, 397.
52. HARRIS, G. I. and EDWARDS, A. G., UK Patent No. 1,305,551, 1973.
53. PHILLIPS, L. N. and MURPHY, D. J. *RAE Tech. Memorandum Mat.*, 1979, 322.

Chapter 5

INITIATOR SYSTEMS FOR UNSATURATED POLYESTER RESINS

V. R. KAMATH and R. B. GALLAGHER

Pennwalt Corporation, Buffalo, New York, USA

SUMMARY

The basic principles of free radical initiators, their important parameters and new developments as they relate to the selection of effective initiator systems for curing unsaturated polyester resins are reviewed.

Peroxide and azo type initiators continue to be the most popular initiators for the fibre reinforced plastics (FRP) industry. Certain peroxides can be decomposed by specific promoters to provide good cures at ambient temperatures. The mechanism of activation and particularly the detrimental effects of excess promoter are described. The results of studies relating the structure of ketone peroxides to their reactivity are reviewed.

The kinetics of thermal decomposition of initiators are related to important processing parameters such as shelf-life, rate of cure and temperature of cure. For easy reference, specific initiators used for curing in the $100°–160°C$ temperature range are tabulated along with their $10h$ $t\frac{1}{2}$ temperatures. The advantages of peroxyketals over popular peroxyesters, and their sensitivity to acidic fillers are discussed. The advantages of peroxide blends in terms of fast cure systems and especially the long shelf-life of blends containing unique azo compounds are described. Results of studies directed towards the reduction of residual styrene in cured resin are reviewed.

In addition to azo and peroxide compounds, other sources of free radicals are reviewed including novel carbon–carbon initiator and photoinitiator systems based on both UV and visible light.

121

5.1. INTRODUCTION

The fibre reinforced plastics (FRP) industry continues to enjoy steady growth in spite of cyclical economic movement. Over the period 1967–1976 the total yearly production grew from 544 million pounds (247 000 tonnes) to 1·86 billion pounds (844 000 tonnes), and is projected to enjoy significant growth in the future. Contributing to such growth is the development of new and improved initiators for unsaturated polyester resin curing.

Unsaturated polyester resins consist of an alkyd polymer containing vinyl unsaturation dissolved in a vinyl monomer, most commonly styrene. Curing or crosslinking of these liquid resins into useful solid thermosets occurs readily by a chain addition type of copolymerisation reaction. Free radicals capable of initiating the cure reaction can be generated in the resin system by the decomposition of organic initiators. Other methods such as photoinitiation and thermal initiation have also been used to some extent in specialised areas of application.

Organic initiators can be defined as compounds which, when subjected to heat, decompose to yield highly reactive free radical species. Organic peroxides and azo compounds are commonly used as initiators, but are often incorrectly referred to as catalysts. In the traditional sense catalysts are substances which are not consumed as they speed (promote) a chemical reaction. In the case of unsaturated polyesters, the initiating molecule, i.e. peroxide or azo, decomposes during the cure reaction. Therefore, the term 'initiator' is more appropriate and as such will be used throughout the discussion.

The phenomenal growth in the plastics industry during the past 10 to 20 years has led to a considerable research effort in the area of free radical initiators. As a result, a number of new initiators as well as improved initiator formulations have been developed. To meet the current demands of highly diverse applications and processing techniques, more than 50 different organic peroxide and azo compounds in over 75 different formulations are produced and offered commercially.

In the past, the accepted practice has been for the compounder or fabricator to depend solely on the resin and/or initiator producer to select the proper initiator for his application. However, in the rapidly advancing technology of today's polyester industry it has become imperative that the user fully understands the properties and characteristics of initiators. With such knowledge, he will not only be able to use initiators more efficiently in existing processes but will also be able to select effective initiators for newly developing processes. To fill this need, basic principles of free radical initiators, their important parameters and new developments as they relate to selection of the most effective initiator system for specific processes and applications, are reviewed.

5.2. ORGANIC INITIATORS

The chemical bond in an organic compound can cleave either symmetrically or asymmetrically, as shown below.

$$X—Y \rightarrow X\cdot + \cdot Y$$
$$X—Y \rightarrow X^+ + {}^-{:}Y$$

In the first case neutral radicals result and in the second case the result is charged species called ions. The term free radical refers to an atom or group of atoms with unpaired electrons. Such radicals are generally reactive species which can react very rapidly to initiate free radical chain reactions.

By supplying sufficient energy, in the form of heat or radiation, chemical bonds can be broken in a symmetrical manner to produce free radicals. In order to serve as a convenient source of free radicals by thermal homolysis, a potential initiator must contain bonds which will break at relatively low temperatures. The energy required to break a chemical bond is called bond energy and is usually expressed as kJ mole^{-1} (kcal mole^{-1}). To break a normal carbon–carbon bond having a bond energy of 380 kJ mole^{-1} (90 kcal mole^{-1}), temperatures of 350 to 550 °C are required to attain sufficient thermal excitation. Several types of compounds have bond energies of 105–150 kJ mole^{-1} (25–35 kcal mole^{-1}). These bonds can be broken in the temperature range of 50 to 150 °C, i.e. convenient temperatures for processing polyester resins.[1]

By far the most common initiators used in the curing of unsaturated polyester resins are organic peroxides (R—O—O—R') and aliphatic azo compounds (R—N≡N—R') since they decompose thermally to produce free radicals at convenient temperatures. Peroxides decompose by initial cleavage of the oxygen–oxygen bond to produce two free radicals.

$$R—O—O—R' \xrightarrow{\Delta} RO\cdot + \cdot OR'$$

Azo compounds, on the other hand, decompose by simultaneous cleavage of two carbon–nitrogen bonds to produce nitrogen and two alkyl radicals.

$$R—N{=}N—R' \xrightarrow{\Delta} R\cdot + N_2 + \cdot R'$$

TABLE 5.1
COMMERCIAL INITIATOR CLASSIFICATION

Type	Structure	$10\,h\ t\frac{1}{2}$ range ($^\circ$C)
Diacyl peroxide	$\underset{\displaystyle R\overset{\displaystyle O}{\overset{\|}{C}}}{}$—OO—$\overset{\displaystyle O}{\overset{\|}{C}}$R	20–75
Dialkyl peroxydicarbonate	RO$\overset{\displaystyle O}{\overset{\|}{C}}$—OO—$\overset{\displaystyle O}{\overset{\|}{C}}$OR	49–51
tert-Alkyl peroxyester	R$\overset{\displaystyle O}{\overset{\|}{C}}$—OO—R′	49–107
OO-tert-Alkyl,O-alkyl monoperoxycarbonate	R—OO—$\overset{\displaystyle O}{\overset{\|}{C}}$OR′	90–100
Di-(tert-alkyl) peroxyketal	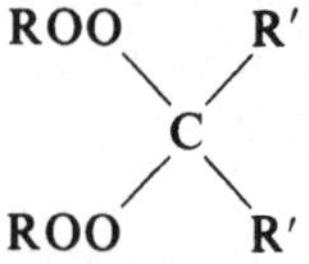 ROO, R′ ╲C╱ ROO, R′	92–115
Di-tert-alkyl peroxide	R—OO—R′	117–133
tert-Alkyl hydroperoxide	R—OO—H	133–172
Ketone peroxide	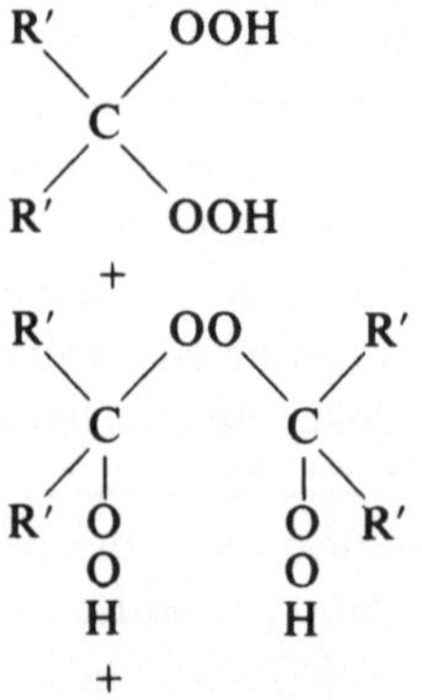 R′ OOH ╲C╱ R′ OOH + R′ OO R′ ╲C╱ ╲C╱ R′ O O R′ ╲ ╲ O O ╲ ╲ H H + other structures	
Symmetrical azonitrile	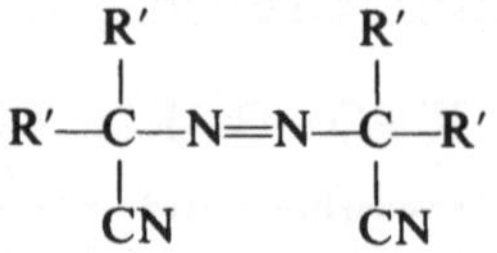 R′—C—N=N—C—R′ with R′ above and CN below each C	50–65
Asymmetrical azonitrile	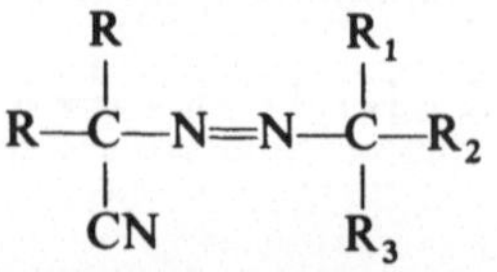 R—C—N=N—C—R₂ with R above first C, CN below; R_1 above second C, R_3 below	55–96

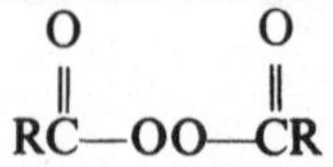

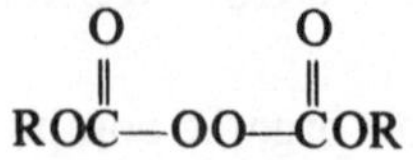

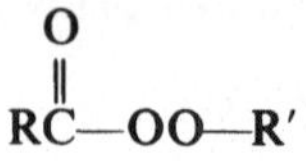

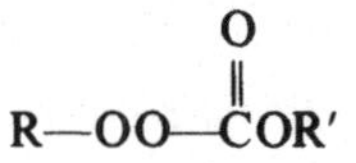

Certain classes of organic peroxides can also be decomposed by specific promoters at temperatures well below their normal thermal decomposition temperature. In this case radicals are produced by an oxidation–reduction (redox) mechanism which represents an asymmetrical decomposition of the peroxide. This phenomenon is utilised not only in curing unsaturated polyester resins at room temperature but also in accelerating their cure at elevated temperatures. Azo compounds, due to their high degree of chemical inertness, are ordinarily not suited for use in promoted cure systems.

The ten major different initiator types or classes that are used commercially to cure unsaturated polyester resins are shown in Table 5.1.

The particular initiator selected for a specific resin formulation will depend primarily on the shelf-life which is required of the formulation

TABLE 5.2

REDOX INITIATORS FOR AMBIENT TEMPERATURE CURE SYSTEMS

Chemical name	Structure
Dibenzoyl peroxide	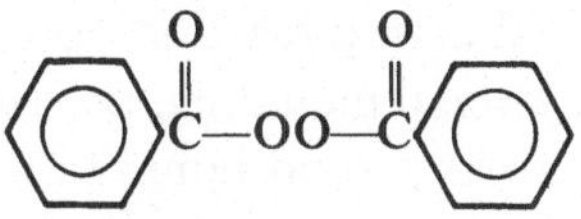
Cumene hydroperoxide	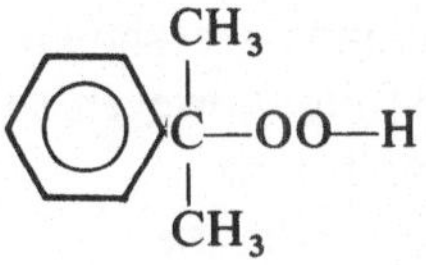
2,4-Pentanedione peroxide	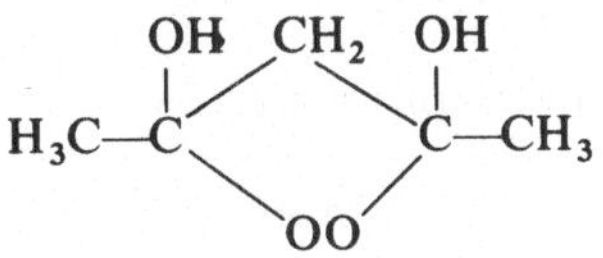
Methyl ethyl ketone peroxide (mixture of structures)	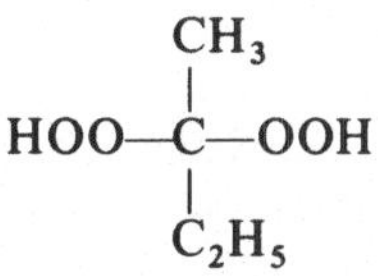

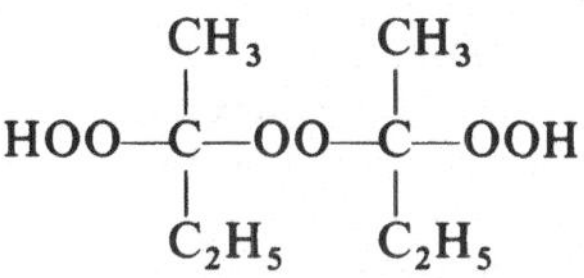

as a function of the moulding process. Shelf-life is defined as the useful life time of a resin/initiator mixture. After an initiator has been added to a resin system the curing process starts as soon as the initiator breaks down into active free radicals. At some early stage in the curing process the resin reaches a gelled state. At this stage the crosslinked network will have formed sufficiently such that flow of the resin is no longer possible. The time taken to reach the point at which the resin has lost its useful flow properties is defined as the shelf-life of the resin/initiator system. Naturally, other factors such as rate of cure desired, temperature of cure, thickness of moulded parts, effect of fillers and initiator cost will enter into selection of the most cost effective initiator system.

Initiators which decompose at low temperatures, either thermally or through the action of promoters, may be selected for those processes which require little or no shelf-life. Contact moulding processes such as hand lay-up, spray-up and resin injection make use of initiators (Table 5.2) which can be activated with promoters at ambient temperatures. After the initiator has been added to the resin these processes require only sufficient time for the resin to wet out the glass reinforcement before the desired curing reaction occurs. Processes such as continuous pultrusion, rotational moulding, pressure bag moulding, etc., require shelf-lives ranging from several hours to a few days. For these processes one may select initiators which decompose at low to intermediate temperatures. Specific compounds are shown in Table 5.3. These initiators will give the desired shelf-life; however, some may require storage under controlled temperature conditions.

Certain processes make use of compound formulations which require

TABLE 5.3

INITIATORS FOR ELEVATED TEMPERATURE CURE SYSTEMS: $10 \text{ h } t_{\frac{1}{2}}$, $80 \,^{\circ}\text{C}$

Name	$10 h \text{ } t_{\frac{1}{2}}$ $0.2 M$ (benzene) ($^{\circ}C$)	Approximate moulding range ($^{\circ}C$)
Di-2-phenoxyethyl peroxydicarbonate	41	70–120
Bis(4-t-butylcyclohexyl) peroxydicarbonate	42	70–120
2,5-Di(2-ethylhexanoylperoxy)-2,5-dimethyl-hexane	67	85–125
Dibenzoyl peroxide	73	90–130
t-Butyl peroxy-2-ethylhexanoate	73	90–130
2-t-Butylazo-2-cyanopropane	79[a]	100–140

[a] Trichlorobenzene solvent.

TABLE 5.4
INITIATORS FOR ELEVATED TEMPERATURE CURE SYSTEMS: $10\,h$, $t\frac{1}{2} > 80\,°C$

Name	$10\,h\ t\frac{1}{2}$ $0\cdot2\,M\ (benzene)$ $(°C)$	Approximate moulding range $(°C)$
2-t-Butylazo-2-cyanobutane	82^a	100–145
1,1-Di(t-butylperoxy)-3,3,5-trimethylcyclohexane	92	130–160
1,1-Di(t-butylperoxy)cyclohexane	93	130–160
1-t-Butylazo-1-cyanocyclohexane	96^b	135–165
OO-t-Butyl, O-isopropyl monoperoxycarbonate	99	130–160
t-Butyl peroxybenzoate	105	135–165
Ethyl 3,3-di(t-butylperoxy)butyrate	111	140–175
Dicumyl peroxide	115	140–175

^{a} Mineral spirits solvent.
^{b} Trichlorobenzene solvent.

a long shelf-life ranging from one week to several months. These include sheet moulding compound (SMC), bulk moulding compound (BMC) and the many recent variations of this form of compound. For these processes one must select initiators which show a high degree of thermal and chemical stability. Specific examples of suitable initiators are shown in Table 5.4.

Very often moulders find it advantageous to combine a thermally stable initiator with a small concentration of a less stable initiator. Such a combination allows faster cure times with some sacrifice in shelf-life. A detailed discussion of recent developments in this area is included in this chapter.

5.3.   INITIATION BY REDOX MECHANISMS

The technique of producing radicals by electron-transfer oxidation–reduction reactions is frequently used since peroxides which have high thermal stability may be used to provide a source of free radicals at low temperatures. The mechanism is based on the reaction of certain classes of peroxides with specific promoters. The terms 'promoter' and 'accelerator' are often used interchangeably to refer to compounds which are added to a resin system to speed the decomposition of the peroxide into free radicals. The terms are different, however, in that promoters can function alone while accelerators are only used in combination with promoters.

Peroxides which are readily susceptible to activation include ketone peroxides, hydroperoxides and diacyl peroxides (see Table 5.2). Ketone peroxides and hydroperoxides both contain the —OOH grouping and their decomposition is commonly promoted by transition metals such as cobalt. For enhanced activity, doubly-promoted systems incorporating an amine such as dimethylaniline (DMA) or diethylaniline (DEA) in combination with the cobalt are often employed. In this case the amine compound acts as an accelerator.

Methyl ethyl ketone peroxides (MEKP) are by far the most widely used organic peroxides for room temperature curing applications. Over 90 % of the spray-up processes are conducted by combining polyester resin and MEKP in the spray gun at the mould site. The reasons for the popularity of ketone peroxides are many. They are low in cost and, since they are liquids, they can be easily and accurately metered. They are readily soluble in polyester resin and are available in a range of activities. Finally, MEK peroxides can be handled safely when the manufacturer's recommendations are followed.[2]

Ketone peroxides are normally promoted with transition metals such as cobalt. A simplified redox mechanism for this reaction is shown below.

(A) $R{-}OOH + Co^{+2} \rightarrow RO\cdot + OH^- + Co^{+3}$

(B) $R{-}OOH + Co^{+3} \rightarrow ROO\cdot + H^+Co^{+2}$

(C) $RO\cdot + Co^{+2} \rightarrow RO^- + Co^{+3}$

Equations (A) and (B) show that either oxidation state of the transition metal can decompose the peroxide. Transition metals can also react with free radicals to convert them to ions as shown in eqn. (C). The significance of this reaction is that a free radical is destroyed and, therefore, can no longer initiate curing reactions. This is why there is an optimum level of transition metal that should be used for promoting efficient cures. This effect is illustrated in Table 5.5 where cobalt naphthenate promoter levels were increased with three different levels of methyl ethyl ketone peroxide in a typical orthophthalic type polyester resin. The data show that, at all three peroxide levels, the cure (Barcol hardness) begins to deteriorate at excessively high levels of cobalt. The data also illustrate that if one wants faster cures without sacrificing cure characteristics, the best approach is to increase peroxide concentration.[3]

A significant development has recently occurred which will affect all

TABLE 5.5

ROOM TEMPERATURE CURING OF POLYESTER RESIN: EFFECT OF PEROXIDE
AND PROMOTER CONCENTRATION

Cobalt naphthenate (6·0% cobalt) (phr)	0·5% MEKP		1·0% MEKP		1·5% MEKP	
	Gel (min)	Barcol hardness	Gel (min)	Barcol hardness	Gel (min)	Barcol hardness
0·025	79·4	45–50	39·2	45–50	27·5	45–50
0·050	46·0	45–50	19·9	45–50	14·3	45–50
0·100	22·1	45–50	12·0	45–50	8·9	45–50
0·200	15·7	45–50	8·6	45–50	6·4	45–50
1·000	8·3	45–50	4·2	45–50	3·4	45–50
2·000	7·2	40–45	4·0	40–45	3·3	40–45
3·000	6·0	35–40	3·6	35–40	2·9	35–40

Gel times are measured at 30 °C in general purpose orthophthalic resin using the SPI exotherm procedure.

MEKP manufacturers as well as end users in the United States. The US Department of Transportation (DOT) has initiated regulations which will prohibit the shipment of MEKPs which contain more than 9% active oxygen. Active oxygen may be defined as the extra oxygen (bound to oxygen) in peroxy compounds as compared to their nonperoxy analogues. The major reason for this move is to provide an extra margin of safety during shipment and storage of ketone peroxides. Previously the most commonly used MEKP formulation was a clear solution of ketone peroxides in dimethyl phthalate with about 11% active oxygen. This type of product is widely used for hand lay-up and spray-up applications since its activity provides suitable working times. In accordance with the DOT regulations, MEKP producers will dilute these products such that they will contain less than 9% active oxygen in their commercial form.

What does this mean to the fabricator who normally uses 1 part peroxide per hundred parts of resin (1 phr) on a weight basis? Slower gel and cure times will be observed since the diluted products contain less active oxygen (9% v. 11%). To obtain equivalent activity the fabricator can either increase the level of peroxide and/or promoter in the resin system or select a more reactive MEKP formulation.

The cure rate is directly related to the concentration of both the peroxide and the promoter. It is clearly shown in Table 5.6 that a 25% increase in the concentration of the 9% MEKP (1·25 phr) will provide reactivity

TABLE 5.6

POLYESTER RESIN CURE DATA: 11 % V. 9 % ACTIVE OXYGEN MEK PEROXIDES

	% Active oxygen	Concentration (phr)	30°C Cure data		
			Gel (min)	Cure (min)	Peak (°C)
MEK peroxide A	11	1·00	27·2	34·3	171
	9	1·00	32·5	41·9	167
	9	1·25	25·8	39·3	167
	9	1·50	21·7	29·2	176
MEK peroxide B	11	1·00	14·3	23·1	173
	9	0·75	28·2	40·8	169
	9	1·00	18·7	28·5	169
	9	1·25	14·3	23·2	175

Cure activity measured in general purpose orthophthalic resin containing 0·065% of 6% cobalt promoter, using the SPI exotherm procedure.

equivalent to that previously obtained with 1·0 phr of the 11% MEKP. Alternatively, the concentration of peroxide can be maintained at the 1 phr level while the concentration of promoter is increased.

A third approach would be to select a more reactive MEKP formulation. As shown in Table 5.6, peroxide B is inherently more reactive than peroxide A at equivalent concentrations. Thus 0·75 phr of peroxide B—9 % will give cure activity which is approximately equivalent to that obtained with 1·25 phr of peroxide A—9 %.

Although the active oxygen content of MEK peroxides does affect their cure activity, other factors such as chemical structure and presence of certain diluents can also cause significant effects. A number of studies have been directed towards establishing the relationship between the chemical structure of methyl ethyl ketone peroxides and their reactivity in various resin systems. Ketone peroxides are generally manufactured by reacting a ketone, $R—C(=O)—R$, with hydrogen peroxide (HOOH) under acidic conditions. This results in a mixture of peroxy structures. The structures of the major peroxide constituents of commercial MEKPs are shown below.

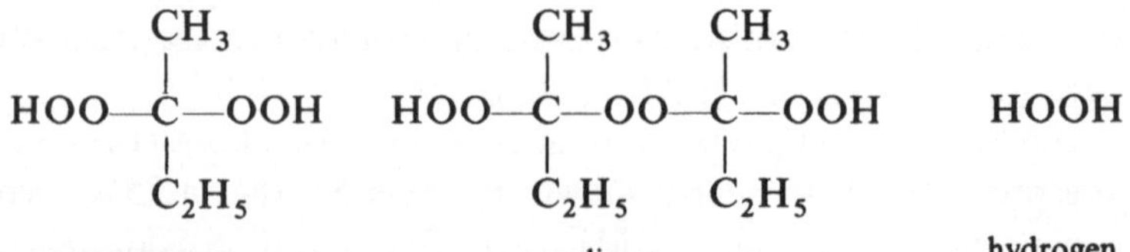

$$\underset{\text{monomer}}{HOO—\overset{\overset{\textstyle CH_3}{|}}{\underset{\underset{\textstyle C_2H_5}{|}}{C}}—OOH} \qquad \underset{\text{dimer}}{HOO—\overset{\overset{\textstyle CH_3}{|}}{\underset{\underset{\textstyle C_2H_5}{|}}{C}}—OO—\overset{\overset{\textstyle CH_3}{|}}{\underset{\underset{\textstyle C_2H_5}{|}}{C}}—OOH} \qquad \underset{\substack{\text{hydrogen}\\\text{peroxide}}}{HOOH}$$

Recent studies[4,5] have shown that the cure characteristics of commercial MEK peroxides are highly dependent upon the presence and the amount of each of these components. As shown in Table 5.7, peroxides A and B, which contain a relatively low level of dimeric species, give significantly faster cures in conventional orthophthalic and isophthalic resins. In vinyl ester resins however, peroxide C, which contains a relatively high level of dimeric species, gives significantly faster cures than peroxides A and B. These relationships hold true for resin systems which are singly promoted with cobalt or doubly promoted with cobalt and dimethylaniline.

TABLE 5.7

MEK PEROXIDE: EFFECT OF COMPOSITION AND STRUCTURE ON CURE ACTIVITY IN THREE RESIN TYPES

1 phr MEK peroxide	% dimer	% H_2O_2	30°C Resin cure time (min)		Vinyl ester
			Orthophthalic	Isophthalic	
A	6·4	0·3	41	34	50
B	6·2	1·9	29	31	56
C	19·3	1·3	43	24	33

Peroxides contain 9% active oxygen.
Cure times were measured using the standard SPI exotherm procedure.

Comparison of the performances of peroxides A and B illustrates that the gellation and cure of conventional polyester resins is augmented by the presence of hydrogen peroxide while, on the other hand, hydrogen peroxide contributes little or nothing to the cure of vinyl ester type resins.

The curing of highly reactive vinyl ester resins must be done carefully using standard MEKP systems. Thick laminates must be built up slowly to avoid high exotherms which can cause warping and cracking. It was recently shown[4] that an MEKP with a high proportion of monomeric species is capable of curing thick vinyl ester parts with lower exotherms and avoids splitting and warping which is characteristic of cures by MEKPs which have a high proportion of the dimeric product. Cumene hydroperoxide has also been found to give good cures with low exotherms in highly reactive vinyl ester resins.[6] These initiators can be used to increase productivity since thick flanges or thick flat laminates can often be fabricated in one step with minimal warpage when they are used in place of standard MEKP systems.

Due to the hazards involved in their manufacture, all commercial MEK peroxides are manufactured in the presence of solvents such as dimethyl phthalate. Special ketone peroxide formulations are also available for specific applications. Fire resistant formulations are desirable because of their inherent safety features which often qualify a fabricator for reduced fire insurance rates. Low assay or half strength formulations are desirable for spray-up applications since they can be more accurately metered.

Diacyl peroxides, especially benzoyl peroxide (BPO), are also frequently used in redox type promoted systems. In this case the promoter is most often a tertiary aromatic amine, such as dimethylaniline. A simplified reaction mechanism is illustrated below.

$$\begin{bmatrix} & \underset{\substack{| \\ CH_3}}{\overset{CH_3}{\underset{|}{C_6H_5-N^+}}} -\overset{O}{\overset{\|}{OC}}-C_6H_5 \end{bmatrix} \quad {}^-\overset{O}{\overset{\|}{OC}}C_6H_5 \rightarrow$$

Benzoyl peroxide   Dimethylaniline

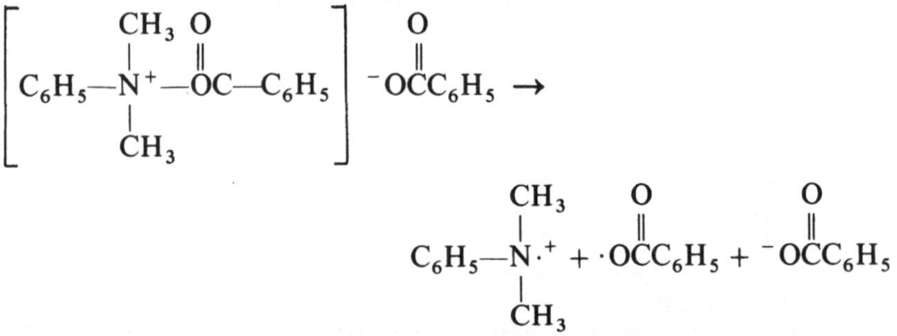

$$C_6H_5-\underset{\substack{| \\ CH_3}}{\overset{CH_3}{\overset{|}{N}}}{}^{\cdot+} + \cdot O\overset{O}{\overset{\|}{C}}C_6H_5 + {}^-O\overset{O}{\overset{\|}{C}}C_6H_5$$

Here again, as in the case of ketone peroxides, overpromotion is detrimental to the cured properties of the polyester resin.

Although not as widely used as the ketone peroxides, BPO is used extensively for applications such as autobody and marine repair kits and for the fast growing mine bolt adhesive industry.

Nonseparating paste and pumpable liquid formulations of BPO have been developed for safer and easier handling and to improve the solubility of BPO in the resin. Typical products consist of 25–55 % BPO dispersed in a plasticiser. Fire resistant formulations which do not require yellow precautionary labelling are also available.

5.4. INITIATION BY THERMAL DECOMPOSITION OF PEROXIDE AND AZO COMPOUNDS

5.4.1. Rate of Decomposition—Half-Life

Peroxides and azos decompose over a range of temperatures. The rate at which they decompose increases with increasing temperature. Half-life

$(t_{\frac{1}{2}})$ is a convenient means of expressing the rate of decomposition at a particular temperature and is defined as the time required to decompose 50 % of the initiator.

Generally, the 10 h half-life temperature is used as a reference point. This is the temperature at which 50 % of the initiator has decomposed in 10 h. As a rule, 10 h $t_{\frac{1}{2}}$ temperatures are universally accepted as the best means of comparing the relative reactivity of different initiators. At any given temperature the most active initiator would be the one with the lowest 10 h $t_{\frac{1}{2}}$ temperature.

The half-life of an initiator is experimentally determined by measuring its concentration as a function of time at a constant temperature. This is usually done in dilute solution using solvents such as n-decane, mineral spirits, etc. Specific analytical procedures vary for each initiator. In the case of peroxides, these generally involve the liberation of iodine from sodium iodide under controlled conditions and then titration of the liberated iodine with standard sodium thiosulphate. Infrared and vapour phase chromatography may also be used. Gas evolution techniques are particularly suited to azo compounds.

When using the 10 h half-life temperature or the kinetic measurements for the decomposition of an initiator, it is important to consider the limitations of the data. The rate of decomposition of peroxide initiators varies markedly with the solvent used. Thus, while for the comparison of kinetic data for the decomposition of various initiators in a given solvent it is useful to place them in order of reactivity, the reported kinetics may not always apply under actual polymerisation conditions.

In general the decomposition kinetics of azo compounds show little solvent or concentration dependence, i.e. they are much less susceptible to any form of induced decomposition. Hence, they may be used advantageously in unsaturated polyester formulations containing fillers, pigments, etc., which can interfere with the efficient decomposition of peroxide initiators.

5.4.2. Decomposition Kinetics

The thermal decomposition of most peroxides in inert solvents has been found to follow first order kinetics, i.e. the rate is directly proportional to the initiator concentration.

$$\text{Rate} = -\frac{d[I]}{dt} = k[I]$$

The term k is the first order rate constant and may be determined from the slope of the straight line obtained by plotting the logarithm of residual peroxide concentration (in moles of peroxide group/litre of solution) as a function of time. By integration between limits it is possible to define a half-life for the decomposition at a particular temperature.

$$t_{\frac{1}{2}} = \frac{0 \cdot 693}{k}$$

The activation energy (E_a) may be obtained by studying the variation of reaction rate with temperature. By measuring k at different temperatures the value of E_a can be obtained from the standard Arrhenius equation,

$$k = A e^{-E_a/RT}$$

which expresses the rate constant in terms of two parameters: activation energy and a pre-exponential term A which is related to the probability of the reaction occurring. The activation energy may be determined from the slope of the line obtained by plotting log k as a function of the reciprocal of absolute temperature. In general, the higher the activation energy, the greater the change in rate of decomposition/unit change in temperature. Thus, initiators with a lower activation energy will give a more uniform rate of decomposition over a broader temperature range. In practical terms this means that these initiators can be used over a wide temperature range without substantial loss in efficiency.

5.5. SELECTING INITIATORS FOR CURING AT ELEVATED TEMPERATURES

Common initiator types used in curing polyester resins in the temperature range of 90° to 160°C are shown in Tables 10.3 and 10.4. Processing techniques which utilise these compounds include pultrusion, compression, injection, transfer moulding, etc. These processes generally require a shelf-life ranging from several days to several months at ambient temperatures.

    As described earlier, the primary consideration in selecting an initiator is generally the shelf-life which is required as a function of the moulding process. Other important parameters for initiator selection include the rate of cure desired, the temperature of cure and part thickness.

    Rate of cure is a function of initiator half-life, initiator concentration and reaction temperature. Data in Table 5.8 show that at a given mould temperature the cure time decreases with decreasing 10 h $t_{\frac{1}{2}}$ temperature. In general, one can select a moulding temperature which will be within

TABLE 5.8

EFFECT OF INITIATOR HALF-LIFE ON CURE CHARACTERISTICS AT 121 °C

Initiator	$10h\ t_{\frac{1}{2}}$ 0·2 M (benzene) (°C)	Cure time (min)	Peak temperature (°C)
2,5-Di(2-ethylhexanoylperoxy)-2,5-imethyl-hexane	67	2·1	186
2-t-Butylazo-2-cyanopropane	79[a]	2·7	212
1,1-Di(t-butylperoxy)cyclohexane	95	3·4	216
t-Butyl peroxybenzoate	105	4·7	230

[a] Trichlorobenzene solvent.

Initiators compared on equal equivalents corresponding to 1·0 wt. % t-butyl peroxybenzoate.

Cure time and peak exotherm temperatures were measured according to the SPI exotherm procedure.

the optimum range for a given initiator by adding 40° to the 10 h $t_{\frac{1}{2}}$ temperature, in °C. Using this as a starting point, it is then usually necessary to experimentally determine the optimum process temperature for the specific moulding compound. For any given resin/initiator system there is an optimum temperature at which the resin can efficiently utilise the free radicals formed by the initiator. When radicals are generated too fast they tend to recombine to form inactive products or they can terminate growing polymer chains (primary radical termination) resulting in lower molecular weight polymers and a loss in physical properties. This is illustrated in Table 5.9 where a temperature of 121 °C is near optimum for t-butyl peroctoate. Above 121 °C cure times are not shortened and cure characteristics begin to deteriorate, while below 121 °C too much time is required to complete the cure.

Thickness of the part being moulded can play an important role in selecting an initiator. As part thickness increases, heat transfer becomes slower and cure time will usually increase because it takes longer to reach reaction temperature, especially in the centre of the part. High-temperature initiators coupled with reduced heat transfer produce higher peak exotherms which can lead to cracking or warping of thick sections. Use of a low-temperature initiator will result in reduced cure times, however, for optimum efficiency one must be careful not to exceed the optimum temperature range. High mould temperatures may also cause pregel at the mould surface which can cause rejected parts. The best answer is to use higher concentrations of a low-temperature initiator

TABLE 5.9

EFFECT OF TEMPERATURE ON INITIATOR EFFICIENCY

Bath temperature (°C)	1·66% t-Butyl peroxy-2-ethylhexanoate[a]		
	Gel time (min)	Cure time (min)	Peak exotherm (°C)
107	2·2	3·3	195
121	1·2	2·4	198
135	1·0	2·4	191
149	1·0	2·4	180

Cure activity measured in general purpose ortho-phthalic resin using the SPI exotherm procedure.

[a] Often referred to as t-butyl peroctoate.

at reduced mould temperatures. This is particularly true, for example, in the pultrusion process.

Pultrusion is the only process for converting continuously and automatically glass reinforcement and resin into finished products and, thus, it is enjoying steady growth and increasing importance in the FRP industry. Although the process can be conducted over a wide range of temperatures, temperatures around 100 °C are most popular. This is especially true when moulding profiles with thick cross sections are used, since lower exotherms are obtained at lower moulding temperatures. Initiators which give acceptable cure rates at these low temperatures are those with a 10 h $t_{\frac{1}{2}}$ below 80 °C (Table 5.3). Effective initiators include benzoyl peroxide, 2,5-dimethyl-2,5-di(2-ethylhexanoyl-peroxy) hexane and bis(4-t-butyl cyclohexyl) peroxydicarbonate. The latter two, in their commercial form, require shipment and storage at controlled temperatures.

Di-2-phenoxyethyl peroxydicarbonate is a relatively new initiator which is now being test marketed in the form of a powdered solid. This initiator is stable at ambient temperatures below 38 °C and gives faster cures than benzoyl peroxide.

Although each of the above initiators give acceptable cures, their performance is enhanced when a low concentration of a higher-temperature initiator such as t-butyl peroxybenzoate is added to the formulation.

For moulding at temperatures around 150 °C, t-butyl peroxybenzoate has been considered for years to be the 'work horse' of the industry. It offered the moulder acceptable reactivity and shelf-life, all at reasonable cost. However, to keep pace with increasing product demand from many application areas, the moulder found it necessary to increase the rate of production. One way of doing this is to reduce mould cure time. Shorter cure times can be obtained by several methods including higher moulding temperatures, using faster (less stable) peroxides or by using initiator blends.

Raising the mould temperature is not always feasible, nor always economical because of equipment limitations or due to the high cost and limited availability of energy.

A faster initiator, i.e. one which decomposes faster at a given temperature, can be used to reduce cure time. As noted in Section 5.4.1 the $10\,h\ t_{\frac{1}{2}}$ temperature is often used to compare the relative reactivity of initiators. The initiator with the lower $10\,h\ t_{\frac{1}{2}}$ temperature is considered the faster initiator. Using fast initiators to obtain short cure times is possible, however, shelf-life is significantly reduced. This can be a serious limitation in SMC and BMC applications where precatalysed moulding compounds are often aged before use. One type of initiator which has been used successfully to produce faster cures and at the same time give good shelf-life is the peroxyketal group.[7]

Peroxyketals are diperoxides that decompose to generate alkoxy radicals upon cleavage of the peroxy bond.

$$
\begin{array}{c}
R'OO \diagdown \quad \diagup R_1 \\
C \quad \rightarrow 2R'O \cdot + \text{other fragments} \\
R'OO \diagup \quad \diagdown R_2
\end{array}
$$

The mechanism of decomposition is not fully understood, however, chemical analysis indicates that other decomposition products include $R_1 \cdot$, $R_2 \cdot$ and CO_2.[8] As with other classes of peroxide, the nature of the R' groups and also the R_1 and R_2 groups plays a role in determining activity. When R_1 and R_2 are the same, t-butyl peroxyketals are the most stable, followed by t-amyl, t-octyl and t-cumyl peroxyketals, in descending order of stability. The effect on $10\,h\ t_{\frac{1}{2}}$ temperature resulting from a change in R_1 and R_2 groups is illustrated in Table 5.10. The first three compounds show a decreasing stability as the R_2 group increases in electron donating capability, i.e. from $C_2H_5OC(O)CH_2-$ to C_2H_5- to $(CH_3)_2CHCH_2-$. The cyclohexyl peroxides show a more significant lowering of stability which may be due to a combination of steric strain relief and electronic effects.[3]

TABLE 5.10

COMMERCIAL DIPEROXYKETALS: EFFECT OF STRUCTURE ON THE TEN HOUR HALF-LIFE
TEMPERATURE

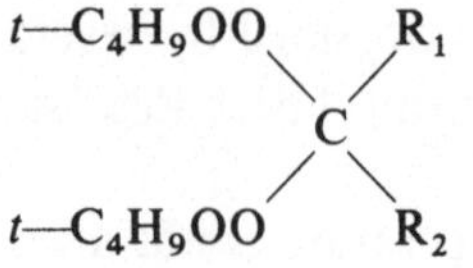

$$t\text{—}C_4H_9OO \diagdown_{\displaystyle C}\diagup R_1$$
$$t\text{—}C_4H_9OO \diagup \diagdown R_2$$

Name	R_1	R_2	$10\,h\ t_{\frac{1}{2}}$ $0{\cdot}2\,M\,(benzene)$ $(^\circ C)$
Ethyl 3,3-di(t-butylperoxy)-butane	—CH$_3$	—CH$_2\overset{\displaystyle O}{\overset{\displaystyle \|}{C}}OC_2H_5$	111
2,2-Di(t-butylperoxy)-butane	—CH$_3$	—C$_2$H$_5$	104
2,2-Di(t-butylperoxy)-4-methylpentane	—CH$_3$	—CH$_2$CH(CH$_3$)$_2$	101
1,1-Di(t-butylperoxy)-cyclohexane	—(CH$_2$)$_5$—		95
1,1-Di(t-butylperoxy)-3,3,5-trimethylcyclohexane	—CH$_2$CH(CH$_3$)C(CH$_3$)$_2$CH$_2$—		92

The decomposition rate of peroxyketals can be accelerated by using certain quaternary ammonium salts.[9] This technique represents a practical means for accelerating the cure of polyester resins while retaining serviceable shelf-life of the premix. Based on the known acid sensitivity of peroxyketals it is also possible to effectively accelerate their decomposition at elevated temperatures by using small concentrations of mineral acids in the resin formulation.[3]

The peroxyketals have certain limitations which can make them inefficient under specific conditions. The effect of resin formulation on the efficiency of peroxyketals has been studied.[10] This study revealed that, unlike t-butyl peroxybenzoate, the shelf-lives and cure activities of moulding compounds precatalysed with peroxyketals are related to filler pH. The higher the pH, the longer the shelf-life and the shorter the cure time. Although filler particle size has no effect on shelf-life, the cure activity of clay filled formulations is inversely related to particle size, i.e. the smaller the particle size, the longer the cure time. The effects of acidic fillers are minimal in SMC/BMC formulations due to the neutral-ising effect of strongly basic thickening agents, such as $Mg(OH)_2$.

Peroxyketals like other single initiator systems can be used efficiently only over a relatively narrow temperature range and often the moulder cannot achieve the desired cure time by using them alone. A number of studies have shown that blends or mixtures of initiators provide versa-tility which cannot be achieved by single initiators.

5.6. INITIATOR BLENDS

When a high-temperature initiator (high 10 h $t_{\frac{1}{2}}$ temperature) is blended with a small concentration (10–25 wt %) of a low-temperature initiator (lower 10 h $t_{\frac{1}{2}}$ temperature), faster cures are obtained often with some sacrifice in shelf-life. In addition to offering reductions in cure time, it has been reported that the depth of sink marks in SMC mouldings can be significantly reduced by using peroxide blends.[11] Recent work with novel azo/peroxide blends shows that faster cures are possible and shelf-life can be improved.[12] A major advantage of azo initiators, com-pared to peroxides, is that they are not subject to induced decomposition by impurities or transition metals. Therefore, azos exhibit markedly superior shelf-life as compared with peroxides of similar thermal stability.

In the past, azo initiators found limited acceptance for polyester mould-ing since, under certain moulding conditions, they reportedly led to surface porosity problems. Most recent work with these initiators has shown that surface porosity is not a problem when they are used as the minor component in initiator blends.

The activity characteristics and the advantages of initiator blends are illustrated in Table 5.11. Adding a low concentration of t-butyl peroctoate to either t-butyl peroxybenzoate or 1,1-di(t-butylperoxy)cyclohexane re-sults in significantly faster cure times, however, shelf-life is reduced in each case. The advantage of azo compounds in blends can be seen by comparing blends 4 and 5. Blend 5, containing the azo initiator, exhibits faster cure time and, at the same time, longer shelf-life.

5.7. RESIDUAL STYRENE

The influence of initiator curing systems on the residual styrene content of cured polyester resins has been the subject of several recent

TABLE 5.11

INITIATOR BLENDS: CURE ACTIVITY AND SHELF-LIFE DATA

No.	1 phr Initiator blend		Cure activity at 121°C			32°C Shelf-life (days)
	Wt. fraction 'A' component	Wt. fraction 'B' component	Gel (min)	Cure (min)	Peak (°C)	
1	—	1·0 t-Butyl peroxybenzoate	4·2	5·6	217	40
2	0·17 t-Butyl peroxy-2-ethylhexanoate	0·83 t-Butyl peroxybenzoate	2·3	3·1	209	12
3	—	1·0 1,1-Di(t-butylperoxy)-cyclohexane[a]	2·7	3·7	209	81
4	0·10 t-Butyl peroxy-2-ethylhexanoate	0·90 1,1-Di(t-butylperoxy)-cyclohexane[a]	2·2	3·1	204	40
5	0·10 2-t-Butylazo-2-cyano-4-methylpentane	0·90 1,1-Di(t-butylperoxy)-cyclohexane[a]	1·9	2·8	199	56

[a] 80% in phthalate solvent.

Cure activity and shelf-life measured in general purpose orthophthalic resin using modified SPI exotherm procedure.

Shelf-life measured using 100 g of resin in a glass jar stored in a constant temperature oven.

studies.[4,13-15] Theoretically, the copolymerisation of the styrene monomer with the unsaturation in the resin molecule will cease when all of the monomer has reacted. In actual practice, however, this state is seldom achieved unless special measures are taken. The quantity of residual styrene in a cured resin is recognised as being an important parameter since optimum physical properties are achieved only at the lowest values. This is important in all applications, and is particularly important for high chemical resistance and food handling applications.

The technique most often used to determine residual styrene in cured mouldings involves extraction of the finely divided sample with methylene chloride followed by analysis of the extract by gas chromatography.[16] Since sample size, extraction time and extraction temperature can all affect results, the exact procedure that one follows must be carefully evaluated to establish the accuracy of results. Experience indicates that for best reproducibility, the sample must be ground to a fine powder and extracted with methylene chloride for at least 7 days at 23 °C.

In polyester resins cured at room temperature, all current studies indicate that relatively high levels of residual styrene are found in laminates prepared using standard cure systems. It has recently been reported[4] that after extensive evaluations no MEKP–promoter-resin systems were found which would give acceptably low residual styrene levels via room temperature cures. The lowest residual styrene levels are obtained when high levels of peroxide are used, when high exotherms are obtained or when post cure techniques are employed.

At the usual initiator concentrations of 1 wt %, reported residual styrene levels are in the range of 5–11%. These levels tend to decrease with time, however, after six months at 20 °C levels are generally greater than 1·0 wt %. Peroxide concentrations as high as 10 wt % will give residual styrene levels of <0·01%, however, such initiator concentrations are not practical. Increasing the level of promoter is far less effective and, in the case of benzoyl peroxide, can have adverse effects.[13,15]

Ketone peroxide/cobalt systems give the lowest residual styrene levels when a post cure treatment is used. For example, an initial residual styrene content of 5·25% is reduced to <0·01% after a post cure at 80 °C for 8 h.

In those cases where post cure is practical, the optimum time and temperature must be experimentally determined for each resin. As a general rule, however, it has been reported that 8 h at, or slightly above, the heat distortion temperature of the specific resin will reduce the residual styrene to acceptably low levels.[13]

In polyester resins moulded at elevated temperatures, lowest residual styrene levels are obtained with peroxyketals[13,14] and medium reactivity peroxyesters such as t-butyl peroxyoctoate.[13] At any given moulding temperature one can decrease residual styrene levels by increasing the initiator concentration and/or increasing the moulding time. In BMC formulations containing t-butyl peroxyoctoate moulded at 140 °C, residual styrene decreased from 0·3% to 0·2% when the initiator concentration was increased from 0·5% to 1·0%.

The type of radical formed by the initiator can have a significant effect on the level of residual styrene. For example, benzoyl peroxide gives relatively high residual styrene values compared to t-butyl peroxyoctoate, an initiator with a similar 10 h $t_{\frac{1}{2}}$ temperature. t-Butyl peroxybenzoate (tBPB) produces radicals similar to those produced by benzoyl peroxide and is also found to give relatively high residual styrene values. For example, in a BMC formulation moulded at 140 °C and containing 1·0% tBPB, the residual styrene was found to be 0·45%.

Certain organometallic promoters are commercially available which can reduce cure times when used with peresters and also significantly reduce residual styrene. Promoter 301,† for example, at a concentration of 1·0%, reduced residual styrene from 0·45% to <0·1% in a formulation initiated with t-butyl peroxybenzoate. These pieces were moulded at 140 °C for 2 min.[13] Cerium compounds are also reported to give significant reductions in residual styrene when added to formulations containing peroxyesters.[17]

5.8. OTHER SOURCES OF FREE RADICALS

5.8.1. Carbon–Carbon Initiators

Azo and peroxide compounds are convenient, low cost sources of free radicals. However, considerable work has been done to develop other sources of free radicals to cure polyester resins. A majority of them are designed for specialised applications where cost considerations are not very critical. These include a new class of compounds, commonly known as carbon–carbon initiators. They are tetra substituted dibenzyl compounds and can be represented by the general structure:

† Air Products and Chemicals, Inc. USA.

where R_1 and R_2 are bulky substituents. The carbon–carbon bond here is highly strained and, thus, undergoes homolytic scission readily at low temperatures to yield two free radicals. The decomposition temperature of these compounds is strongly influenced by the nature of the substituents.

A series of tetraphenyl ethane compounds were recently investigated[18] as potential initiators for curing polyester resins. Results of this study indicate that compounds such as 1,2-dimethoxy-1,1,2,2-tetraphenyl ethane (DMTPE) and 1,2-dichloro-1,1,2,2-tetraphenyl ethane (DCTPE) will effectively initiate the cure reaction at temperatures in the range of 100–130 °C. The properties of the cured resin, i.e. flexural strength, impact strength and hardness were found to be equivalent to those obtained with benzoyl peroxide. Although these compounds tend to be more stable than peroxides, they have certain disadvantages which have, to date, limited their acceptance on a commercial scale. They are solids which can pose dispersion and resin solubility problems. In addition, their chemical structure does not lend itself to activation and, thus, they are limited to elevated temperature curing systems. These disadvantages along with their high cost will probably limit commercial utilisation.

5.8.2. Photoinitiators

The unimolecular decomposition of photoinitiators as a source of free radicals has certain inherent advantages. Considerable effort has been directed towards development of commercially viable systems using ultraviolet and visible light as sources of energy. These systems are especially important in the field of coatings where benzoin ether compounds are popular UV photoinitiators. A recent review[19] of these systems indicates that a number of new photoinitiators are being developed to keep pace with projected growth in these applications.

Visible light curing of unsaturated polyesters using photoinitiators is a commercially feasible method which has recently been described.[20] The use of visible light offers the benefit of eliminating the hazards associated with high energy UV radiation and thus increases the practicality of the system for the FRP industry. Another major advantage compared to UV technology is that thick parts can be cured quite easily.

Various photoinitiator–amine reducing agent combinations have been evaluated in visible light activated systems.[21] The amine compound is selected such that it will reduce the photoinitiator in its excited state only. Of the various amines evaluated, shortest gel times were obtained with *l*-methyl imidazole. In terms of photoinitiators, faster gel times (e.g. 30 to 60 s) were obtained with benzoin ether as compared to benzil (~ 3 min). This study also identified vinyl ester resins, bisphenol A/fumaric acid polyesters and isophthalic polyesters as having clear visible-light-curing specificity. The technology is projected as being suitable for electronic materials processing including potting, encapsulation and prepreg processing applications.

REFERENCES

1. PRYOR, W. A. *Free Radicals*, 1966, McGraw-Hill Inc., New York.
2. GALLAGHER, R. B. and KAMATH, V. R. *Plastics Design and Processing*, 1978, **18**(7), 38.
3. SHEPPARD, C. S. and KAMATH, V. R. *Polym. Eng. Sci.*, 1979, **19**, 597.
4. CASSONI, J. P., HARPELL, G. A., WANG, P. C. and ZUPA, A. H. *32nd SPI Reinforced Plastics/Composites Conf.*, 1977, Paper 3-E.
5. THOMAS, A., JACYSZYN, O., SCHMITT, W. and KOLCZYNSKI, J. *32nd SPI Reinforced Plastics/Composites Conf.*, 1977, Paper 3-B.
6. MAXSTADT, A. *Kunststoffe*, 1979, **69**, 266.
7. THOMAS, A., ROSKOTT, L., GROENENDAAL, A. A. M. and KOLCZYNSKI, J. R. *33rd SPI Reinforced Plastics/Composites Conf.*, 1978, paper 5-E.
8. BUKATA, S. W., ZABROCKI, L. L., MCLAUGHLIN, M. F., KOLCZYNSKI, J. R. and MAGELI, O. L. *Ind. Eng. Chem. Prod. Res. Develop.*, 1964, **3**, 261.
9. US Patent No. 4,032,596 issued to Air Products and Chemicals Inc., 28 June, 1977.
10. CASSONI, J. P., GALLAGHER, R. B. and KAMATH, V. R. *34th SPI Reinforced Plastics/Conposites Conf.*, 1979, Paper 14-F.
11. AMPTHOR, F. J. *Plastics World*, 1978, **36**(5), 48.
12. KAMENS, E. R., GALLAGHER, R. B. and KAMATH, V. R. *34th SPI Reinforced Plastics/Composites Conf.*, 1979, Paper 16-B.
13. ROSKOTT, L. and GROENENDAAL, A. A. M. *33rd SPI Reinforced Plastics/Composites Conf.*, 1978, Paper 5-B.
14. IZZARD, K. J. and NEWTON, G. P. *33rd SPI Reinforced Plastics/Composites Conf.*, 1978, Paper 5-C.
15. VARCO, P. *30th SPI Reinforced Plastics/Composites Conf.*, 1975, Paper 6-C.
16. SHAPRAS, P. and CLOVER, G. C. *Anal. Chem.*, 1964, **36**, 2282.
17. German Patent No. 2,815,924 issued to Akzo G.m.b.H., 19 April, 1977.
18. BRAUN, D. and QUELLA, F. *Kunstoffe*, 1979, **69**, 100.
19. DELZENNE, G. A. *Makromol. Chem.*, 1979, Suppl. 2, 169.
20. DIXON, B. G., LONGENECKER, D. M. and GRETH, G. G. *32nd SPI Reinforced Plastics/Composites Conf.*, 1977, Paper 5-D.
21. KOLEK, R. L. and HAMMILL, J. L. *Plastics Compounding*, 1979, **2**(6), 52.

Chapter 6

HIGH-TEMPERATURE PROPERTIES OF THERMALLY STABLE RESINS

G. J. Knight

Royal Aircraft Establishment, Farnborough, Hants, UK

SUMMARY

This chapter deals with the chemistry and the thermal and mechanical properties of a number of resin systems known to possess good high-temperature properties. The resins examined include the poly(aryl ether sulphones), poly(phenylene sulphide), phenol–formaldehyde and melamine–formaldehyde resins, silicones, and the polyimides. In the Introduction the term thermal stability is discussed, together with factors to be considered when looking for a heat resistant polymer.

6.1. INTRODUCTION

6.1.1. The Search for New Resins

Mainly due to the demands of the aerospace industry, there has been a considerable amount of effort put into the search for temperature resistant materials during the past 15 to 20 years. One of the first synthetic polymeric materials prepared, phenol–formaldehyde resin, is, in fact, reasonably thermally stable and is still used for such applications as brake linings. Where good electrical properties are required together with temperature resistance then melamine–formaldehyde or silicone resins are used. Although these materials have served their purpose for many years, and the silicones, for example, are suitable for prolonged service at temperatures up to 250 °C, the new demands for structural composites to be used in aircraft and missiles have stimulated the search for stronger and yet more thermally resistant resins.

6.1.2. The Meaning of 'Thermally Stable'

Before going on to consider the chemistry and properties of the various resin systems commercially available today one should consider exactly what is meant by the term 'thermally stable'. It has been known for an author, describing the synthesis of a potentially thermally stable polymer, to make the statement that: 'the polymer was heated strongly and showed no signs of decomposition', no mention being made of the temperature nor of the length of time the sample was heated. One of the first techniques for assessing thermal stability systematically was that of thermogravimetric analysis (TGA). This is a very useful tool and helps to indicate the relative order of stability of various polymers, either in inert or

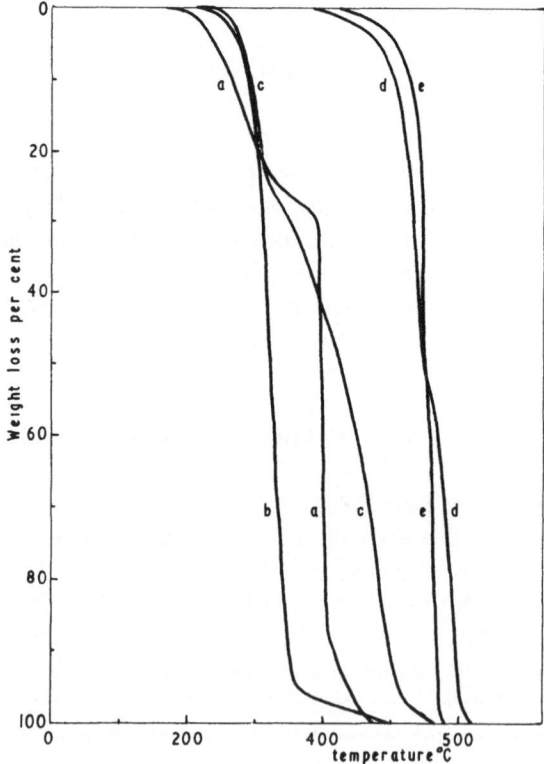

FIG. 6.1.   Thermogravimetric analysis in air of (a) polyethylene, (b) polystyrene, (c) epoxy resin, (d) polyethersulphone and (e) Kapton.

oxidising atmospheres, as illustrated in Fig. 6.1. This illustrates the results given by various common polymers when the temperature is raised by 2 °C/min. To a certain extent the results depend on the heating rate; the higher the heating rates, the greater the apparent stability. (In this chapter where a figure illustrates the weight loss behaviour of a polymer the work has been performed on a Du Pont 951 Thermogravimetric Analyzer attached to a 990 Thermal Analyzer, with a gas flow rate of 50 ml/min and a heating rate of 2 °C/min.) The temperature of decomposition, T_D, the temperature at which the weight loss curve departs from the baseline is subject to errors due to loss of volatiles from the polymer. A commonly quoted value, therefore, is the temperature for 10 % weight loss. Whatever value is quoted it must be emphasised that it does not indicate the maximum temperature at which a material may be used. It is not possible to give a single value for this temperature as it depends on many variables including, for example, the local environment, the length of time at a temperature and the type of mechanical stress to which the polymer is subjected. In a missile to be used only once, with a lifetime at temperature measured in minutes, much higher temperatures are permissible than in an aircraft structure, where lifetimes of 10 000 to 20 000 h would be expected. Where an accurate assessment of the life of a component is required the only way available at present is to model the conditions and expose the article for the required time. Attempts to shorten the test procedure by raising the temperature and then extrapolating the results have to be treated with caution, as the rise in temperature can provide the activation energy for reactions that do not take place at lower temperatures. The Arrhenius equation for the rate constant of a reaction is written

$$k = A_e^{-E_a/RT}$$

where k is the rate constant, A the Arrhenius constant, E_a the activation energy for the reaction, R the gas constant and T is the temperature in degrees Kelvin. By taking logs the equation becomes

$$\ln k = -\frac{E_a}{RT} + \ln A$$

Isothermal weight loss experiments can be conducted at various temperatures, then, if the reaction causing weight loss follows simple reaction kinetic relationships, a plot of log (rate of weight loss) v. $1/T$ should give a straight line. Such a plot is shown in Fig. 6.2 for a polysulphone. The experiments were performed over the temperature range 383 °C to

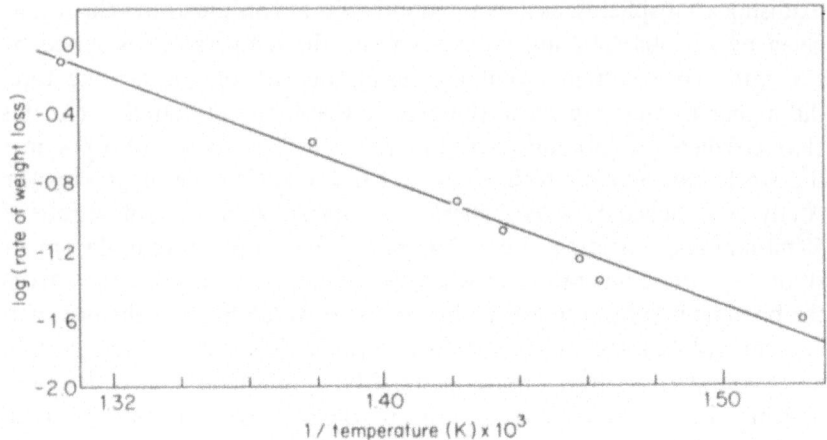

FIG. 6.2. Arrhenius plot of log (rate of weight loss) at a constant temperature v. 1/temperature (K) for Udel P1700 polysulphone. Isothermal weight loss in air.

493 °C. By extrapolation, a weight loss of 0·5 %/year would be expected at 200 °C, but without knowing the actual mechanisms of the reactions causing weight loss it is not possible to be certain of this prediction. Isothermal weight loss experiments, like TGA, are better considered as another means of comparing different polymers and of discovering more about the chemistry of the compounds rather than of predicting their useful life.

When used as a load bearing material the glass transition temperature

TABLE 6.1

THE GLASS TRANSITION TEMPERATURES T_g AND DECOMPOSITION TEMPERATURES T_D OF THE POLYMERS ILLUSTRATED IN FIG. 6.1

Polymer	T_g °C	T_D °C
Polyethylene	−175[1]	200
Polystyrene	100[1]	240
ICI Polyethersulphone 200P	230[2]	410
Epoxy resin[a]	210[3]	250
Kapton type polyimide[b]	325[4]	430

[a] The diglycidyl ether of bisphenol A hardened by 4,4′ diaminodiphenyl sulphone, post cured for 4 h at 200 °C.
[b] The polypyromellitimide of 4,4′diaminodiphenyl ether.

(T_g) of a polymer gives a better indication of the upper temperature limit for use. For example, polystyrene can be seen from Fig. 6.1 to be stable up to about 240 °C, but its T_g is 100 °C[1] and its melting point is about 240 °C.[1] Examples of the T_gs of various polymers are given in Table 6.1 together with the decomposition temperatures, T_D, in air.

From Table 6.1 it can be seen that the thermoplastic materials soften at temperatures well below their decomposition temperatures. A thermoset resin can show a T_g but this represents the increased mobility of segments within the crosslinked network, the overall structure remains fixed and decomposition precedes any softening or melting. The value of 325 °C quoted as the T_g of the polyimide is an estimate and it is thought that crosslinking reactions take place at temperatures below this.[4]

6.1.3. Fabrication

As mentioned above, this chapter is concerned with resins used as reinforced plastics. A reinforced plastic can be considered as a composite material consisting of continuous or discrete fillers bonded by a resin matrix, each component contributing to the properties of the whole. The reinforcement is the more important factor in the mechanical performance of the composite provided that there is good adhesion between the fibres and the matrix. Thermoplastic resins being fusible and often also soluble lend themselves to easy processing; no chemical reactions take place during fabrication. An important factor with respect to thermoplastics is that the heat distortion temperature of a polymer can be raised by the incorporation of a suitable filler (e.g. Nylon 6.6, resin alone has a heat distortion temperature of 65 °C but with 40 % glass fibre it rises to 260 °C).[5] Other interactions are known between the resin and the reinforcement but these will be discussed below with the resin properties.

The thermoset materials have to be fabricated at some intermediate prepolymer stage as, in the final crosslinked state, they are intractable. Also most of the resins which show temperature resistance and are suitable for laminating do not have properties that enable castings to be prepared, e.g. they are very brittle and the cure produces samples with voids and cracks. This means that few, if any, property measurements have been made on the resins alone; the majority of values quoted in the literature relate solely to composites.

The fact that the processor is, in effect, synthesising the final polymer as he fabricates the product needs to be borne in mind. The curing reaction may require the addition of a crosslinking agent and catalysts as well as the application of heat. The mechanical properties of the final polymer will depend upon the chemical reactions that govern the structure of the main chain and of the crosslink chain, and upon the number of crosslinks. It is the variations which are possible that account for the many resin systems that can be made. An increase in the crosslink density of a resin causes a concomitant increase in modulus and hardness, and in the resistance to creep, heat, and chemical reagents. At the same time there is a decrease in impact strength, elongation at break and reversible extensibility.

Chemically the crosslinking process can be brought about either through addition or a condensation process. This is of practical importance as, in the addition process, no volatiles are evolved. This means that processing is simpler and, therefore, cheaper, lower pressures need to be used, void formation is less likely and thicker sections are easier to make. Theoretically, after fabrication all the functional groups in the original resin will have been used up but in practice this is rarely the case. Hence, further cure may be possible when the article is exposed to elevated temperatures causing changes in the mechanical properties. Also with respect to electrical properties the number and type of polar groups in the resin are of importance, for example OH, CO and COO groups in the final structure may give rise to permanent dipoles in the material.

6.1.4. The Effect of Moisture

A further factor which needs to be considered when a composite is to be used at elevated temperatures is the effect of moisture. It has been shown[6-9] that water has a marked effect on the mechanical properties of epoxy resins. It acts as a plasticiser, lowering the T_g of the resin by as much as 20°C for every 1% water content. As well as fostering research into epoxy resin systems with low water pick-up it has also focused attention on the fact that the high T_gs of the temperature resistant resins makes them attractive candidates as the matrix for composites to be subjected to severe environmental exposure.[8,10-13] It can be seen, therefore, that each resin system has its own advantages and disadvantages, and the subsequent sections describing the various polymers will attempt to draw attention to these.

6.2. THERMOPLASTIC RESINS

6.2.1. Introduction

A major factor in the development of reinforced thermoplastics is ease of processing; this includes the ability to rework scrap materials. The introduction of carbon fibre reinforcement together with the high-temperature resins, poly(arylene ether sulphones) and poly(phenylene sulphide) has opened up even further fields of application. The two poly-imide resins, Du Pont's NR-150 and Upjohn's Polyimide 2080, are thermoplastic materials and are considered in Section 6.4.

6.2.2. The Poly(Arylene Ether Sulphones)

The poly(arylene ether sulphones) are thermoplastic polymers, some of which may be used under stress at temperatures up to 150 °C. Synthetic routes to these polymers were discovered independently in the laboratories of the 3M Corporation[14] and Union Carbide Corporation[15] in the USA and in the Plastics Division of ICI[16] UK. All three companies now market different polysulphone plastics and the structures of these materials and their T_gs are shown in Table 6.2.

Poly(phenylene sulphone) itself decomposes before melting, at > 500 °C, and the improvement in tractability of the polymers has been brought about by the introduction of the aryl ether linkages. As can be seen from the table, the more there are of these, the lower the T_g of the polymer.

There are two synthetic routes to these polymers:

(1) Reaction of a dihalo aryl compound with a diphenoxide of an alkali metal.[15]

$$Hal—\langle O \rangle—SO_2—\langle O \rangle—Hal + MO—Ar—OM \rightarrow$$

$$\left[\langle O \rangle—SO_2—\langle O \rangle—O—Ar—O \right]_n + 2MHal$$

The sulphone group plays an essential part in the reaction as it activates the atoms to attack by the phenoxides.

(2) A polysulphonylation process in which sulphone linkages are formed by reaction of arylsulphonyl chlorides with aromatic nuclei

$$H—Ar—H + ClSO_2—Ar'—SO_2Cl \rightarrow [Ar—SO_2—Ar'—SO_2] + 2HCl$$

The reaction is performed in the presence of catalytic amounts of a Friedel–Crafts reagent such as $FeCl_3$, $SbCl_5$ or $InCl_3$.[2,17] The ArH_2 can be a diphenyl, diphenyl ether or naphthalene but not a benzophenone or a diphenyl sulphone as the electron withdrawing substituents reduce the ability of the ring to provide an electron pair for bond formation.

The polymerisation can be carried out in the melt but it is preferable

TABLE 6.2
POLY(ARYLENE ETHER SULPHONES)

Polymer	Structure	T_g °C
Astrel 360 (Carborundum Co.)	Mainly 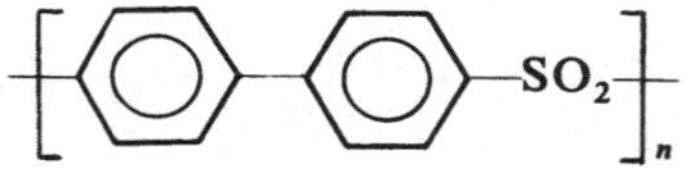 $\left[\!-\!C_6H_4\!-\!C_6H_4\!-\!SO_2\!-\!\right]_n$ also some $\left[\!-\!C_6H_4\!-\!O\!-\!C_6H_4\!-\!SO_2\!-\!\right]_n$	285
Udel P1700 (Union Carbide)	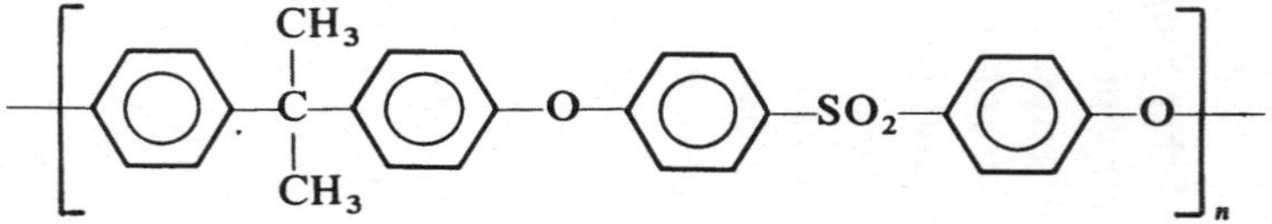 $\left[\!-\!C_6H_4\!-\!C(CH_3)_2\!-\!C_6H_4\!-\!O\!-\!C_6H_4\!-\!SO_2\!-\!C_6H_4\!-\!O\!-\!\right]_n$	190
Radel (Union Carbide)	Not known	204[a]
Poly(ether sulphone) 200P (ICI Ltd)	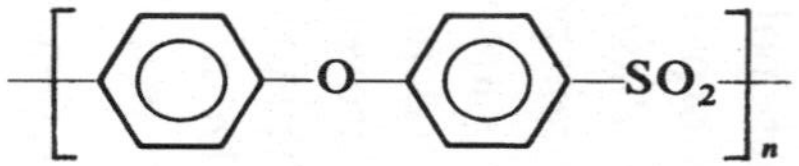 $\left[\!-\!C_6H_4\!-\!O\!-\!C_6H_4\!-\!SO_2\!-\!\right]_n$	230

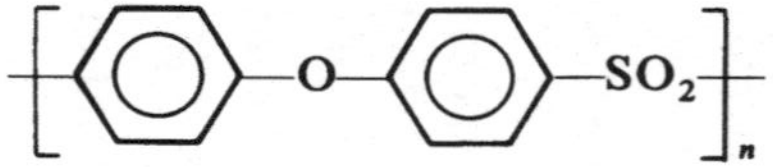

[a] Heat deflection temperature at 1·82 MPa, ASTM D648.

to use an inert solvent. Melt polymerisation requires very high temperatures to provide sufficient mobility for the growing polymer chains. The preferred solvents are nitrobenzene or dimethyl sulphone; these are capable of dissolving both the starting materials and the resultant polymer. They are also compatible with the strong Lewis acid catalysts and stable to electrophilic attack by the sulphonyl chloride.

6.2.3. Thermal Stability of the Polysulphones

The results of thermogravimetric analysis of the polysulphones are illustrated in Figs. 6.3 to 6.6; the experiments were performed in air and in nitrogen at a rate of temperature rise of 2 °C/min. In addition a plot of the first derivative is shown. It can be readily seen that all the polymers show very high thermal and thermo-oxidative stability, the majority of the degradation occurring between 450 °C and 550 °C in each case. The

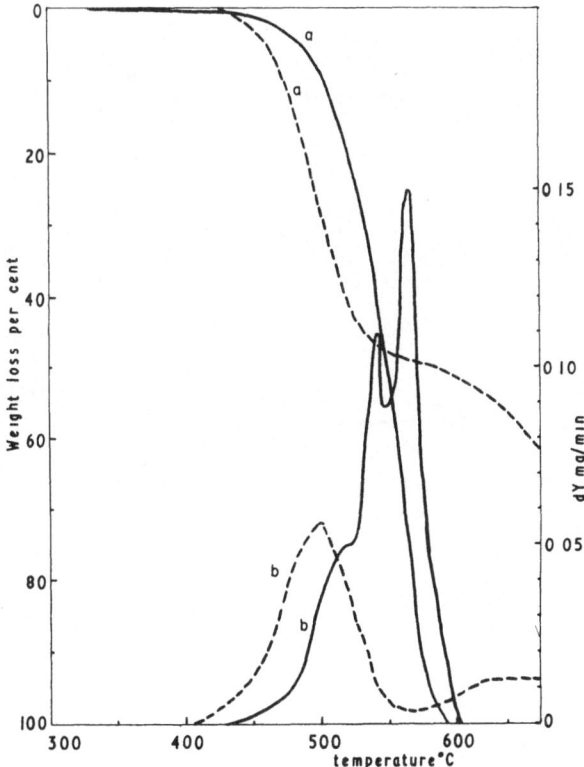

FIG. 6.3.   Astrel 360—(a) weight loss and (b) rate of weight loss (dY). ——— in air; ———— in nitrogen.

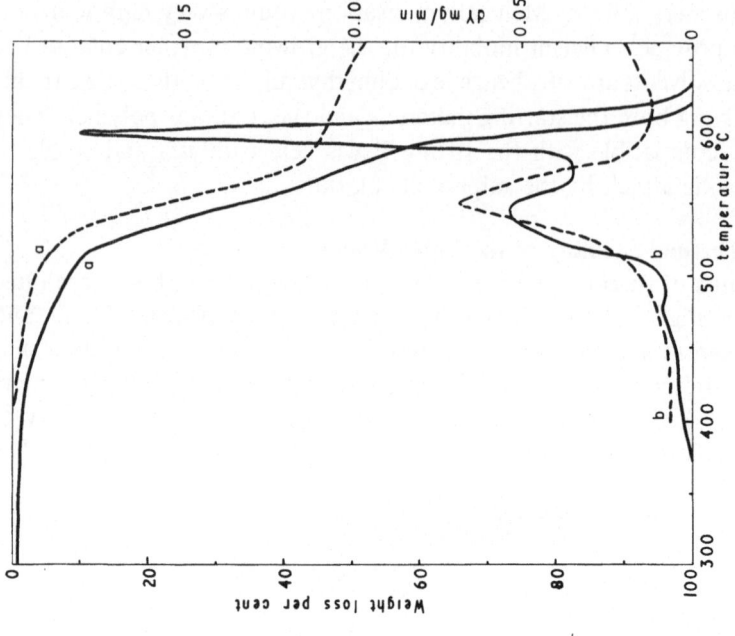

FIG. 6.5. Radel—(a) weight loss and (b) rate of weight loss (dY). —— in air; - - - - in nitrogen.

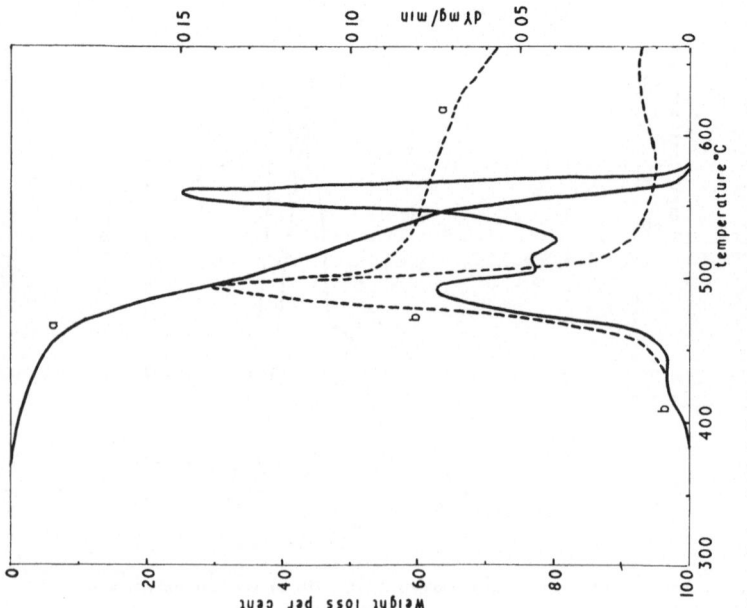

FIG. 6.4. Udel P1700—(a) weight loss and (b) rate of weight loss (dY). —— in air; - - - - in nitrogen.

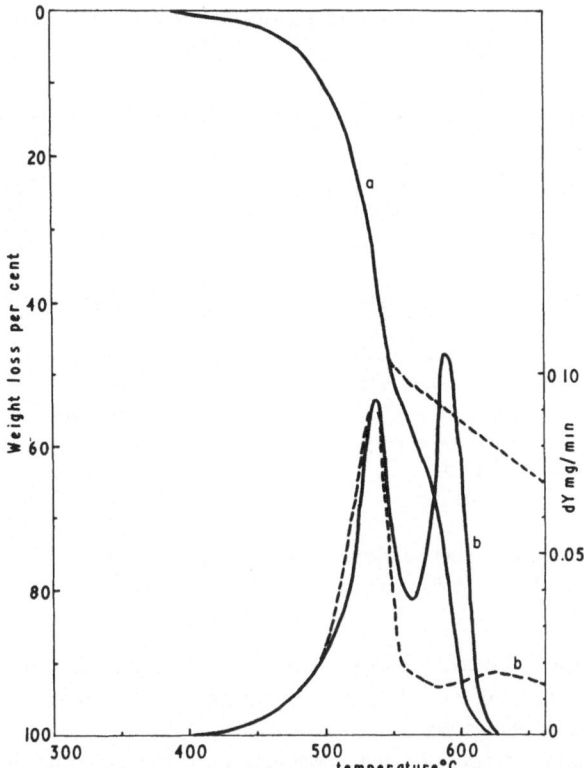

FIG. 6.6.   Poly(ether sulphone) 200P—(a) weight loss and (b) rate of weight loss (dY). ——— in air; ---- in nitrogen.

dY (rate of weight loss) plots show very clearly that in air the degradation is a multistage process, in contrast, in nitrogen there is essentially a single stage reaction with the formation of a fairly stable char representing about 40% of the starting material. The derivative plots also indicate that the primary mode of breakdown is thermal rather than oxidative.

The activation energy for thermal degradation of Udel P1700 in an argon atmosphere has been given as 293 kJ/mole.[18] As a result of isothermal experiments the values shown in Table 6.3 have been evaluated.[19]

It has been shown[18,19] that the primary thermal breakdown route of poly(arylene ether sulphones) *in vacuo* is by cleavage of the C–S bonds to release SO_2 gas with a concomitant crosslinking process within the polymer.

As a class the polysulphones are very stable chemically and they all show good stability with acids and alkalis, in addition to the good oxidative stability described above. Resistance to hydrolysis is high but hot water promotes stress cracking. Solvent resistance is good towards the non-polar hydrocarbons, alcohols, Freons, silicone oils etc., but chlorinated hydrocarbons, ketones and other polar solvents tend to swell or, in specific cases, dissolve the polymers. Unfilled material has been shown to have

TABLE 6.3

ACTIVATION ENERGY E_a FOR DEGRADATION IN AIR
AND NITROGEN

Sample	$E_a \, kJ/mole$	
	Air	Nitrogen
Astrel 360	172 ± 10	270 ± 12
Udel P1700	137 ± 11	270 ± 12
Radel	78 ± 4	310 ± 4
Poly(ether sulphone) 200P	183 ± 10	311 ± 8

poor resistance to ultraviolet radiation,[21-24] but with high pigment loadings the weathering stability can be improved to an acceptable level for outdoor applications.[17]

A summary of the physical properties of the polymer is shown in Table 6.4.

The polysulphones have been shown to have outstandingly good creep properties. For example, a creep specimen of poly(ether sulphone) at a stress of 50 MPa endured for more than two years, developed only slight crazing and showed no inclination towards runaway creep.[25] The combustion characteristics and low smoke emission of these plastics make them suitable, for example, for applications in electrical apparatus and in the interior of aircraft.

When used in combination with glass or carbon fibres these materials offer many attractive properties. Much effort has gone into developing them in industrial, automotive and aerospace applications,[26] e.g. in the space shuttle[27] and the YC-14 aircraft.[28] Unfortunately the susceptibility of the resins to attack by halogenated solvents, for example dichloromethane which is commonly used as a paint stripper, would seem, at present, to preclude their use in aircraft structures.[10] Research is being conducted on modification of the chemical structure of the resin to reduce the effect of solvents.[28]

TABLE 6.4

PHYSICAL PROPERTIES OF POLY(ARYLENE ETHER SULPHONES)

Properties	Unit	Astrel 360	Udel	Radel	200P
Density	g/cm^3	1·36	1·24	1·29	1·37
Tensile strength	MPa	91	70	72	84
Elongation	%	13	50 to 100	7	30 to 80
Tensile modulus	GPa	2·60	2·60	2·14	2·44
Flexural strength	MPa	121	108	86	129
Flexural modulus	GPa	2·78	2·60	2·28	2·57
Rockwell hardness	—	M110	M69, R120	—	M88
Volume resistivity	ohm.cm	3×10^{16}	5×10^{16}	9×10^{16}	10^{17}
Power factor at 10^6 Hz	—	0·0035	0·0010	0·0076	0·0035
Service temperature	°C	260	150 to 170	175	200
Heat distortion temperature at 1·82 MPa (264 psi)	°C	274	174	204	202
Coefficient of linear expansion	cm/cm/°C	$4·7 \times 10^{-5}$	$5·4 \times 10^{-5}$	$5·5 \times 10^{-5}$	$5·5 \times 10^{-5}$
Flammability ASTM D635	—	—	Self extinguishing	Self extinguishing	—
Water absorption (23°C, 24 h)	%	1·8	0·02	1·1	0·43

6.2.4. Poly(Phenylene Sulphide)

Poly(phenylene sulphide) resins were reported as early as 1948.[29] The polymer being obtained by heating either an alkali metal sulphide or an alkali metal salt and sulphur mixture with a dihalo-arylene compound. The proposed reactions are:

$$2Cl-\!\!\left\langle\bigcirc\right\rangle\!\!-Cl + 3Na_2CO_3 + 4S \rightarrow \left[\!\!\left\langle\bigcirc\right\rangle\!\!-S\!\!-\!\!\left\langle\bigcirc\right\rangle\!\!-S\right]$$

$$+ Na_2S_2O_3 + 4NaCl + 3CO_2 \qquad (1)$$

$$3Cl-\!\!\left\langle\bigcirc\right\rangle\!\!-Cl + 4Na_2CO_3 + 4S \rightarrow$$

$$\qquad\qquad\qquad\qquad\qquad\qquad (2)$$

$$\left[\!\!\left\langle\bigcirc\right\rangle\!\!-S\!\!-\!\!\left\langle\bigcirc\right\rangle\!\!-S\!\!-\!\!\left\langle\bigcirc\right\rangle\!\!-S\right]$$

$$+ Na_2SO_4 + 6NaCl + 4CO_2$$

The product was described as a thermoplastic of high thermal stability, soluble in molten sulphur and with a molecular weight greater than 9000. Phillips Petroleum Company developed a large scale synthetic method of condensing *para*-dichlorobenzene with sodium sulphide[30] and were able to offer trial quantities of the Ryton plastic in 1968.

The following characteristic properties are quoted for the material: high strength, toughness and rigidity, high service temperature, good resistance to chemicals, hydrolysis and oxidation, low water absorption, good processability, low shrinkage, good dimensional stability and good fire performance.

An illustration of the high thermal stability is given in Fig. 6.7 where the TGA curves are shown for experiments conducted in air and in nitrogen. From this it can be seen that the initial reactions causing weight loss are purely thermal and that oxygen causes a certain amount of crosslinking to take place. It has been shown that the thermal stability can be increased by a form of post cure which causes branching and chain extension to take place.[31] Heating *in vacuo* at 250–260 °C showed that the volatile degradation products were dimeric and trimeric chain fragments, dibenzothiophene and possibly thianthrene.[32] The polymer shows excellent resistance to high energy radiation.[33]

TABLE 6.5

PHYSICAL PROPERTIES OF POLY(PHENYLENE SULPHIDE) AND ITS COMPOSITES WITH GLASS AND CARBON FIBRE

Property	ASTM Method	Units	Poly(phenylene sulphide)	With 30% glass fibre	With 30% carbon fibre
Density		g/cm^3	1·34	1·56	1·45
Water absorption in 24 h	D570	%	0·2	0·04	0·04
Tensile strength	D638	MPa	74	138	186
Tensile elongation		%	3–4	3–4	2–3
Flexural strength	D790	MPa	138	200	234
Flexural modulus	D790	GPa	4·13	11·02	16·88
Heat distortion temperature	D648	°C	137	260	260
Coefficient of thermal expansion		cm/cm/°C × 10^{-5}	5·4	2·34	1·08
Surface resistivity		ohm/cm	10^{16}	10^{16}	1–3
Flammability	UL Subj. 94		94 V-O	94 V-O	94 V-O

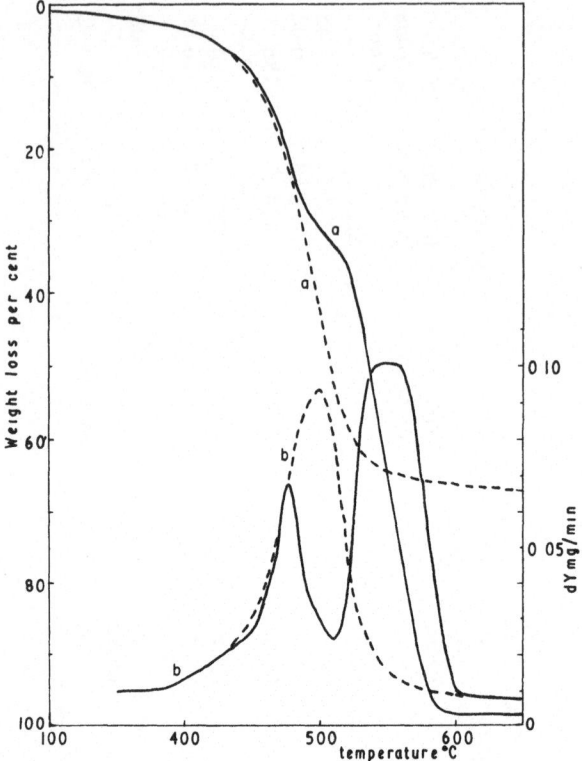

FIG. 6.7.   Poly(phenylene sulphide)—(a) weight loss and (b) rate of weight loss (dY). ——— in air; ---- in nitrogen.

The resin adheres well to filler materials[33] and both glass and carbon fibre reinforced poly(phenylene sulphide) exhibit good mechanical properties.[5,26,33] Examples of typical properties are listed in Table 6.5.

6.3.   THERMOSETS

6.3.1. Introduction

The thermoset resins consist of three dimensional crosslinked networks that do not show softening or melting points, hence there are more of them, compared with thermoplastics, that can be used at elevated temperatures. This section deals with those materials that show potential at temperatures of 200 °C or higher; the Friedel–Crafts resins are not included in this chapter as they are dealt with in Chapter 4.

6.3.2. Phenol–Formaldehyde Resins

The phenol–formaldehyde resins were the first wholly synthetic polymers to be utilised, first going into production in 1910. Many papers and books have been written on the subject so it will be mentioned only briefly here. The formation of resinous products by the reaction of phenols and aldehydes has been known for a long time, and was studied especially by Baeyer[34] in 1872. There was, however, little interest in resins at that time. The classical work of Baekeland[35] opened the way to the large scale manufacture of articles made from these resins.

The preparation of phenol–formaldehyde resins is described in Chapter 1. Briefly, the resins are prepared by the reaction of phenol (or a mixture of phenols) with formaldehyde. Cresols, xylenols and resorcinol are also used, but to a much lesser extent than phenol itself. Several types of low molecular weight prepolymer are produced commercially, but these can conveniently be divided into the so called resols and novolacs.

Resols are produced when phenol and excess formaldehyde are reacted under alkaline conditions to give a complex mixture of mono and polynuclear phenols with methylol substituent groups.

The novolacs are produced when formaldehyde and excess phenol are reacted under acid conditions. This results in polynuclear phenols linked by methylene groups, as shown:

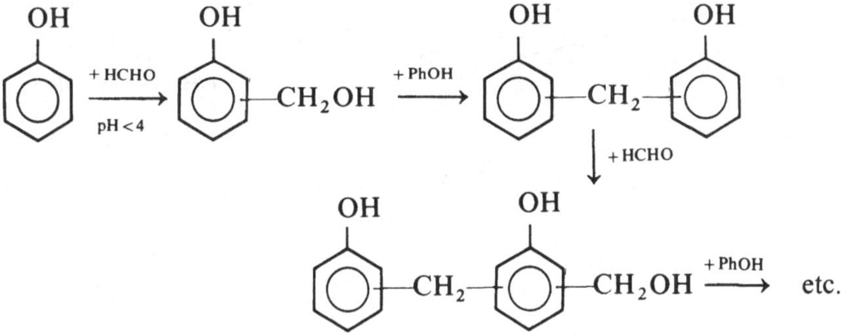

Thus novolacs are mixtures of isomeric phenols of various chain lengths but with an average of 5–6 benzene rings/molecule and a range of from 2 to 13 as shown by molecular weight determination. Some unreacted phenol and water are also usually present. Unlike resols, novolacs contain no reactive methylol groups and, therefore, do not condense on heating. In order to obtain cured material they need a suitable crosslinking agent, such as hexamethylene tetramine or paraformaldehyde. The mechanism of the crosslinking process is not fully understood but the final structure is not very different from that obtained with the resols. In both crosslinking reactions volatiles are evolved and, therefore, all manipulations have to be performed under pressure to prevent the formation of voids.

6.3.3. Thermal Degradation of Phenol–Formaldehyde Resins

Pyrolysis of phenolic resins always gives higher homologues of the phenols originally used.[36][37] The probable decomposition products of a novolac are:

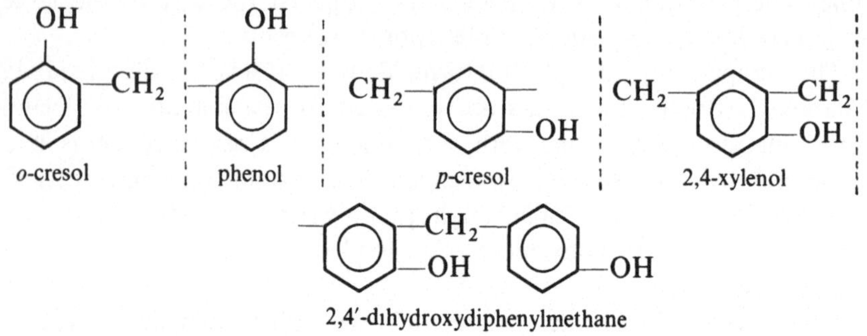

o-cresol   phenol   p-cresol   2,4-xylenol

2,4'-dihydroxydiphenylmethane

It has been shown that when heated to 400 °C in air the initial degradation reactions are:[38][39]

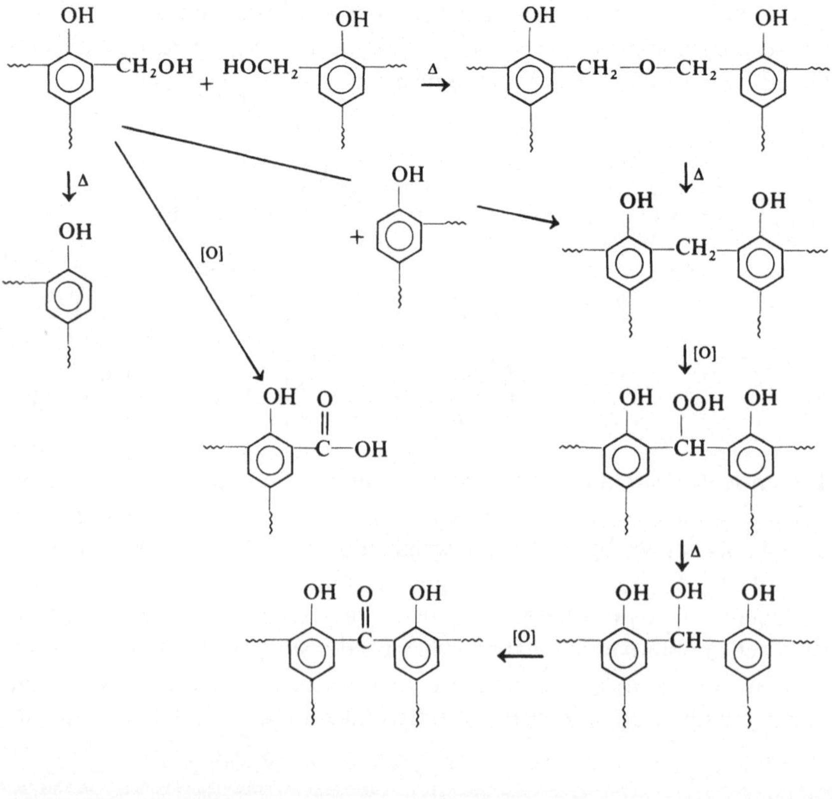

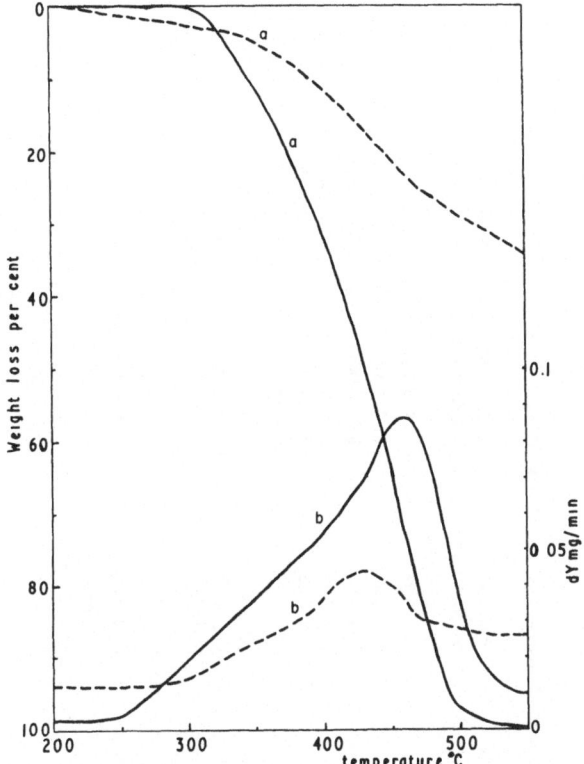

FIG. 6.8.   Phenol–formaldehyde resin—(a) weight loss and (b) rate of weight loss (dY). ———— in air; – – – – in nitrogen.

An illustration of the TGA of a phenol–formaldehyde resin in air and nitrogen is shown in Fig. 6.8. This shows the good thermal stability of the resin. Electrical properties are not very good because of the presence of polar groups, and the resins show relatively poor tracking resistance under conditions of high humidity. In respect of chemical resistance they are practically unaffected by water or organic solvents. They are resistant to all except strong oxidising acids, but are readily attacked by alkalis. Resins based on cresols have better acid and alkali resistance. Some typical properties of phenol–formaldehyde composites are given in Table 6.6.

6.3.4. Melamine–Formaldehyde Resins

The production of melamine–formaldehyde resins began in the mid 1930s, the first patent being granted in 1935.[40] Melamine is a cyclic trimer

TABLE 6.6

TYPICAL PROPERTIES OF RESIN GLASS CLOTH LAMINATES

Property	Units	ASTM test method	Phenol formaldehyde	Melamine formaldehyde	Silicone	Furan[a]
Laminating temperature	°C	—	135-175	135-150	160-250	RT-80
Laminating pressure	MPa	—	0·1-14	7-13	0·2-14	0 upwards
Tensile strength	MPa	D638	60-350	175-500	70-265	80
Tensile modulus	GPa	D638	8-17	14-17	10-14	6
Flexural strength	MPa	D790	110-560	245-595	70-265	153
Compressive strength	MPa	D695	235-520	170-590	170-320	60
Impact strength Izod	kJ/m^2	D256	20-95	25-80	25-70	—
Water absorption in 24 h	%	D570	0·12-2·70	0·20-2·50	0·07-0·65	1·3
Heat resistance, continuous	°C	—	120-260	150	200-370	200-250
Dielectric strength 3 mm thickness	kV/mm	D149	12-28	8-24	7-19	4·7
Dielectric constant at 1 MHz	—	D150	3·7-6·6	6·0-9·0	3·7-4·3	—
Dissipation factor at 1 MHz	—	D150	0·005-0·050	0·011-0·025	0·005-0·010	—
Arc resistance	s	D495	Tracks	175-200	150-250	—

[a] Chopped strand glass cloth laminate.

of cyanamide. Like urea it forms colourless resins with formaldehyde, but the melamine resins are superior to those from urea in heat and water resistance.

The melamine reacts with formaldehyde under slightly alkaline conditions giving methylol derivatives with up to six methylol groups/molecule.

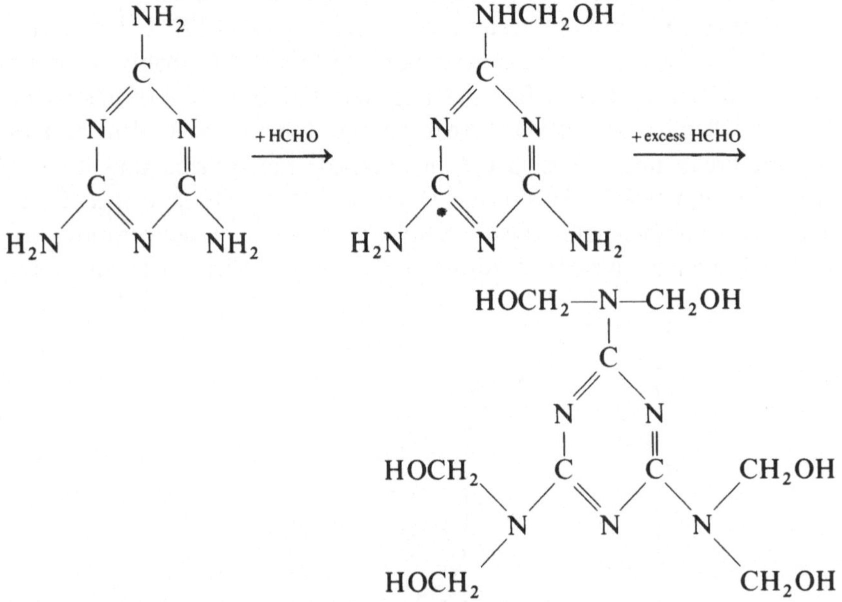

$$\text{melamine} \xrightarrow{+\,\text{HCHO}} \text{monomethylol melamine} \xrightarrow{+\,\text{excess HCHO}} \text{hexamethylol melamine}$$

The triazine ring structures bear NH_2, $NHCH_2OH$, $HOCH_2-N-CH_2OH$ and CH_2OH substituents as shown.

The presence of all possible methylolmelamines in a reaction mixture has been demonstrated.[41]

On heating, methylolmelamines condense to form resinous products. The rate of resinification is strongly dependent on pH, but for practical purposes sufficient crosslinking occurs under the action of heat alone. At temperatures of 150°C the crosslinking takes place by the reaction of the methylol groups, water and formaldehyde being evolved. A cured resin may be represented as

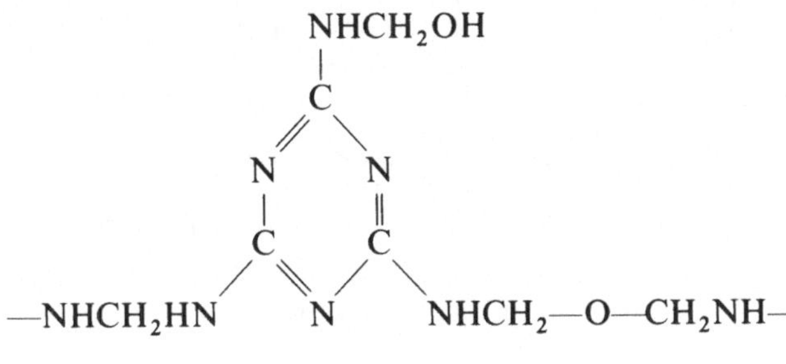

A cured triazine network bearing $NHCH_2OH$, $-NHCH_2HN-$, and $NHCH_2-O-CH_2NH-$ linkages.

There will be some reacted methylol groups, some methylene and some ether linkages. All three groups are present in a normal cured resin though not necessarily in equal amounts; it is generally assumed that ether links predominate over methylene links.

6.3.5. Thermal Stability of Melamine–Formaldehyde Resins

The methylol groups have been shown to decompose at 140 °C.[42,43] formaldehyde being given off. The next bond to break is probably the ether bond, followed by loss of formaldehyde and the formation of methylene bridges.[43] Under inert atmosphere it is believed that the methylene bridges do not break down until 380 °C is reached; the triazine ring itself is stable up to 400 °C.[43] This means in practice that so long as inert fillers are used and the material is carefully post cured at temperatures up to 200 °C, then the melamine–formaldehyde resins do have a useful service

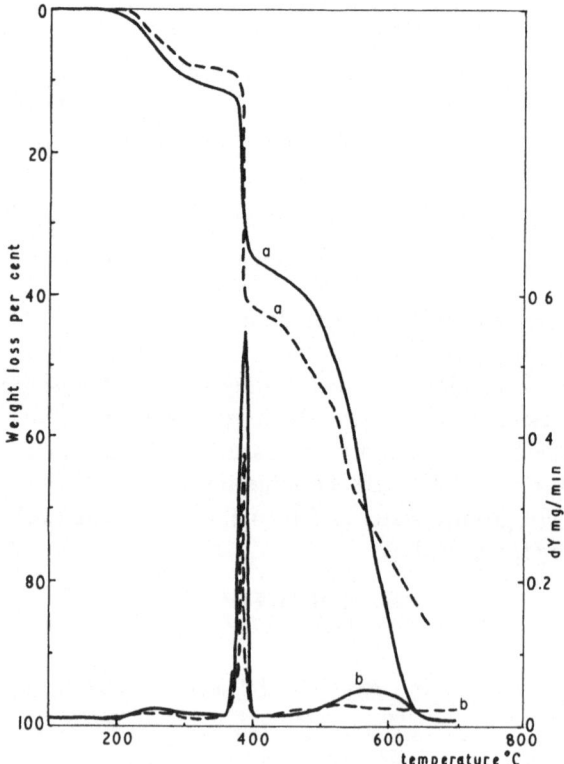

FIG. 6.9.   Melamine resin—(a) weight loss and (b) rate of weight loss (dY).
———— in air; – – – – in nitrogen.

life at elevated temperatures. The TGA curves for a melamine resin cured at 200 °C are illustrated in Fig. 6.9; this shows that the resin breaks down in a similar manner in air or in nitrogen, initial weight loss occurring at about 180 °C in air and 210 °C in nitrogen. It can be clearly seen that there are three phases to the degradation probably corresponding to the breakdown of ether links between 200 and 300 °C, methylene links at 370 °C to 380 °C, followed by the residual char decomposing between 400 and 700 °C. Some typical properties of these resins are listed in Table 6.6. Laminated with glass cloth, they have been used as arc barriers, switchboard panels, circuit-breaker parts and for other high duty electrical purposes.

6.3.6. Silicone Resins

The polyorganosiloxanes are polymers in which the backbone is comprised of alternate silicon and oxygen atoms, the silicon atoms being also joined to organic groups.[44,45] A representation of the structure would be

$$
\begin{array}{cccc}
& & \xi & \\
& & | & \\
\mathrm{R} & & \mathrm{O} & \\
| & & | & \\
\text{\textasciitilde\textasciitilde}\mathrm{Si}\text{---}\mathrm{O}\text{---}\mathrm{Si}\text{---}\mathrm{O}\text{---}\text{\textasciitilde\textasciitilde} & \\
| & & | & \\
\mathrm{O} & & \mathrm{R} & \\
| & & | & \mathrm{R} \\
& & \mathrm{R} & | \\
\text{\textasciitilde\textasciitilde}\mathrm{Si}\text{---}\mathrm{O}\text{---}\mathrm{Si}\text{---}\mathrm{O}\text{---}\mathrm{Si}\text{---}\text{\textasciitilde\textasciitilde} \\
| & | & | \\
\mathrm{R} & \mathrm{R} & \mathrm{O} \\
& & | \\
& & \xi
\end{array}
$$

The resins are manufactured by first formulating an appropriate blend of monoethyl, dimethyl, monophenyl, diphenyl, methylphenyl, monovinyl and methylvinylchlorosilanes with silicon tetrachloride. The final properties of the resin depend as much on the processing and cure conditions as on the original composition, but it is possible to generalise and say that trichlorosilanes produce hard resins, immiscible in other organic polymers, whereas dichlorosilanes increase softness and flexibility and phenylsilanes give resins that are more miscible in organic polymers, that are less brittle and have superior thermal resistance. The chlorosilane blend, commonly dissolved in inert solvents which modify the rate of reaction, is then mixed with water. Because of the difference in the rates of hydrolysis of the various chlorosilanes, problems can arise in producing polymers incorporating a proper balance of all the constituents. It is possible to control this by the choice of reaction solvent or, if necessary, by sequential addition of the chlorosilanes. After hydrolysis the resin incorporates a certain proportion of silanol groups, these serve to initiate the final cure of the resin to a fully crosslinked structure. This final cure is achieved by processing at elevated temperatures either with small amounts of residual acid left in the resin as catalyst or by incorporation of metal soaps such as tin or zinc octoate, cobalt naphthenate or amines such as triethanolamine.

6.3.7. Thermal Stability of Silicone Resins

Detailed reviews of the thermal stability of the silicones have been written.[46,47] Overall the silicones can be regarded as providing materials

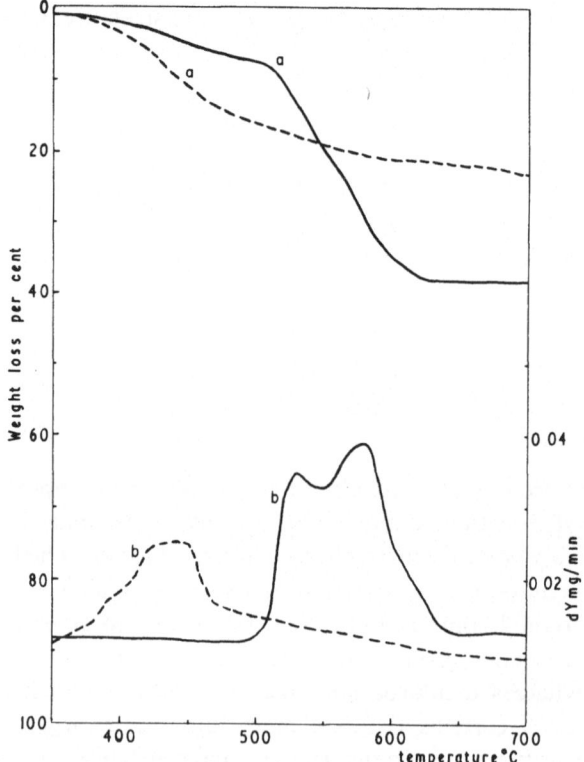

FIG. 6.10.   Silicone resin MS 840—(a) weight loss and (b) rate of weight loss (dY).
——— in air; – – – – in nitrogen.

with good high-temperature resistance as illustrated in Fig. 6.10, the TGA curves for a silicone resin in air and in nitrogen. It is known that branched-chain polysiloxanes are stable at temperatures in excess of 400 °C and tetramethylsilane vapour is stable to above 600 °C showing the stability of the Si–C bond in non-oxidative conditions. Oxidative attack takes place on the organic substituents, for example the methyl groups giving rise to evolution of formaldehyde, water, carbon monoxide and crosslinking through siloxane bridges. Thus the resin finally forms a silica residue. The incorporation of phenyl groups improves both the thermal and oxidative stability of a resin.

The resins are used as coatings, for laminating and, to a lesser extent, as moulding compounds. Glass fibre laminates are used where the combined properties of good electrical insulation and heat resistance are required. The mechanical properties of the laminates, given in Table 6.6, are inferior to those of epoxy or polyester resins. In sealed equipment the gradual loss of volatiles from these resins can cause build up of non-conducting silicone deposits on contacts or commutator brushes, but this has been partially overcome by a special design of the brushes.

6.3.8. Furan Resins

The basic chemistry and properties of the early systems were reviewed by Dunlop and Peters[48] in 1953. The intermediate used in the production of furan resins are derived entirely from natural resources by the steam distillation of such food by-products as corn cobs and oat husks. The furfural and furfuryl alcohol so produced may be resinified by the action of heat in the presence of an acid catalyst. The reaction is a condensation one with the elimination of water to form a linear polymer.

The polymer is normally neutralised and further diluted with monomeric furfuryl alcohol or furfural to give a low viscosity resin suitable for use by the fabricator. Once stabilised the resins may be stored many months without change in activity or viscosity.

The final reaction stage, again in the presence of an acid catalyst, produces a highly crosslinked infusible resin (the mechanism is uncertain). Control of this reaction by suitable catalysts led to the development of laminates with good mechanical properties.[49-53]

The densely crosslinked structure and absence of polar groups gives the resins a good resistance to all chemicals except concentrated sulphuric and hydrofluoric acids, strong caustic soda and oxidising media. In addition they have high heat distortion temperatures and good thermal stability. They are also suitable for applications where fire retardancy and low smoke generation are required. On heating, the furan resins form a very stable char with little loss of volatile material. For example, weight loss in air starts at 300 °C, 10 % weight loss occurs at 380 °C and 50 % weight loss occurs at 530 °C.[51] Some examples of the properties of glass fibre laminates are given in Table 6.6 and an illustration of the strength retention at temperature is given in Fig. 6.11.

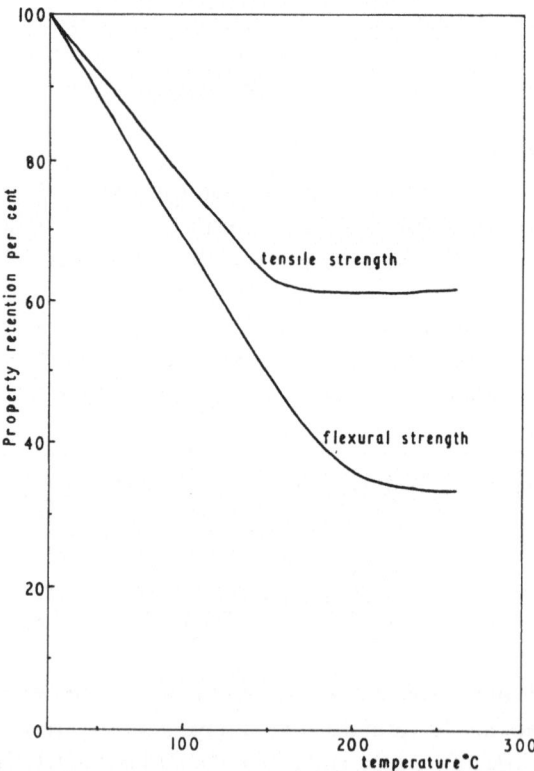

FIG. 6.11.   Property retention of Furan resin chopped glass fibre laminates with varying temperature.

6.3.9. Polyimide Resins—Introduction

The polyimides comprise the most thermally stable family of organic resins currently made. They are available in several forms suitable for different purposes. Films and varnishes for electrical uses were introduced commercially in 1961, to be followed later by moulding powders, laminating resins and metal-to-metal adhesives. Fibres and foams have also been evaluated.

The aromatic imide structure may be represented as

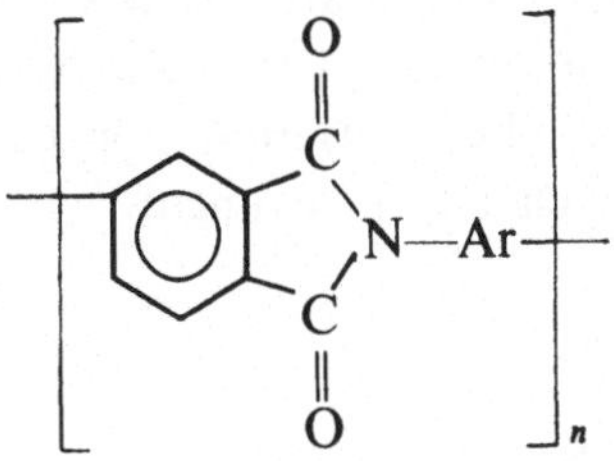

and it is very stable both thermally and oxidatively. Polymers of this basic structure have very high melting points, often above their decomposition temperatures, and the history of the polyimides is the development of processable resin systems.

6.3.10. Condensation Type Polyimides

The first polyimides to be developed were prepared from aliphatic diamines by a melt fusion of the salt formed from the diamine and tetra-acid, or diamine and diacid/diester.[54]

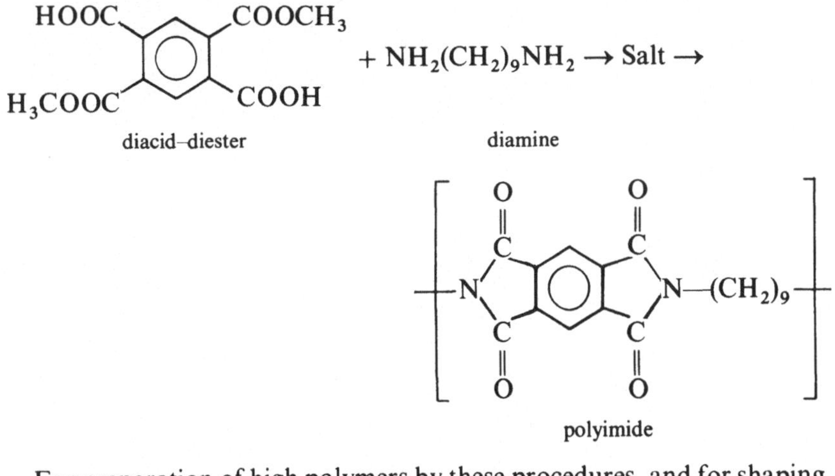

diacid–diester   diamine

$$+ NH_2(CH_2)_9NH_2 \rightarrow Salt \rightarrow$$

polyimide

For preparation of high polymers by these procedures, and for shaping, the polyimide had to be fusible. As a result, the synthesis of useful polypyromellitimides required the use of aliphatic diamines that possessed nine or more carbons in a normal chain, or seven in a branched chain.[55] These long chain aliphatic groups make the polymer susceptible to thermo-oxidative degradation.

The method of preparation of fully aromatic polyimides was developed about twenty years ago.[56-58] It involves the synthesis of a soluble polyamic acid precursor and a final cyclisation step to the required polyimide.

The dianhydride and diamine are reacted at ambient temperatures in polar solvents such as dimethyl formamide, dimethyl acetamide or N-methyl-2-pyrrolidone. The cyclisation step can be accomplished by heating or by a suitable chemical treatment.

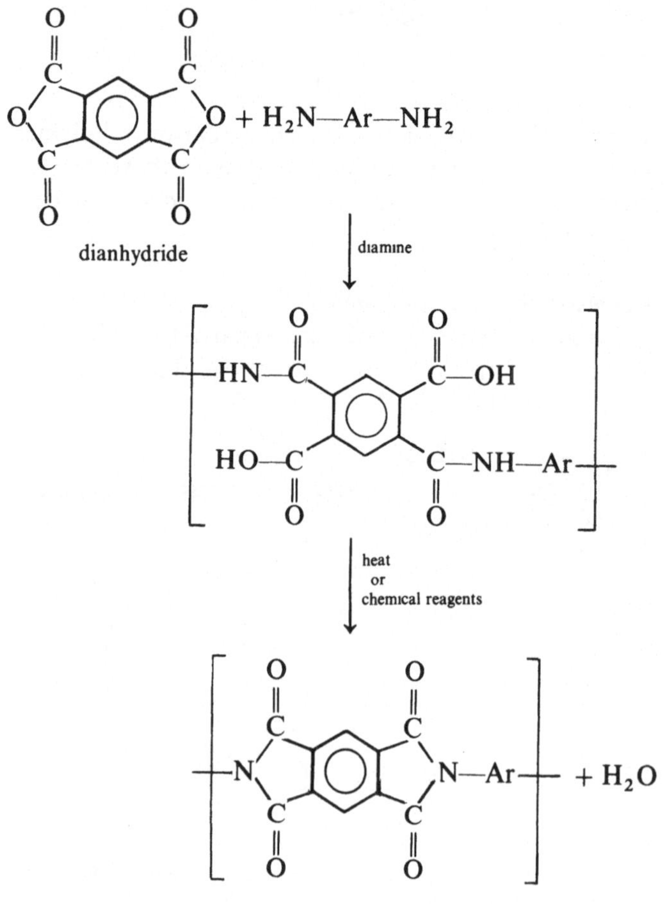

Pure reagents and solvents are necessary for the synthesis of high molecular weight polymers and the ratio of reactants used and the method of mixing are also critical factors.[59,60] The intermediate polyamic acid solutions are unstable, necessitating storage under dry refrigerated conditions. The polyimides finally produced are usually insoluble and infusible and since the temperatures for fusion of the intermediate and the cyclisation reaction are similar, precipitation effects markedly influence the flow characteristics and make for difficulties in moulding or laminating. Matters are not helped by the evolution of water during the process. All these factors limit the size and thickness of articles which can be made. Nevertheless, despite these disadvantages the condensation type polyimides are being used very successfully. Du Pont's Pyralin and Monsanto's Skybond resins are examples of the condensation type of polyimide. The exact constitution of these resins is not known but essentially they consist of the polyamic acids derived from either pyromellitic dianhydride, benzophenone tetracarboxylic dianhydride or diesters of these acids with diamines such as 4,4'-diaminodiphenyl methane or diaminodiphenyl ether. By using the esters the stability of the polyamic acid intermediate is improved, giving better shelf lives for the resin solutions at room temperature. The cure reaction is also slower giving better resin flow and allowing the use of vacuum bag autoclave rather than the comparatively expensive heated press. However, the volatiles generated during the cyclisation reaction, ethanol or methanol, can give rise to voids in the final product.

6.3.11. Thermal Degradation of Condensation Polyimides

The condensation polyimides are very stable as can be seen from Figs. 6.12 and 6.13, the weight loss curves in air and nitrogen for Kapton film and Skybond 700. These polymers are thought to have the following structures:

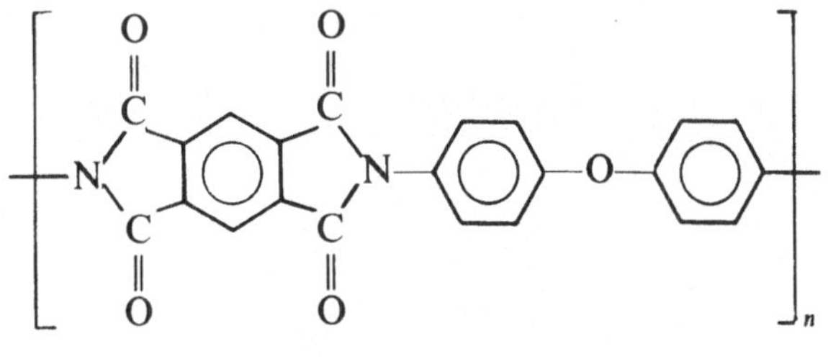

Kapton film from Du Pont

174   G. J. KNIGHT

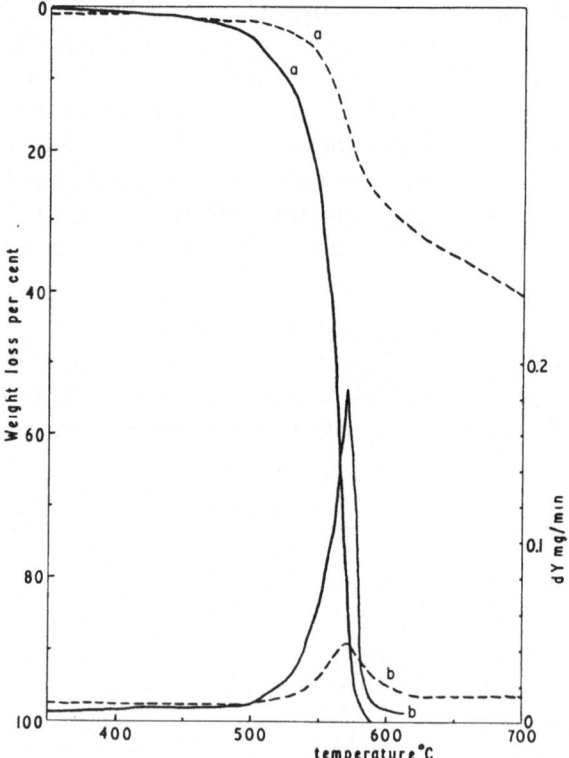

Skybond 700 from Monsanto

The figures show that the polymers are equally stable in air or in nitrogen, weight loss not occurring until over 450 °C, the dY plots indicate that the maximum rate of weight loss occurs at about 570 °C. Pyrolysis experiments in vacuum have shown that the principal volatile decomposition products from these structures are carbon monoxide 59 % and carbon

FIG. 6.12.   Kapton film—(a) weight loss and (b) rate of weight loss (dY). ——— in air; – – – – in nitrogen.

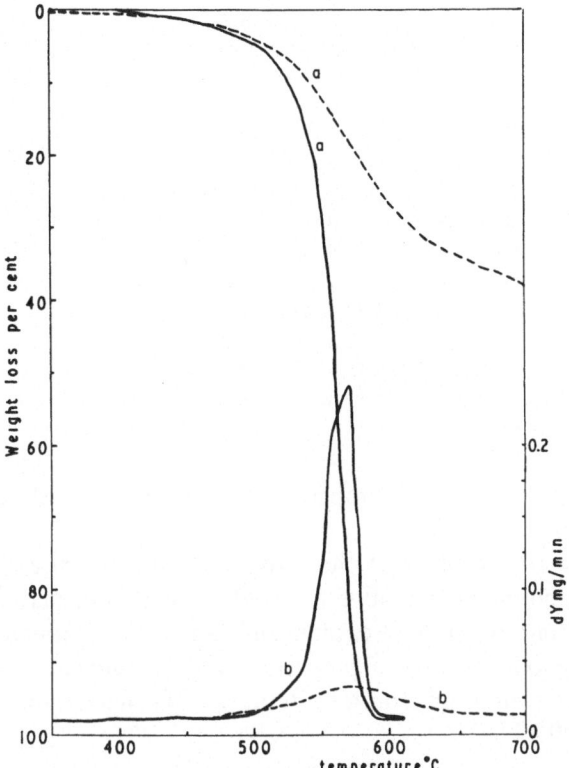

FIG. 6.13.   Skybond 700—(a) weight loss and (b) rate of weight loss (dY). ——— in air; – – – – in nitrogen.

dioxide 35%.[61-63] Pyrolysis of model compounds supports the view that the carbon monoxide derives from the decomposition of the imide ring, and the carbon dioxide is formed from the isoimide structure resulting from a rearrangement reaction.[64,65] Other experiments have shown that in air the reactions involved in weight loss are accelerated by moisture[66] and it has also been shown that considerable crosslinking and degradation takes place in the Kapton film before any weight loss is observed.[67,68] In fact after five hours at 400 °C, and only 2–3% weight loss, 50% of the diphenyl ether units had reacted and 30% of the pyromellitimide units had undergone modification.[69]

6.3.12. Processing and Properties of Condensation Polyimides
The processing conditions recommended for producing glass cloth laminates of Monsanto's Skybond 703 are shown in Table 6.7 and examples

TABLE 6.7

TYPICAL PRESS CURE AND POST CURE CYCLE FOR
SKYBOND 703[70]

B-stage prepreg at 121–149 °C
Preheat press to 204 °C
Prepreg into press–kiss contact for 1 min
Increase pressure to 1·72 MPa, leave for 30 min
Cool in press to 65 °C

Post cure to obtain optimum hot strength
   raise to 204 °C hold for 4 h
   raise to 260 °C hold for 4 h
   raise to 315 °C hold for 4 h
   raise to 343 °C hold for 4 h

---

of typical properties and property retention at elevated temperatures are shown in Table 6.8.

It will be realised that at the recommended post cure temperatures a certain amount of oxidative crosslinking is taking place. Work has also been done to prepare carbon and boron fibre laminates of these resins and good retention of properties has been shown at temperatures up to 315 °C, although for long term use the maximum temperature is nearer 250 °C.[71-74]

TABLE 6.8

TYPICAL PROPERTIES OF POLYIMIDE/
GLASS CLOTH LAMINATE AS PRE-
PARED IN TABLE 6.7

Resin content 30–35 %
Void content   ~5 %

Flexural strength
  at room temperature—489–545 MPa

  after   500 h at 260 °C[a]—455 MPa
  after 1 000 h at 260 °C[a]—407 MPa

  after   100 h at 315 °C[a]—427 MPa
  after   100 h at 343 °C[a]—172 MPa
  after   100 h at 371 °C[a]— 48 MPa

---

[a] Values given for elevated temperature ageing are the values that were obtained on the test specimens at the exposure temperature.

6.3.13. Addition Type Polyimides: Polyaminobismaleimides

One method of incorporating the imide structure into a polymer without the problems caused by volatile evolution is to make use of the activated maleimide double bonds. For example, 1,4-dimaleimidobenzene when irradiated at 240 °C yields a highly crosslinked polymer structure.[75]

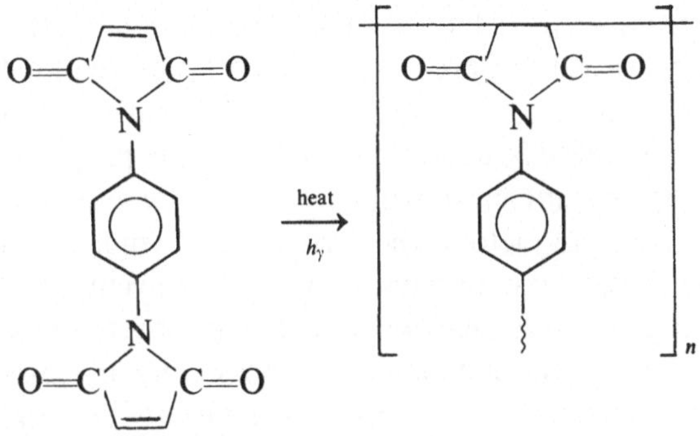

Di(4-maleimidophenyl) methane will form a polymer in the same way when heated.[76] Being so highly crosslinked these materials are very intractable and are very brittle. It is also known that amines will react with a maleimide group.[77] By combining these two reactions workers at Rhone–Poulenc have shown that a proper ratio of bismaleimide to aromatic diamine (e.g. diaminodiphenylmethane, DDM) will give polyimides with good physical properties as well as good thermal stability.[78]

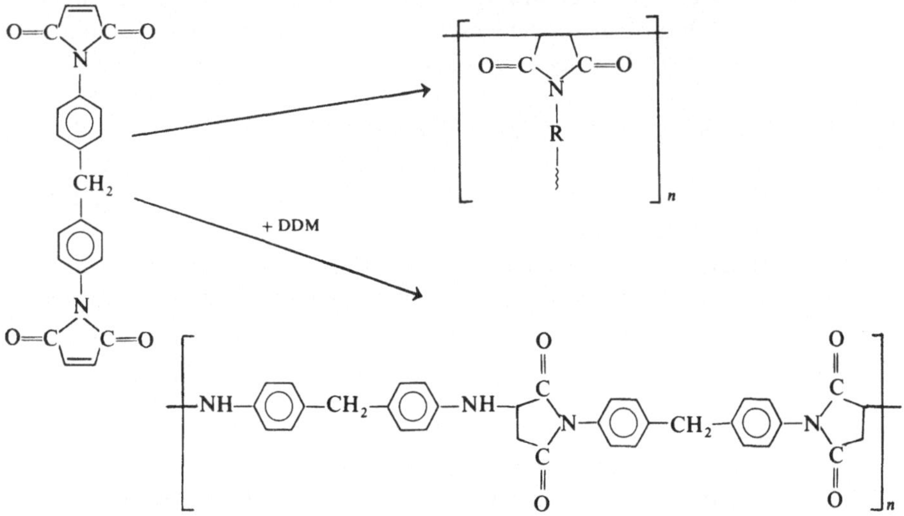

These resins are available from Rhone–Poulenc in forms suitable for moulding (Kinel) or for laminating (Kerimid 601). The latter is supplied as a yellow powder which is dissolved in N-methylpyrrolidone for preparing prepregs. Kerimid 601 is readily soluble in many organic solvents but an extensive study of polymer–solvent interactions has shown that many of the solvents produce laminates with poor thermo-oxidative stability, for example 1,2-dichloroethane affords laminates that blow apart on post cure.[79] For ease of processing, prepregs are required with drape and tack, this means leaving 10% or more solvent in the resin, which can cause considerable degradation in final laminate properties. The exact reasons for this are not clear from the literature but it is possibly a combination of several factors; the formation of voids in the composite by the volatile solvent, the incorporation of the solvent into the polymer structure during the cure reaction and the plasticisation effect of the excess solvent left in the laminate. The possibility of solvent–polymer interaction has been noted with respect to the effect of dimethylformamide on poly-N,N'-(4,4'-diphenyl ether)pyromellitimide.[66]

6.3.14. Thermal Degradation of Kerimid 601
An illustration of the thermal stability of the resin is shown in Fig. 6.14, the TGA curves in air and in nitrogen. Initial breakdown occurs at about 300 °C in air and at 320 °C in nitrogen, 10% weight loss being at 370 °C in air and at 385 °C in nitrogen. The dY plots show that the degradation is a two stage process in air with the first peak coinciding with the nitrogen curve at 410 °C. These values show how the changes in structure of the polymer considerably reduce the overall thermal stability when compared to the fully aromatic condensation type polyimides.

6.3.15. Processing and Properties of Kerimid 601
An example of the suggested processing conditions necessary to produce glass cloth laminates is shown in Table 6.9 and some typical mechanical properties of such laminates are shown in Table 6.10.

The strength retention at temperature is illustrated in Fig. 6.15 which shows that the polymer has good strength retention at temperatures up to 250 °C, but that for long-term use 180 to 200 °C is the upper limit.

6.3.16. Kerimid 353
In the search for easier processing a very interesting resin system is that developed by Technochemie GmbH and marketed by Rhone–Poulenc

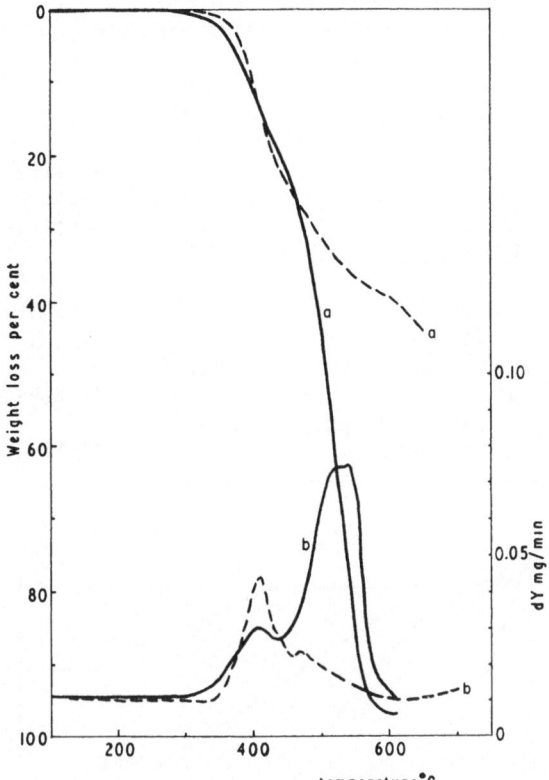

FIG. 6.14.   Kerimid 601—(a) weight loss and (b) rate of weight loss (dY). ——— in air; – – – – in nitrogen.

TABLE 6.9

PROCESSING CONDITIONS FOR KERIMID 601/GLASS CLOTH LAMINATES[80]

Preheat press to 120 °C

Place prepreg in press and pressurise to 1·45–5·52 MPa

Increase temperature progressively to 180 °C over 30 min

Cure for 1 h at 180 °C
(The laminate may be removed hot from the press)

To obtain maximum mechanical properties a post cure of 24 h at 250 °C or 48 h at 200 °C is recommended

TABLE 6.10

SOME TYPICAL MECHANICAL PROPERTIES OF
KERIMID 601/GLASS CLOTH LAMINATES[80]

Property	ASTM test method	Result
Flexural strength	D790	
at 25°C		480 MPa
at 200°C		414 MPa
at 250°C		345 MPa
Flexural modulus	D790	
at 25°C		27·5 GPa
at 200°C		26·2 GPa
at 250°C		22·1 GPa
Tensile strength	D638	
at 25°C		345 MPa
Compressive strength	D695	
at 25°C		345 MPa

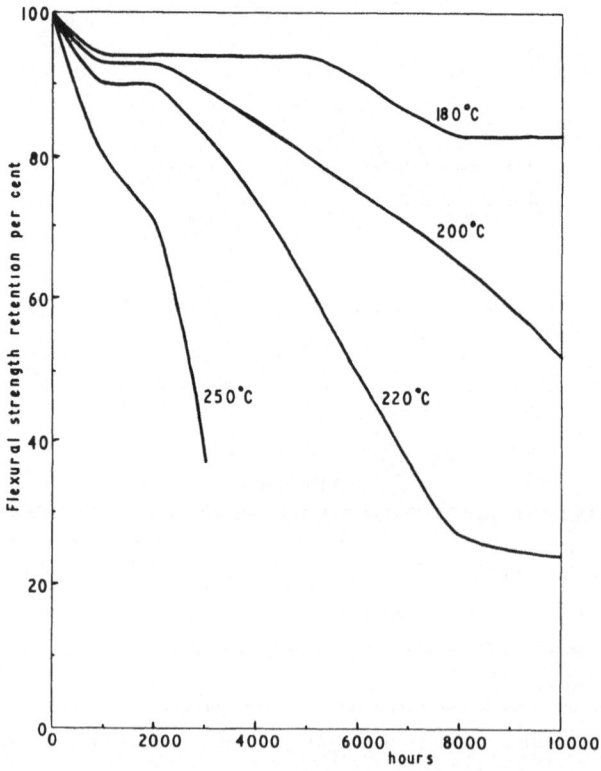

FIG. 6.15. Flexural strength retention of Kerimid 601 glass cloth laminates aged at 180, 200, 220 and 250°C.

as Kerimid 353. It is reported to consist of a mixture of three bismale-imides.[81] These form a eutectic mixture with an overall melting point of 70–125 °C. This is well below the curing temperature of 200–240 °C and means that prepreg can be prepared by means of a filament winding technique using molten resin. As solvent is not used there are no problems with removing solvent during lamination. The prepreg is stiff at room temperature but acquires tack and drape if warmed to 50 °C.

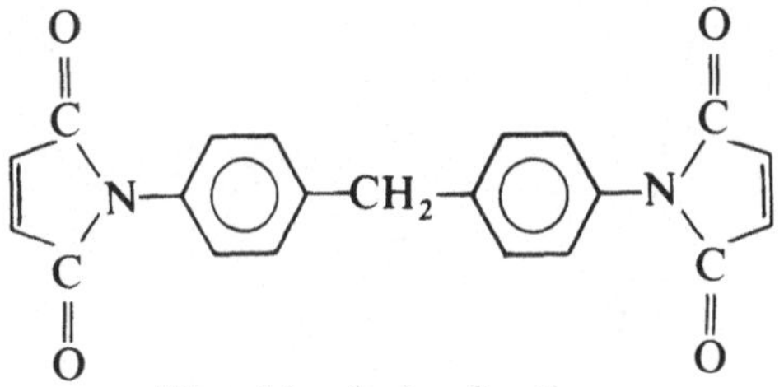

di(4-maleimodophenyl)methane   m.p. 154–156 °C

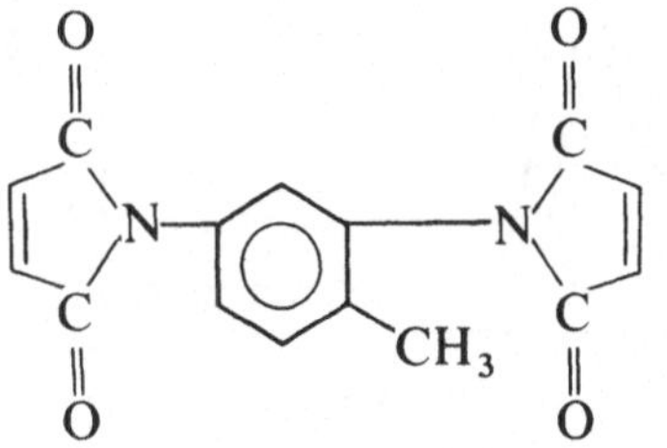

2,4-bismaleimido-toluene   m.p. 172–174 °C

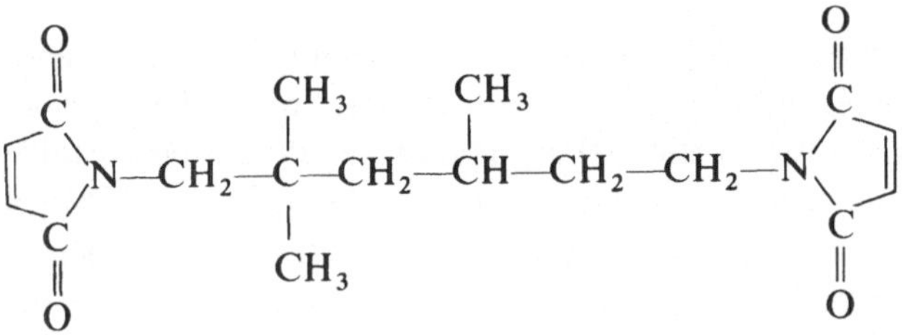

1,6-bismaleimido-2,2,4-trimethylhexane   m.p. 70–130 °C

6.3.17. Thermal Degradation of Kerimid 353

Figure 6.16 shows the TGA curves in air and in nitrogen of a sample of cured Kerimid 353 resin and illustrates the good thermal stability of the resin. Decomposition starts at about 350 °C in air and at about 400 °C in nitrogen, 10 % weight loss being at 420 °C in air and at 460 °C in nitrogen. Unlike Kerimid 601, which is equally stable in air or nitrogen,

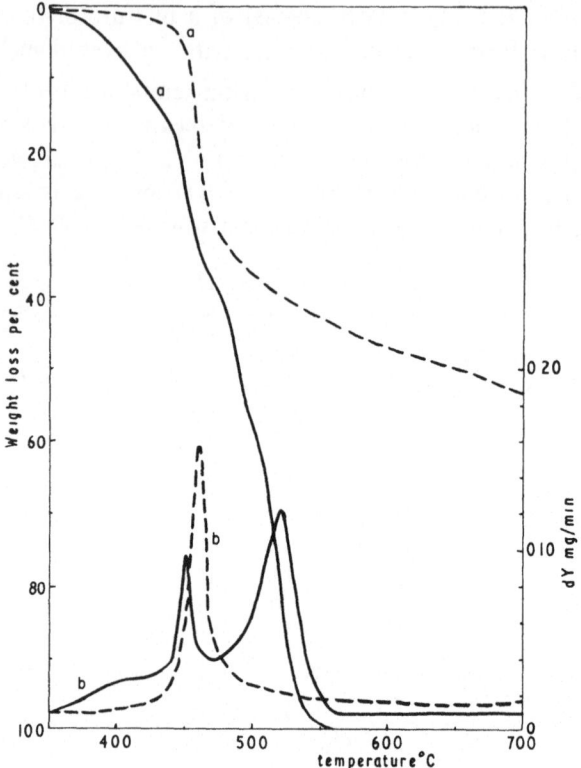

FIG. 6.16.   Kerimid 353—(a) weight loss and (b) rate of weight loss (dY). ———— in
air; ———— in nitrogen.

there is a marked decrease in thermo-oxidative stability. This difference
is not altogether surprising considering the incorporation of the aliphatic
bismaleimide. In fact the surprise is that the polymer should be as stable
as it apparently is. The dY plots also emphasise the difference between
the thermo-oxidative and thermal breakdown, with a three stage reaction
in air and just one period of rapid weight loss in nitrogen before the
development of a stable char.

Thermogravimetric analysis experiments have been reported on the
polymers derived from a series of aliphatic bismaleimides with different
numbers of methylene linking groups, also on several aromatic bismale-
imides, confirming the superior stability of the aromatic compounds.[82]
Pyrolysis mass spectrometry was also done on the same set of polymers
showing that the aliphatic polymers decompose by cleavage of the C–C
bonds in the aliphatic bridge, mainly in the vicinity of the N–C bond.

The aromatic polymers decompose in a similar manner to the condensation type polyimides by breakdown of the imide ring with evolution of CO and CO_2.[82]

6.3.18. Processing and Properties of Kerimid 353

Descriptions of the preparation of unidirectional glass, Kevlar and carbon fibre laminates are given in the literature.[83-85] An example of the autoclave cure of the resin is given in Table 6.11 and the properties achieved by the laminate are shown in Table 6.12.

The values in Table 6.12 illustrate the high strengths that can be achieved with the resin system and the comparative ease of processing. The property retention at 250 °C is also high but it has been shown that appreciable degradation of a carbon fibre/Kerimid 353 laminate takes place during isothermal ageing at 260 °C in air for 125 hours. It was found that coating the laminate with aluminium filled polyimide resin improved the resistance of the laminate to degradation.[84]

TABLE 6.11

RECOMMENDED CYCLE FOR AUTOCLAVE LAMINATION OF GLASS FIBRE KERIMID 353 PREPREG

Lay up prepreg at 45–60 °C
Apply full vacuum for 10 min, then heat to 100 °C whilst maintaining vacuum
Release vacuum, heat to 180 °C, hold for 1 h
Apply 0·3 MPa pressure and maintain at 180 °C for 30 min
Apply 0·7 MPa pressure and heat to 210 °C, hold for 4 h
Cool to room temperature
Post cure at 240 °C for 15 h

TABLE 6.12

PROPERTIES OF UNIDIRECTIONAL KERIMID 353-E GLASS COMPOSITE

Property	At room temperature	At 250 °C
Fibre content	60 %	—
Void content	<2 %	—
Density (g/cm³)	2·1	
Tensile strength (MPa)	1 058	1 028
Tensile modulus (GPa)	41	38
Flexural strength (MPa)	1 215	837
Flexural modulus (GPa)	42	40
Short beam shear (MPa)	76	50

6.3.19. Other Bismaleimide Polyimides

Other addition type polyimides based on bismaleimides are available commercially, examples of these are Kerimid 711, another product of Rhone–Poulenc, and Hexel F178, sold by the Hexel Corporation but only available in the form of prepreg.

Kerimid 711 has also been designed as a filament winding resin and consists of a mixture of components that give the resin a very low melting point of ~60 °C. The TGA of the cured resin is illustrated in Fig. 6.17 which shows the weight loss in air and in nitrogen. The initial weight loss starts at about 300 °C in air and in nitrogen, 10 % weight loss occurring at 390 °C in air and in nitrogen. The lower melting point and the lower stability when compared to Kerimid 353 suggests that Kerimid 711 might contain a greater proportion of aliphatic material. The recommended processing conditions for filament winding are to melt the resin by heating

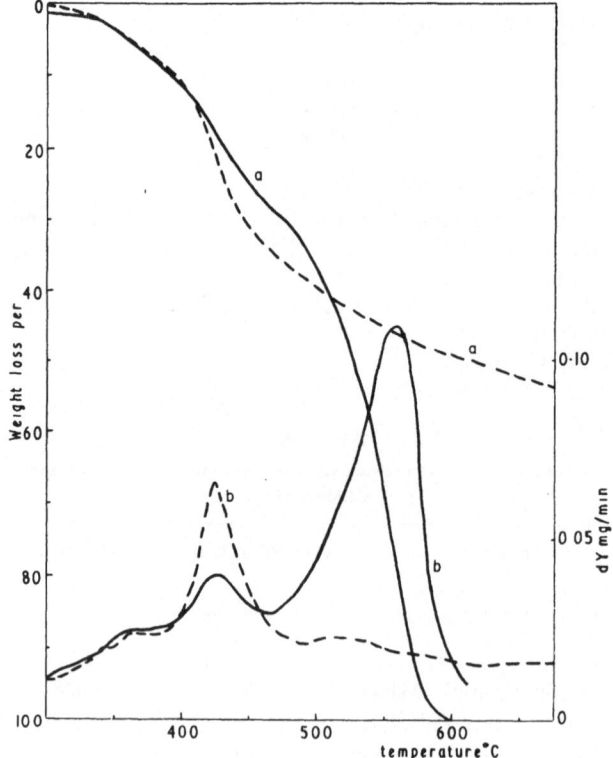

FIG. 6.17.   Kerimid 711—(a) weight loss and (b) rate of weight loss (dY). ——— in air; - - - - in nitrogen.

TABLE 6.13
PROPERTIES OF CURED KERIMID 711 RESIN[80]

Property		At room temperature	At 200°C
Density	(g/cm³)	1·3	—
Glass transition temperature	(°C)	280	—
Flexural strength	(MPa)	108	59
Flexural modulus	(GPa)	2·84	2·26
After 2 000 hours at 200°C			
Flexural strength	(MPa)	108	39
Flexural modulus	(GPa)	2·75	1·96

to 115 °C, and to wind the impregnated fibres onto a mandrel heated to 120–150 °C. The component is maintained at 150 °C for approximately 1 h then the temperature is raised to 200 °C over 2 h and cured for 4 h at 200 °C. The whole is then cooled slowly to room temperature. To achieve maximum mechanical properties a post cure of 15 h at 240°–250 °C is recommended. Some typical properties of the cured resin are given in Table 6.13. From the figures in Table 6.13 it can be seen that 200 °C is probably the upper temperature limit for use.

Hexel F178 can be obtained only in prepreg form. The resin is known to be of the bismaleimide type and it contains triallylisocyanurate.[86]

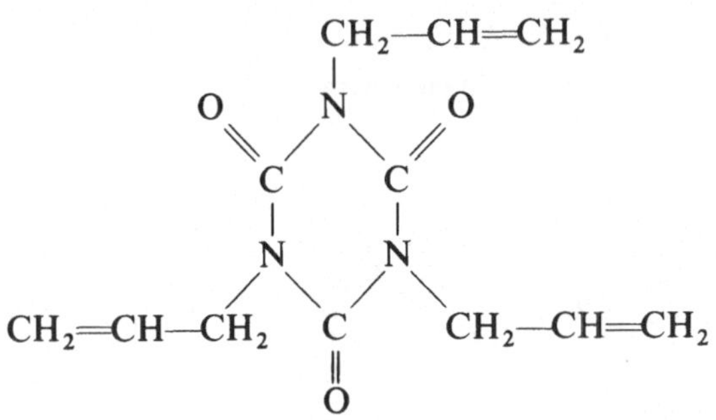

This compound will homopolymerise to give thermally stable materials and presumably it is also able to take part in the free radical polymerisation reactions of the bismaleimide. Its low melting point, 24 °C, means that it helps to make the F178 a solvent free system with good drape but no tack at room temperature; heating to 60–90 °C is sufficient to give good ply-to-ply tack. If extreme drape and tack are required then very small amounts of N-methylpyrrolidone can be added.[87] The suggested autoclave cure is 30 min at 149 °C and 0·4 MPa minimum pressure. A post cure is said not to be required.[87] But a post cure of 64 h at 204 °C is recommended for a carbon fibre reinforced component after an initial cure of 2 h at 177 °C and 0·6 MPa pressure.[86] The cured resin has a density of 1·28 g/cm³ and a glass transition temperature of about 275 °C. Weight loss experiments at 5 °C/min show 3 % weight loss at 375 °C and 7 % at 400 °C.[87]

6.3.20. Norbornene Systems

An alternative method that has been developed for obtaining addition type polyimides is the use of norbornene end groups rather than the maleimides described above. The method of synthesis originally adopted was the preparation of low molecular weight amic acid prepolymers end capped with the reactive norbornene rings.[88]

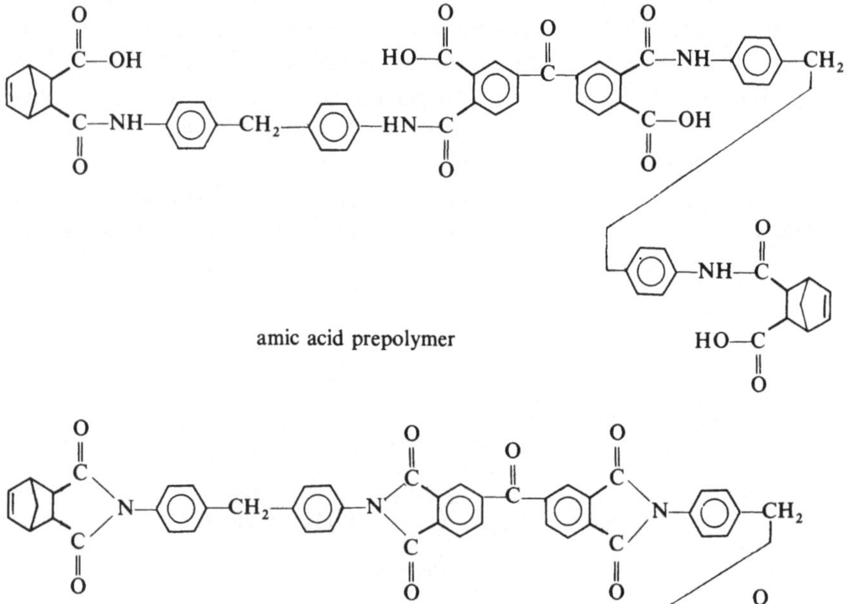

amic acid prepolymer

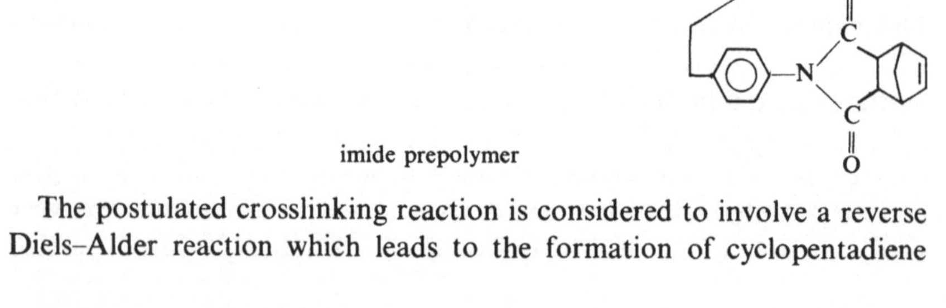

imide prepolymer

The postulated crosslinking reaction is considered to involve a reverse Diels–Alder reaction which leads to the formation of cyclopentadiene and a maleimide structure. These immediately react to form an adduct (I), and this adduct then initiates homopolymerisation of the norbornene species.[89]

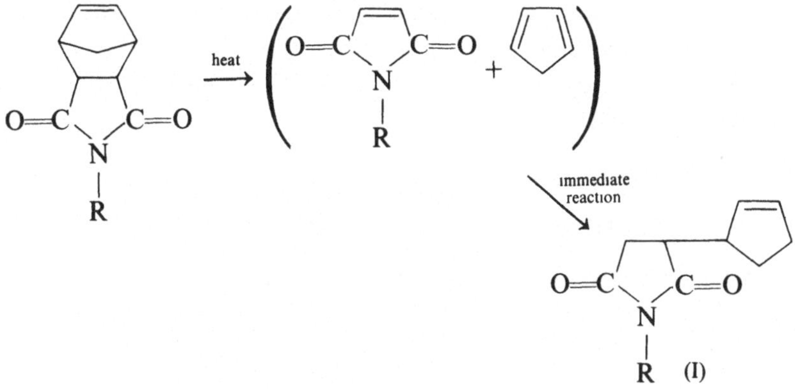

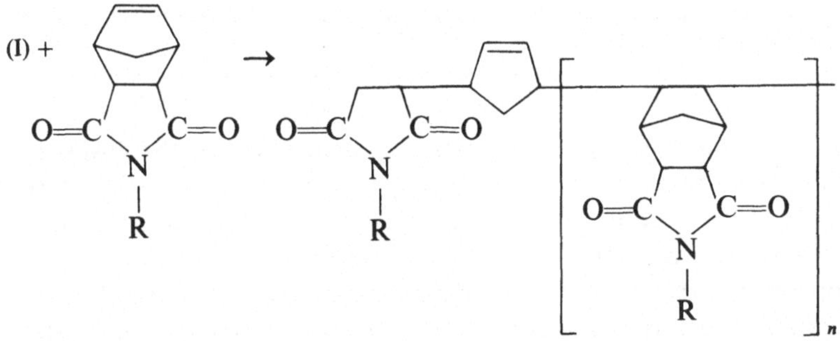

These polymers were successfully developed by TRW Inc. and marketed as P13N resin by Ciba–Geigy Corporation. Further developments were the use of pyromellitic dianhydride in place of benzophenone tetra-carboxylic dianhydride and the incorporation of 4,4'-diaminodiphenyl-sulphone in addition to 4,4'-diaminodiphenylmethane. These changes gave resins with improved long term thermo-oxidative stability, but it was observed that the prepolymer solutions exhibited a very limited shelf life at room temperature.[90] In addition, the preferred solvents, dimethyl formamide or N-methylpyrrolidone pose the same problems during processing as they do with the condensation polyimides. Many of these problems have been overcome by a method of polymerising the monomeric reactants in situ, i.e. the laminate or moulding being manufactured. The basic chemistry of these PMR resins is the same as the P13N type, but in this case the fibres are impregnated using a methanol solution containing the correct proportions of 4,4'-diaminodiphenylmethane (DDM), the dimethylester of benzaphenone tetracarboxylic acid (BTDE) and the monomethyl ester of 5-norbornene-2,3-dicarboxylic acid (NE). Using the molar proportions three parts DDM, two parts BTDE and two parts NE the prepolymer formed has a formulated molecular weight of 1500, hence the designation PMR-15.[91] Other reactant ratios have been examined giving PRM-11, 13 and 19. A recommended cure cycle is shown in Table 6.14.

TABLE 6.14

RECOMMENDED PRESS CURE FOR PMR-15 COMPOSITES[92]

Collimation and impregnation from 50% solids solution
Air or IR dry, oven stage to desired tack and drape
Imidise by heating, in oven or in tool at 120°–200°C for 1–3 h
Place in mould preheated to 230–315°C, dwell for 30 s–10 min
Apply pressure of 1–7 MPa and cure at 315°C for 1 h
To obtain optimum mechanical properties at elevated temperatures it is necessary to post cure at 343°C for 16 h

In addition to the press cure it is possible to use the PMR resins in autoclave procedures and also to produce mouldings with chopped fibre.[92] More recently, improved compositions employing the PMR concept have been developed[93,94] which exhibit significantly improved high-temperature performance while retaining the processing ease of PMR-15. The improved formulations contain p-phenylene diamine, the dimethyl-ester of hexafluoro-isopropylidene-bis-(4-phthalic acid) and the mono-methyl ester of 5-norbornene-2,3-dicarboxylic acid.

Some properties of HM-S carbon fibre/PMR-15 composites are shown in Table 6.15.

TABLE 6.15

PROPERTIES OF HM-S CARBON FIBRE/PMR-15 COMPOSITES[92]

Property	Without post cure		After post cure of 16 h at 343°C
	At room temperature	At 316°C	At 316°C
Flexural strength (MPa)	1 262	483	1 103
Flexural modulus (GPa)	185	104	173
Shear strength (MPa)	59	22	44

The density of the cured PMR-15 is $1.30 \, g/cm^3$ [87] and it has a T_g of about 335°C after the 343°C post cure. Higher temperature post cures can give higher T_gs, but with a concomitant lowering of interlaminar shear strength.[95] This lowering was interpreted as being due to microcracking within the composite, partly due to the residual stress that originates from the large differences between the matrix and fibre coefficients of thermal expansion. Such cracking is observed with epoxy resins and, consequently, can be expected to be greater with the polyimides as they require even higher temperatures for cure. But in this case there appears to be an additional factor in that the cracking was greater with some varieties of carbón fibre. It has also been observed that the carbon fibre can affect the thermo-oxidative degradation of the resin to a considerable extent, as measured on a weight loss basis. For example, PMR-II

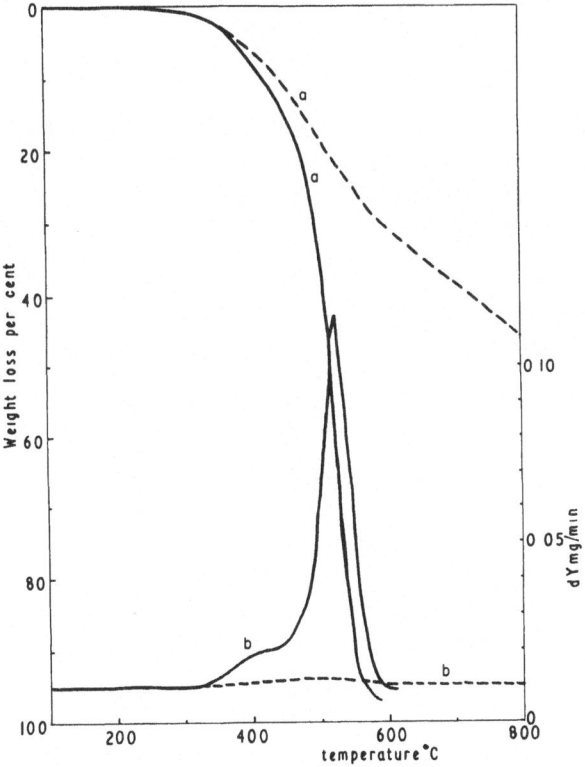

FIG. 6.18.   PMR-15—(a) weight loss and (b) rate of weight loss (dY). ———— in air; ———— in nitrogen.

resin alone loses 13 % by weight after 2100 h at 316 °C, whilst incorporation of HTS-1 carbon fibre causes a weight loss of 24–26 % under the
same conditions. This increase in degradation of the resin was interpreted
as being due to impurities in the carbon fibre—probably sodium salts.[96,97]

The TGA curves for cured PMR-15 are illustrated in Fig. 6.18 which
show the resin to be initially equally stable in air or in nitrogen, initial
weight loss occurring at about 350 °C. Ten per cent weight loss is not
reached until 420 °C in air or 445 °C in nitrogen. This is very similar
to the weight loss curves for Kerimid 601, Fig. 6.14, although PMR-15
gives a much higher residual char in nitrogen.

The bismaleimide and norbornene methods of crosslinking the polyimide do give resins with easier processability. It is possible to fabricate
composites with void contents < 1 % so that the mechanical properties
are as good as those achieved with epoxy resins, but because of the
alicyclic entities introduced into the polymer structure the thermal stability
is reduced compared to that of the condensation type polyimides.

6.3.21. Thermid 600

Another variety of thermosetting polyimide that cures by an addition
polymerisation process has been developed. This is marketed by Gulf
Oil Chemicals and is known as Thermid 600. The structure of the
prepolymer is shown below.

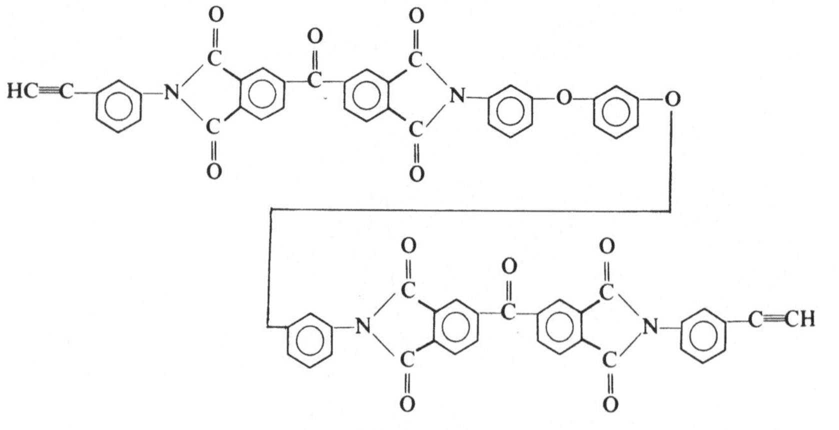

Thermid 600

The compound is prepared from 1 mole 1,3-bis(3-aminophenoxy)benzene,
2 moles benzophenone tetracarboxylic dianhydride and 2 moles 3-amino-
phenylacetylene.[98]

Polymerisation and cure occurs via a process which is believed to involve the trimerisation of the terminal acetylenic groups into aromatic rings. This theory has not been proven, but NMR studies on a model compound, N-(3-ethynylphenyl)phthalimide showed that new aromatic C–H groups were formed as the acetylenic groups disappeared. The high thermal stability of the cured resins also supports the formation of aromatic linking groups.

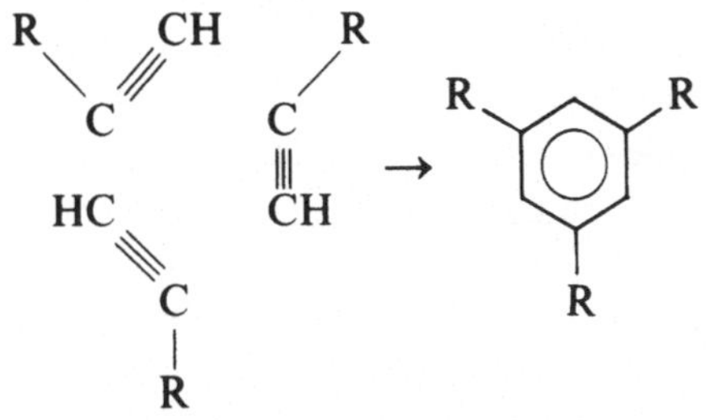

The prepolymers are readily soluble in dimethylformamide or N-methyl-pyrrolidone and these solutions can be used to apply the resin to fibres or fabrics in the manufacture of prepregs. As mentioned above such solvents are very difficult to remove and can cause trouble during processing and degrade the properties of the laminates. To obviate these problems a method of hot melt prepregging has been devised, which provides prepregs that are easily processed into composites.[99] The hot

TABLE 6.16

RECOMMENDED CURE CYCLE FOR GLASS OR CARBON FIBRE/THERMID 600 LAMINATES

Place prepreg into press preheated to 252 °C
Close to contact and maintain for 60 s
'Bump' lay-up (momentarily release the pressure) every 10 s for 30 s
Slowly increase pressure to 1·4 MPa
Cure for 1 to 2 h at 1·4 MPa and 252 °C
Cool under pressure to below 93 °C and remove from press
To obtain optimum high-temperature mechanical properties a post cure is
   recommended either in air or inert atmosphere
      Place laminate in cold oven
      Raise temperature to 232 °C rapidly
      Increase from 232° to 343 °C at 14 °C/h
      Hold at 343 °C for 4 h
      Increase from 343 °C to 371 °C at 7 °C/h
      Hold at 371 °C for 4 h
      Cool slowly in oven to below 93 °C before removing

---

melt process takes place at 310°C and the resin cures rapidly at this temperature. To prevent the cure going too far, the prepreg has to be chilled rapidly as it exits from the hot melt zone.

A further problem is that the resin will cure without melting if the heating rate of the lay-up is insufficient.[99] Heat up rates of the order of 8°C/min are required; such heating rates are readily achieved in a press but require specially modified autoclaves. Allowing for this it is possible to obtain good quality laminates by either press or autoclave cure. It is also possible to use the resins to prepare mouldings and as high temperature adhesives.

A typical cure cycle is shown in Table 6.16, examples of the properties of this resin and its laminates with glass and carbon fibre are shown in Table 6.17.

The TGA curves for a sample of cured Thermid 600 are illustrated

TABLE 6.17

TYPICAL PROPERTIES OF CURED THERMID 600 RESIN AND ITS LAMINATES WITH GLASS[a] AND CARBON FIBRE[b] [100]

Property	Resin	Glass	Carbon
Flexural strength (MPa)			
at room temperature	131	479	1 344
at 316°C	29	310	1 020
Flexural modulus (GPa)			
at room temperature	4·48	32·4	103
at 316°C	—	20·7	83
Short beam shear (MPa)			
at room temperature	—	64	83
at 316°C	—	45	55
Tensile strength (MPa)	83	—	—
Tensile modulus (GPa)	3·93	—	—
Compressive strength (MPa)	172	—	—
Effect of ageing at 316°C			
% weight loss—after 500 h	2·89	—	—
—after 1 000 h	4·04	9·72	10·00
Flexural strength after 1 000 h at 316°C			
at room temperature	92	162	600
at 316°C	18	160	572

[a] Glass laminates are 3·18 mm thick, 7 628 glass cloth W/CS 290 finish. Resin content 35%.
[b] Carbon fibre laminates are 1·59 mm thick, unidirectional, Hercules HTS fibre. Resin content 35%.

in Fig. 6.19, where the weight losses in air and in nitrogen are shown. From this it can be seen that weight loss starts at about 350 °C in both air and nitrogen, 10% weight loss occurring at 460 °C in air and 525 °C in nitrogen. On a weight loss basis this makes Thermid 600 almost as stable in nitrogen as Kapton or Skybond 700, good evidence, as mentioned

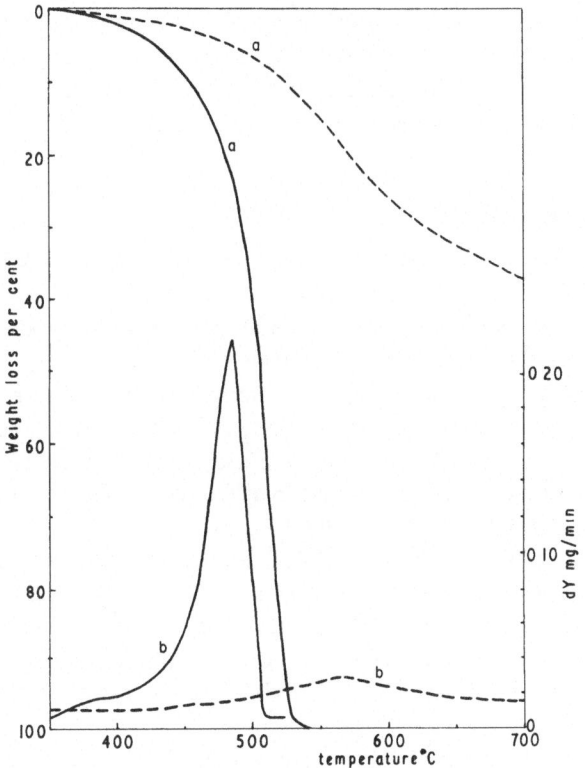

FIG. 6.19.   Thermid 600—(a) weight loss and (b) rate of weight loss (dY). ——— in air; – – – – in nitrogen.

above, for the highly aromatic character of the cured resin. The great reduction in oxidative stability shows that there are some fundamental differences between the resins, possibly related to the difficulty in getting all the acetylenic end groups to react fully during the cure. Any olefinic or acetylenic unsaturation left in the molecule would presumably act as a site for the initiation of thermo-oxidative degradation.

6.4. THERMOPLASTIC POLYIMIDES

6.4.1. Introduction

As has been shown above, the addition type polyimides although easier to process than the fully aromatic condensation type polyimides suffer a marked reduction in their overall thermal stability. An alternative method of improving processability is to introduce flexibilising groups into the polyimide molecule so that after the initial cyclising reaction the application of suitable temperatures and pressures can eliminate any voids formed. Two such products will be described here, one is Du Pont's NR-150 resin system, and the other is Polyimide 2080 marketed by the Upjohn Corporation.

6.4.2. NR-150 Polyimides

A considerable amount of work has gone in to developing these resins for use either as matrix resins for composites, as chopped fibre moulding compounds, or as high-temperature adhesives; several papers have been published on the subject over the past eight years.[101-110] The resins were originally sold as solutions in N-methylpyrrolidone or in N-methyl-pyrrolidone/ethanol mixtures, of a free tetrafunctional acid and a diamine which could be coated onto the reinforcement. These solutions have recently (January 1980) been withdrawn from the market, and Du Pont will now sell only fabricated products.

The basic chemistry of the NR-150 binder solutions and their cured products is shown below. When sold as solutions, they had a shelf life of at least nine months when stored at 4 °C or below.[110] The hexafluoropropylidene group flexibilises the molecule, lowering the T_g of the polyimides to 280–300 °C in the case of NR-150A2 and to 350–371 °C for NR-150B2.

The NR-150B material has a better thermo-oxidative stability than the NR-150A. An illustration of the thermal stability is given in Fig. 6.20 which shows the TGA curves obtained by heating cured NR-150A2 in air and in nitrogen. From this it appears that the polymer is equally stable in air or nitrogen, weight loss starting at about 300 °C in inert and oxidising atmospheres, but 10 % weight loss not being reached until 485 °C in air and 495 °C in nitrogen.

Because of the thermoplastic nature of the resin, prepregs based on NR-150 systems are easier to process than any other aromatic condensation type polyimides. High quality mouldings can be produced using a variety of processes. This processing versatility is indicated by the four methods

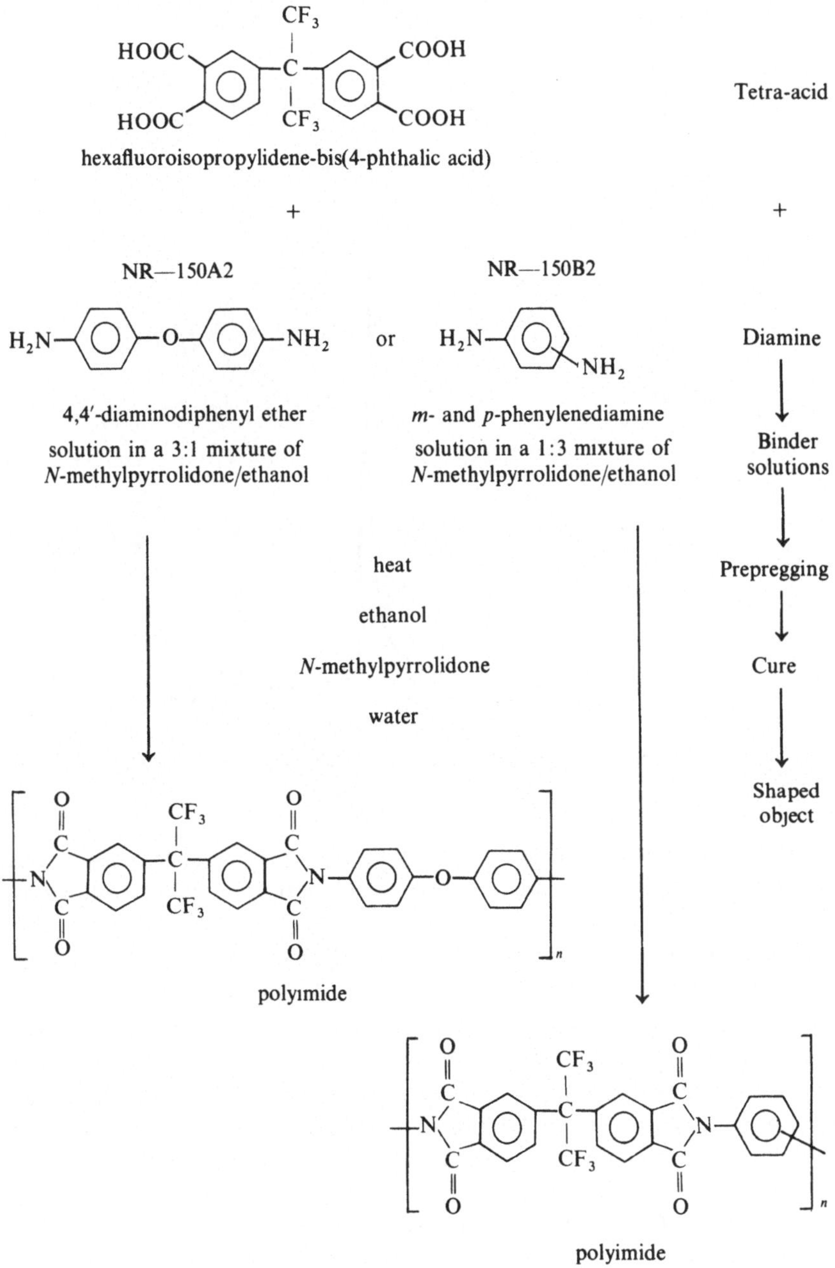

Tetra-acid

hexafluoroisopropylidene-bis(4-phthalic acid)

+

+

NR—150A2   NR—150B2

or

Diamine

4,4'-diaminodiphenyl ether   m- and p-phenylenediamine

solution in a 3:1 mixture of   solution in a 1:3 mixture of

N-methylpyrrolidone/ethanol   N-methylpyrrolidone/ethanol

Binder
solutions

heat

Prepregging

ethanol

N-methylpyrrolidone

Cure

water

Shaped
object

polyimide

polyimide

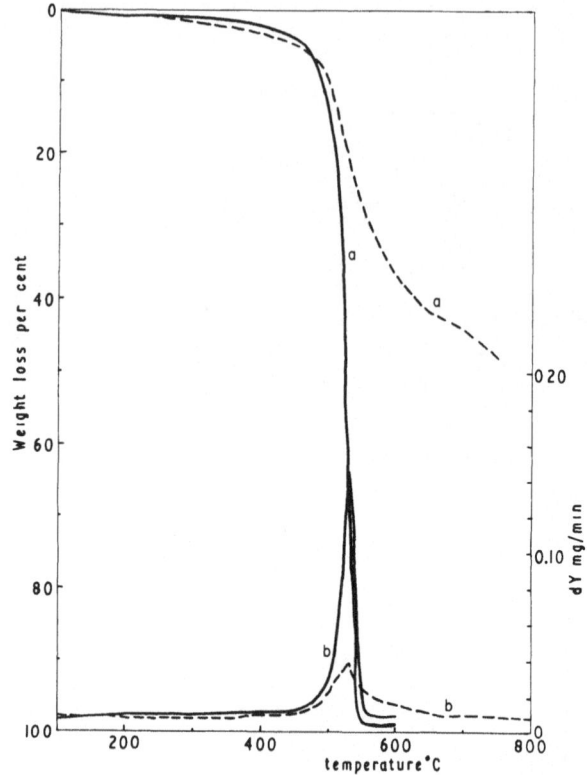

FIG. 6.20. NR-150A2—(a) weight loss and (b) rate of weight loss (dY). ———— in air; – – – – in nitrogen.

illustrated in Table 6.18 and some typical properties of the resins and composites made from them in glass and carbon fibre are shown in Table 6.19.

The effect of long term ageing on the mechanical properties of four different laminates from NR-150 solutions is shown in Table 6.20. It can be seen that there is a minimal effect of exposure on the mechanical properties at elevated temperature for all the laminates, though there is evidence that the room temperature properties are falling. There was no sign of any surface cracking or erosion. The reasons for this minimal reduction in mechanical properties may be associated with the small weight change of the samples after exposure and the glass transition data. The T_g for all samples increased during exposure, presumably as a result of oxidative crosslinking reactions as was seen to occur in Kapton films. Any increase in crosslink density would be seen as an

TABLE 6.18
RECOMMENDED METHODS OF PROCESSING NR-150 POLYIMIDE PREPREGS[111]

Method A:[a] Vacuum bag/autoclave at high temperatures
Precure
   1. Heat from room temperature to 200 °C at 2 °C/min under full vacuum
   2. Hold 30 min at 200 °C under full vacuum
Removal of volatiles and consolidation
   3. Heat to 343 °C over 2 h under full vacuum
Mould
   4. Hold 1 h at 343 °C under full vacuum and 1·4 MPa
   5. Cool under pressure

Method B:[b] Vacuum bag precuring followed by matched die moulding
Precure
   1. Heat from room temperature to 200 °C at 2 °C/min under full vacuum
   2. Hold 30 min at 200 °C under full vacuum
Mould
   3. Preheat matched die to 427 °C and insert precured laminate
   4. Hold 10 min at 427 °C and 17 MPa
   5. Cool to below the T_g under pressure no greater than 1·4 MPa

Method C:[c] Vacuum bag/autoclave moulding at low temperatures followed by vacuum bag/oven post cure
Precure
   1. Heat from room temperature to 175 °C at 2 °C/min under full vacuum
Consolidate and mould
   2. Hold 3 h at 175 °C under full vacuum and 1·4 MPa pressure
Removal of volatiles in oven
   3. Heat from 23 °C to 316 °C over 24–48 h under full vacuum
   4. Cool under full vacuum

Method D:[d] Vacuum bag/oven only
Precure
   1. Heat from room temperature to 200 °C at 2 °C/min under partial vacuum
   2. Hold 30 min at 200 °C under full vacuum
Consolidate and removal of volatiles
   3 Heat from 200 °C to 316 °C over 1 h under full vacuum
   4. Hold 1 h at 316 °C under full vacuum
   5. Heat to 343 °C and hold for 2 h under full vacuum

[a] Produces a 3·18 mm thick laminate using NR-150A2/181 style E-glass fabric. Void content <1%.
[b] Produces a 3·18 mm thick laminate using NR-150B2/181 style E-glass fabric. Void content <1%.
[c] Produces a 1·78 mm thick laminate using NR-150B2/S-2 glass fabric. Void content 3–6%
[d] Produces a 3·18 mm thick laminate using NR-150A2/181 style E-glass fabric. Void content 7–10%.

TABLE 6.19

PROPERTIES OF NR-150 RESINS AND THEIR COMPOSITES WITH GLASS AND CARBON FIBRE[111]

Property	Cured resin		Composites with E-glass fabric		Carbon fibre composites	
					Magnamite AS	Magnamite HMS
	NR-150A2	NR-150B2	NR-150A2	NR-150B2	NR-150A	NR-150B2
Laminate thickness (mm)	—	—	2·54 3·05	2·54 3·05	3·05	3·18
Volume % fibre	—	—	53	52	55	52
Volume % voids	—	—	1	1	1	2
T_g (°C)	280–300	350–371	—	—	257	360
Flexural strength (MPa)						
room temperature	97	117	510	558	1 530	869
250°C					641	—
260°C			407			
288°C				262	—	641
316°C					—	682
343°C					—	496
Flexural modulus (GPa)						
room temperature			23·4	30·3	110	145
250°C					96	—
260°C			28·3			
288°C				25·5	—	124
316°C					—	138
343°C					—	117
Tensile strength (MPa)						
room temperature	110	110			—	—
Short beam shear strength (MPa)						
room temperature			67·6	74·5	120·0	51·0
250°C					63·4	—
260°C			37·2			
288°C				28·3	—	37·2
316°C					—	31·7
343°C					—	31·7

TABLE 6.20

THERMO-OXIDATIVE STABILITY OF NR-150 LAMINATES EXPOSED IN AN AIR OVEN[111]

Composition resin reinforcement	NR-150A 47% quartz fibre		NR-150B2 53% E-glass fabric		NR-150B2 52% undirectional Magnamite HMS carbon fibre		NR-150B2 52% undirectional Magnamite HMS carbon fibre	
Ageing conditions temperature °C	260		316		316		343	
time at temperature (h)	20 000		1 000		3 000		500	
Properties	Control	After exposure	Control	After exposure	Control	After exposure	Control	After exposure
Flexural strength (MPa)								
room temperature	613	—	434	345	869	—	869	—
250°C	365	358^a	—	—	—	—	—	—
316°C	—	—	241	296	682	565	496	517
343°C	—	—	—	—	—	—	—	—
Flexural modulus (GPa)								
room temperature	24·1	—	33·1	27·6	145	—	145	—
250°C	22·1	20·0^a	—	—	—	—	—	—
316°C	—	—	23·4	26·9	138	131	—	—
343°C	—	—	—	—	—	—	117	117
Short beam shear strength (MPa)								
room temperature	93·8	—	68·9	30·3	51·0	—	51·0	—
250°C	29·0	33·8	—	—	—	—	—	—
316°C	—	—	22·1	22·1	31·7	40·7	31·7	40·0
343°C	—	—	—	—	—	—	—	—
Weight loss (%)	—	3·0	—	1·7	—	2·5	—	1·5
T_g (°C)	270	295	343	356	350	360	350	381

^a After 10 000 h at 260°C.

increase in flexural strength but would probably cause a decrease in impact properties.

By extrapolation of this data it can be shown that the time taken to lose 50 % of the original flexural strength measured at the test temperature is approximately 50 000 h at 260 °C, 5000 h at 316 °C and 1500 h at 343 °C. This extrapolation is made for low-void laminates reinforced with fibres such as E-glass, quartz or high modulus carbon fibre. Composites based on the high strength, or type A, carbon fibres would show a lower level of stability due to the poorer oxidative resistance of such fibres.

6.4.3. Polyimide 2080

Polyimide 2080 is unique compared to all the polyimides discussed above in that it is a polymer with a fully imidised structure, which still remains fusible and soluble in solvents such as dimethyl formamide or N-methyl-pyrrolidone. The polymer is prepared[112] by reacting benzophenone tetra-carboxylic dianhydride with toluene di-isocyanate (a mixture of the 2,4- and 2,6-isomer) in dimethyl sulphoxide solution. On completion of the first part of the reaction 4,4'-diisocyanate diphenyl methane is added and the reaction taken to completion. The aim is to produce a block copolymer with the approximate structure shown below.

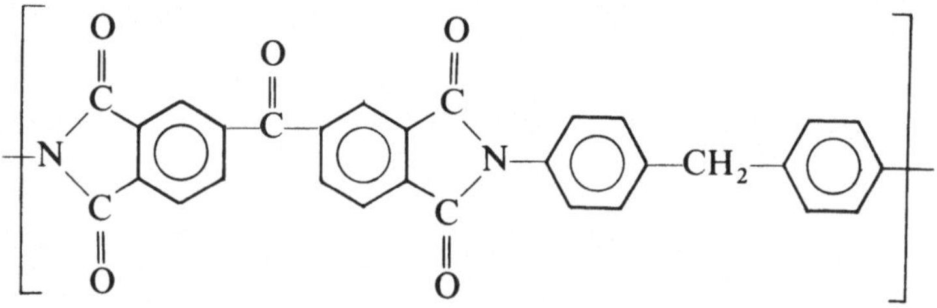

and

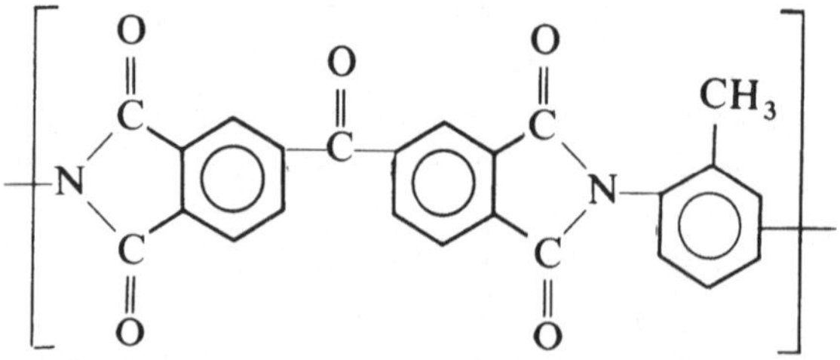

The polymer with this constitution has a T_g of 305 °C. The thermal stability of the polymer is illustrated in Fig. 6.21 which shows the TGA

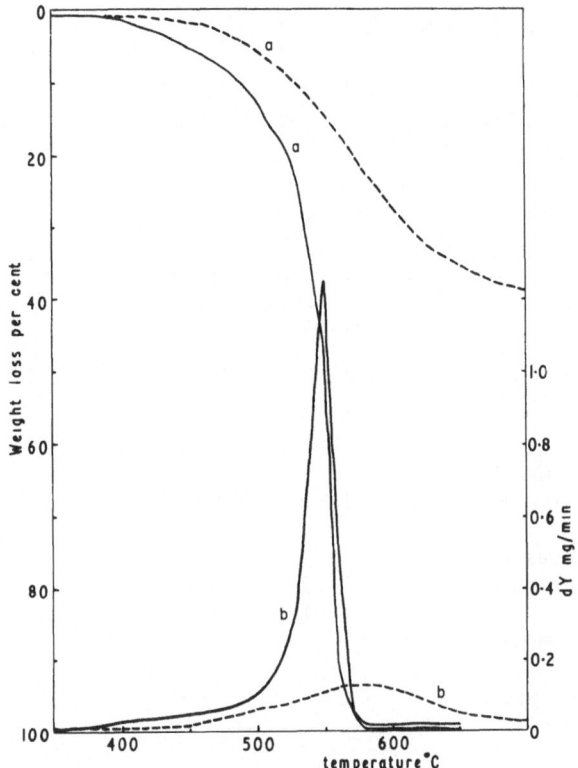

FIG. 6.21.   Polyimide 2080—(a) weight loss and (b) rate of weight loss (dY).
———— in air; —–—– in nitrogen.

curves obtained in air and in nitrogen. The initial weight loss is at 375 °C in air and 410 °C in nitrogen, 10 % weight loss is at 490 °C in air and at 530 °C in nitrogen.

6.4.4. Processing and Properties of Polyimide 2080
It is claimed that it is possible to obtain good quality laminates from Polyimide 2080 using either high pressure platen press, low pressure

TABLE 6.21
PRESS CURE OF POLYIMIDE 2080 LAMINATES

1. 30 min at 250 °C to reduce solvent level to less than 4 %
2. Raise temperature of press to 350 °C and bump repeatedly to release the remaining volatilised solvent
3. Hold for one hour in the press at 350 °C and 2·1 to 3·4 MPa

vacuum bag, or moderate pressure autoclave vacuum bag methods of
processing. No B-staging is required and no water is given off due to
ring closure, curing consists simply of solvent removal. The recommended
processing conditions are given in Table 6.21 and some typical properties
are shown in Table 6.22.

TABLE 6.22
POLYIMIDE 2080, RESIN AND LAMINATE PROPERTIES

Property	Resin[113]	Type 181E glass cloth, 30% resin[113]	Fortafil 5-T unidirectional carbon fibre, 30% resin[114]
Flexural strength (MPa)			
room temperature	198	379	1 027
288 °C	34	303	—
Flexural modulus (GPa)			
room temperature	3·3	27·3	116
288 °C	1·1	24·8	—
Interlaminar shear strength (MPa)			
room temperature	—	15·9	—

The process is essentially compression moulding, utilising the thermoplastic properties of the polymer. As in a conventional moulding process
the laminate can be formed in direct contact with the pressure surfaces,
eliminating the waste associated with bleeder cloths and resin flow.

6.4.5. Polyamide-Imides

Another method of introducing flexibilising groups into a polyimide is
via the ester or amide groupings. The polyester- and polyamide-imides
have been available for some time. Being soluble in polar solvents and
capable of forming films with good mechanical properties, they have
been used as varnishes and wire coating enamels. As with the other
linking groups described above, the overall thermal stability of the polymer
is reduced compared with that of the aromatic polyimides.

In addition to their use as insulating coatings, the polyamide-imides
have been developed for use as thermoplastic moulding compounds, which
may be used unfilled or reinforced with, for example, carbon fibre.[115-117]

The amide-imides are based on the combination of trimellitic anhydride with aromatic diamines.

These materials may be moulded, extruded or machined to produce the required articles. For compression moulding, temperatures of 325–350 °C are pressures of 21–28 MPa are required. To obtain optimum engineering properties the moulded parts require to be heat treated for 50–200 h at temperatures of up to 260 °C.

Examples of the engineering properties of a polyamide-imide, Torlon, and carbon fibre reinforced Torlon are given in Table 6.23.

TABLE 6.23

PROPERTIES OF TORLON RESIN AND RESIN CONTAINING CHOPPED CARBON FIBRE

Property	Torlon 2000	Torlon 4000	Carbon fibre reinforced Torlon	
			25% Fibre	50% Fibre
Tensile strength (MPa)				
room temperature	92	117	172	165
149 °C	81	79	124	103
260 °C	61	28	13	8
Tensile modulus (GPa)				
room temperature	—	—	19	31
149 °C	—	—	10	26
260 °C	—	—	1·6	1·8
Flexural strength (MPa)				
room temperature	161	189	272	299
149 °C	127	133	150	237
260 °C	98	62	66	60
Flexural modulus (GPa)				
room temperature	4·8	3·6	17	24
149 °C	3·7	2·8	11	23
260 °C	3·1	1·9	6	10
Shear strength (MPa)				
room temperature	130	111	109	103

The maximum temperature for use is about 250 °C because of the loss of mechanical strength at that temperature, but the retention of properties with time at temperature is very good, for example, 94% retention of strength after 2000 h at 260 °C.

6.5. PSP RESIN

6.5.1. Synthesis
Another resin capable of forming crosslinked networks without evolution of volatiles is the PSP resin.[118] The oligomeric prepolymer is prepared by condensation of aromatic dialdehydes with methylated derivatives of pyridine, in particular 2,4,6-trimethylpyridine.

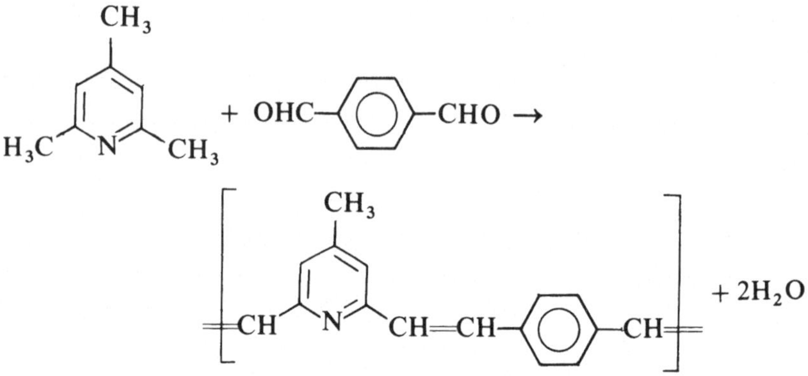

The synthesis can be arrested at various points producing either liquid resins, solids soluble in ethanol or acetone/ethanol mixtures, solids soluble in polar solvents such as dimethylformamide or N-methylpyrrolidone, or materials more suitable for use in injection or compression moulding. The prepolymer has been shown to have a good shelf life, over six months at 5 °C. The resin undergoes initial cure by heating at 200 °C, but to obtain optimum engineering properties a post cure of at least 2 h at 250 °C is required.[119] The resin so obtained has good thermal stability, TGA in air or in argon shows initial weight loss occurring at 300 °C and 10% weight loss at 435 °C in air, or 460 °C in argon. In inert atmosphere the pyrolytic residue is of the order of 65% at 1000 °C.[120]

6.5.2. Processing and Properties of PSP Resin
After impregnation the prepregs are dried at about 100 °C to a volatile content of about 10%, this gives good drape and tack but not too much

TABLE 6.24
RECOMMENDED PRESS CURE CYCLE FOR PSP LAMINATES

1. Place prepreg in mould, raise temperature to 200 °C at about 10 °C/min, hold for 1 h
2. Apply a pressure of 0·5–1·0 MPa, hold for two hours, maintaining the temperature at 200 °C
3. Post cure:
   2 h at 250 °C
   or 16 h at 250 °C
   or 3 h at 300 °C

The first post cure gives material suitable for use at temperatures up to 150 °C; the latter permits the use of the material at temperatures up to 300 °C.

flow when moulding. Such prepregs may then be moulded in a press or subjected to vacuum bag-autoclave cure. A recommended cure cycle is given in Table 6.24.

The materials prepared in this way have lifetimes of >10 000 h at 200 °C, 1000–1500 h at 250 °C and 120–150 h at 300 °C. Some of the properties of cured resin and its laminates with glass and carbon fibre are shown in Table 6.25.[119]

TABLE 6.25
PROPERTIES OF CURED PSP RESIN AND ITS LAMINATES WITH GLASS AND CARBON FIBRE[119]

Property	*Resin*	*Glass fibre E-181 fabric*	*Unidirectional carbon fibre HTS*	*Unidirectional carbon fibre HMS*
Flexural strength (MPa)				
room temperature	100	540	1 700	1 100
200 °C	—	530	—	—
250 °C	—	—	1 250	—
Flexural modulus (GPa)				
room temperature	2·5	30	110	150
200 °C	—	30	—	—
250 °C	—	—	110	—
Shear strength (MPa)				
room temperature	—	48	90	50
200 °C	—	31	—	40
250 °C	—	—	70	—

6.6. FIRE RESISTANCE

One property of these resins that has not been discussed above is their flammability. The high-temperature resins are all fire resistant. The polyimides in particular are very resistant, being self extinguishing and having very good limiting oxygen index figures. Where fire and smoke emission are known to be of consequence then any of the above polyimide resins could be recommended for use.

6.7. CONCLUSIONS

Overall, it can be said that a considerable effort has been expended on the synthesis of polymers with high-temperature capabilities. That the synthetic work has paid off can be seen in the number of commercially available resins especially the polyimides. Some of the latter are capable of service at temperatures over 300 °C and even at 400 °C for very short times. There is one major drawback to all these resins when the construction of large structural items for aircraft is being considered, i.e. the high temperatures and extended times required for post cure; 250 °C for a minimum of 15 h. Where composites are concerned, and especially carbon fibre reinforced composites, the internal stresses set up by the high temperature cure are very severe. Prolonging the time taken over the post cure can help to alleviate the stresses but then the costs of manufacture escalate rapidly.

It would appear that the processor would like a resin with a long shelf life at room temperature, yet capable of being cured at temperatures of no more than 50–100 °C, then with no post cure the structures should show good property retention at temperatures of 300 °C or even higher. It seems unlikely that the polymer chemists will ever develop such a resin system, and the processor will have to learn to come to terms with the exacting processing conditions if he wants to make use of the excellent high temperature properties that can be achieved with carbon fibre laminates.

REFERENCES

1. LEE, W. A. and RUTHERFORD, R. A. Glass transition temperatures of polymers, chapter in *Polymer Handbook* 2nd ed., eds. Brandrup, J. and Immergut, E. H., 1975, Wiley–Interscience, New York.

2. ROSE, J. B. *Polymer*, 1974, **15**, 456.
3. BARTON, J. M. Unpublished work.
4. BARTON, J. M. *Polymer*, 1970, **11**, 212.
5. THEBERGE, J. *Nat. SAMPE Symp.*, 1975, **20**, 72.
6. BROWNING, C. E. *28th SPI Reinforced Plastics/Composites Conf.*, 1973, Paper 15-A, pp. 1–16.
7. BROWNING, C. E., HUSMAN, G. E. and WHITNEY, J. M. *Comp. Mat. Test. and Design, 4th Conf.*, ASTM STP 617, 1977, pp. 481–96.
8. HERTZ, J., NASA-CR-124290, 1973.
9. MORGAN, R. J. and O'NEAL, J. E. *Polym. Plast. Technol., Eng.*, 1978, **10**, 49.
10. HUSMAN, G. E. *Nat. SAMPE Symp.*, 1979, **24**, 21.
11. HANSON, M. P., NASA-TN D-6604, Dec. 1971.
12. HASKINS, J. F., NASA-CP-001 Part 2, Nov. 1976, p. 799.
13. SCOLA, D. A., *Nat. SAMPE Symp.* 1978, **23**, 950.
14. VOGEL, H. A., UK Patent No. 1,060,546, March 1967.
15. JOHNSON, R. N., FARNHAM, A. G., CLENDINNING, R. A., HALE, W. F. and MERRIAM, C. N. *J. Polym. Sci.*, 1967, **5**, Part A1, 2375.
16. JONES, M. E. B., UK Patent No. 979,111, Jan. 1965.
17. VOGEL, H. A. *J. Polym. Sci.*, 1970, **8**, Part A1, 2035.
18. HALE, W. F. *ACS Polymer Preprints*, 1966, **7**, 503.
19. KNIGHT, G. J. Unpublished work.
20. DAVIES, A. *Makromol Chem.*, 1969, **128**, 242.
21. ALVINO, W. M. *J. Appl. Polym. Sci.*, 1971, **15**, 2521.
22. ABDUL-RASOUL, F. *Europ. Polym. J.*, 1977, **13**, 1019.
23. BROWN, J. R. *Polym. Letters*, 1970, **8**, 121.
24. GESNER, B. D. *J. Appl. Polym. Sci.*, 1968, **12**, 1199.
25. GOTHAM, K. V. and TURNER, S. *Polymer*, 1974, **15**, 665.
26. THEBERGE, J. *29th SPI Reinforced Plastics/Composites Conf.*, 1974, Paper 20-D, pp. 1–14.
27. HOUSTON, A. M. *Materials Engineering*, 1975, **81**, 28.
28. HOUSE, E. E. *Nat. SAMPE Symp.*, 1979, **24**, 201.
29. MACALLUM, A. D. *J. Org. Chem.*, 1948, **13**, 154.
30. EDMONDS, J. T. and HILL, H. W, US Patent Nos. 3,354,129, Nov. 1967 and 3,524,835, Aug. 1970.
31. BLACK, R. M., LIST, C. F. and WELLS, R. J. *J. Appl. Chem.*, 1967, **17**, 269.
32. EHLERS, G. F. L., FISCH, K. R. and POWELL, W. R. *J. Polym. Sci.*, 1969, **7**, Part A1, 2955.
33. VERNON JONES, R. *30th SPI Reinforced Plastics/Composites Conf.*, 1975, Paper 12-B, pp. 1–5.
34. BAEYER, A. VON. *Chem. Ber.*, 1872, **5**, 25, 280, 1094.
35. BAEKELAND, L. H. *Ind. Eng. Chem.*, 1909, **1**, 149, 545; 1910, **2**, 478; 1911, **3**, 932; 1912, **4**, 737; 1913, **5**, 506; 1916, **8**, 568; 1925, **17**, 225; 1935, **27**, 538.
36. MEGSON, N. J. L. *Trans. Faraday Soc.*, 1936, **32**, 336.
37. WATERMAN, H. I. and VELDMAN, A. R. *Brit. Plast., Mould. Prod. Trans.*, 1936, **8**, 125, 182.
38. CONLEY, R. T. *ACS Div. Org. Coatings and Plastics Chem.*, 1966, **26**, 138.
39. SHULMAN, G. P. and LOCKTE, H. W. *ACS Div. Org. Coatings and Plastics Chem.*, 1966, **26**, 149.

40. UK Patent No. 455,008 issued to Henkel and Cie, 1935.
41. KOEDA, K. J. *J. Chem. Soc. Japan, Pure Chem. Sec.*, 1954, **75**, 571.
42. LADY, J. H., ADAMS, R. E. and KESSE, I. *J. Appl. Polym. Sci.*, 1960, **4**, 65.
43. MANLEY, T. R. *J. Polym. Sci.*, Symp. No. 42, 1973, 1377.
44. FRIEDEL, C. and CRAFTS, J. M. *Ann.*, 1865, **136**, 203.
45. ROCHOW, E. G., US Patent No. 2,380,995, Aug. 1945.
46. WRIGHT, W. W. and LEE, W. A. The search for thermally stable polymers, chapter in *Progress in High Polymers* Vol. 2, eds. Robb, J. C. and Peaker, F., 1968, Iliffe Books Ltd, London.
47. GOLDOVSKII, E. A. *Kauchuk i Rezina*, 1979, 24.
48. DUNLOP, A. P. and PETERS, F. N. Chapter 18 in *The Furans*, ACS Monograph Series, 1953, Reinhold Publ. Corp., Washington.
49. BOZER, K. B., BROWN, L. H. and WATSON, D. *26th SPI Reinforced Plastics/Composites Conf.*, 1971, Paper 2-C, pp. 1–6.
50. BOZER, K. B. and BROWN, L. H. *27th SPI Reinforced Plastics/Composites Conf.*, 1972, Paper 3-C, pp. 1–5.
51. SELLEY, J. E. *29th SPI Reinforced Plastics/Composites Conf.*, 1974, Paper 23-A, pp. 1–4.
52. RADCLIFFE, A. T. Furane resins, Chapter 5 in *Developments with Thermosetting Plastics*, eds. Whelan, A. and Brydson, J. A., 1975, Applied Science Publishers Ltd, London.
53. DOWNING, P. A. *Chem. Eng. (London)*, 1978, **331**, 272.
54. EDWARDS, W. M. and ROBINSON, I. M. US Patent Nos. 2,710,853, 1955 and 2,867,609, 1959.
55. EDWARDS, W. M. and ROBINSON, I. M., US Patent No. 2,880,230, 1959.
56. IDRIS JONES, J., OCHYNSKI, F. W. and RACKLEY, F. A. *Chem. Ind.*, 1962, 1686.
57. EDWARDS, W. M., US Patent Nos. 3,179,614, 1965 and 3,179,634, 1965.
58. ENDREY, A. L. Canadian Patent No. 659,328, 1963 and US Patent Nos. 3,179,631, 1965 and 3,179,633, 1965.
59. BOWER, G. M. and FROST, L. W. *J. Polym. Sci.*, 1963, **1**, Part A-1, 3135.
60. DINE-HART, R. A. and WRIGHT, W. W. *J. Appl. Polym. Sci.*, 1967, **11**, 609.
61. BRUCK, S. D. *Polymer Preprints*, 1964, **5**, 148.
62. HEACOCK, J. F. and BERR, C. E. *SPE Trans.*, 1965, **5**, 105.
63. JOHNSTON, T. H. and GAULIN, C. A. *J. Macromol. Sci., Chem. A.*, 1969, **3**, 1161.
64. COTTER, J. L. and DINE-HART, R. A. *Chem. Commun.*, 1966, 809.
65. COTTER, J. L. and DINE-HART, R. A. *Organic Mass Spectrometry*, 1968, **1**, 915.
66. DINE-HART, R. A. *Brit. Polym. J.*, 1971, **3**, 163.
67. DINE-HART, R. A., PARKER, D. B. V. and WRIGHT, W. W. *Brit. Polym. J.*, 1971, **3**, 222.
68. DINE-HART, R. A., PARKER, D. B. V. and WRIGHT, W. W. *Brit. Polym. J.*, 1971, **3**, 226.
69. DINE-HART, R. A., PARKER, D. B. V. and WRIGHT, W. W. *Brit. Polym. J.*, 1971, **3**, 235.
70. *Skybond 703, heat resistant resin*, Monsanto Co. Plastics Products and Resins division, technical bulletin No. 6236, Jan. 1970.
71. PETKER, I. *Nat. SAMPE Symp.*, 1976, **21**, 37.

72. PIKE, R. A., MAYNARD, C. P. *Nat. SAMPE Tech. Conf.*, 1969, 331.
73. STUCKEY, J. M. *SAMPE Quarterly*, 1972, **3**, 22.
74. WOLKOWITZ, W. and POVEROMO, L. M. *28th SPI Reinforced Plastics/ Composites Conf.*, 1973, Paper 15-C.
75. IVANOV, V. S., KOPTEVA, I. P. and LEVANDO, L. K., USSR Patent No. 164,678, Aug. 1964.
76. SAMBETH, J. and GRUNDSCHOBER, F. French Patent No. 1,455,514, 12 Nov., 1965.
77. KOVACIC, P., US Patent No. 2,818,407, Dec. 1957.
78. BARGAIN, M., COMBET, A. and GROSJEAN, P., US Patent No. 3,562,222, 1971.
79. DARMORY, F. P. *Nat. SAMPE Symp.*, 1973, **18**, 317.
80. *Kerimid, high temperature polyimide resin*, technical data sheet, Rhone–Poulenc UK, London.
81. PETROVICKI, H., STENZENBERGER, H. D. and HARNIER, A. VON. *Proc. 32nd SPE Tech. Conf.*, 1974, 88.
82. STENZENBERGER, H. D. *J. Polym. Sci., Polym. Chem. Ed.*, 1976, **14**, 2911.
83. ALVAREZ, R. T. *Nat. SAMPE Symp.*, 1975, **20**, 253.
84. JONES, J. S. *Nat. SAMPE Symp.*, 1976, **21**, 438.
85. STENZENBERGER, H. D. *Appl. Polym. Symp.*, 1973, **22**, 77.
86. HARRUFF, P. W. *Nat. SAMPE Tech. Conf.*, 1979, **11**, 1.
87. LORENSEN, L. E. *SAMPE Quarterly*, 1975, **6**, 1.
88. BURNS, E. A., LUBOWITZ, H. R. and JONES, J. F., NASA-CR-72460, Oct. 1968.
89. SERAFINI, T. T. *Appl. Polym. Symp.*, 1973, **22**, 89.
90. SERAFINI, T. T. *J. Appl. Polym. Sci.*, 1972, **16**, 905.
91. CAVANO, P. J. and WINTERS, W. E., NASA-CR-135,113, Dec. 1976.
92. WINTERS, W. E. *Nat. SAMPE Symp.*, 1975, **20**, 629.
93. DELVIGS, P., ALSTON, W. B. and VANNUCCI, R. D. *Nat. SAMPE Symp.*, 1979, **24**, 1053.
94. WINTERS, W. E. *Nat. SAMPE Tech. Conf.*, 1978, **10**, 661.
95. PETKER, I. *Nat. SAMPE Symp.*, 1978, **23**, 775.
96. ALSTON, W. B. *Nat. SAMPE Symp.*, 1979, **24**, 72.
97. MCMAHON, P. E. *Nat. SAMPE Symp.*, 1978, **23**, 150.
98. BILOW, N. *Nat. SAMPE Symp.*, 1975, **20**, 618.
99. APONYI, T. J. *Nat. SAMPE Symp.*, 1978, **23**, 763.
100. *Thermid 600*, technical data sheet, Gulf Oil Chemicals Co., Houston, Texas.
101. GIBBS, H. H. *Nat. SAMPE Symp.*, 1972, **17**, Section III-B-Six, pp. 1–9.
102. GIBBS, H. H. *Nat. SAMPE Symp.*, 1973, **18**, 292.
103. GIBBS, H. H. *29th SPI Reinforced Plastics/Composites Conf.*, 1974, Paper 11-D, pp. 1–15.
104. GIBBS, H. H. *Adv. Chem. Ser.*, 1975, **142**, 442.
105. GIBBS, H. H. *Nat. SAMPE Symp.*, 1975, **20**, 212.
106. GIBBS, H. H. *Nat. SAMPE Symp.*, 1976, **21**, 592.
107. GIBBS, H. H. *Nat. SAMPE Tech. Conf.*, 1978, **10**, 211.
108. GIBBS, H. H. *Nat. SAMPE Symp.*, 1978, **23**, 806.
109. GIBBS, H. H. *Nat. SAMPE Symp.*, 1979, **24**, 11.
110. HOEGGER, E. F. *Nat. SAMPE Symp.*, 1979, **24**, 533.
111. *Du Pont NR-150 polyimide precursor solutions*, technical data sheet, E. I. Du Pont de Nemours and Co., Wilmington, Delaware.

112. ALBERINO, L. M., US Patent No. 3,708,458, Jan. 1973.
113. RAUSCH, K. W., FARRISSEY, W. J. and SAYIGH, A. A. R. *28th SPI Reinforced Plastics/Composites Conf.*, 1973, Paper 11-E.
114. FARRISSEY, W. J., ALBERINO, L. M., RAUSCH, K. W. and SAYIGH, A. A. R. *Nat. SAMPE Tech. Conf.*, 1972, **4**, 29.
115. CHEN, Y. T. *Nat. SAMPE Symp.*, 1978, **23**, 826.
116. ANON. *Kunststoffe*, 1977, **67**, 17.
117. WALKER, R. H. *Nat. SAMPE Symp.*, 1974, **19**, 186.
118. ROPARS, M. and BLOCH, B. French Patent No. 2,261,296, 1974 and US Patent No. 3,994,862, 1976.
119. BLOCH, B. *Nat. SAMPE Symp.*, 1978, **23**, 836.
120. ROPARS, M. and BLOCH, B. *La recherch aerospatiale*, 1977(2), 103.

Chapter 7

STRUCTURE–PROPERTY RELATIONSHIPS AND THE ENVIRONMENTAL SENSITIVITY OF EPOXIES

ROGER J. MORGAN

Lawrence Livermore Laboratory, University of California, USA

SUMMARY

The structure, deformation and failure process, and the mechanical properties, of amine-cured epoxies are discussed, together with the question of how fabrication and environmental factors can affect resin behaviour.

The nature of the chemical reactions that produce amine-cured epoxide networks, and the chemical and physical features controlling these reactions, are described, together with the physical structural parameters which determine the mechanical response of epoxies in terms of their crosslinked network morphologies and microvoid characteristics. The author discusses critically the mechanisms responsible for heterogeneous crosslinked morphologies, and the techniques for detecting them.

Evidence is presented for the deformation of the epoxies in terms of crazing and shear banding, and consideration is given to the ability of these crosslinked glasses to undergo flow, in terms of incomplete network formation, network morphology, and bond breakage.

Modification of the structure–property relationships by combinations of stress, thermal environment and humidity is discussed.

7.1. INTRODUCTION

The increasing use of high-performance light-weight fibrous composites, particularly in the aircraft and automobile industries, has led to a need to predict the lifetimes of these materials in service environments. The

211

durability of the epoxy matrices (primarily amine-cured epoxy thermosets) used in these high-performance composites and of the overall composite has been a cause for concern. A number of studies have indicated that the combined effects of stress, sorbed moisture, and thermal exposure can cause significant changes in the mechanical response of the composite.[1,2] The structural and mechanical integrity of both the epoxy matrix and the fibre–matrix interfacial region can be modified by these environmental factors.

Predicting the durability of the epoxies in service environments with confidence requires knowing the structure, modes of deformation and failure, mechanical-response relationships, and the possible modification of such relationships by fabrication procedures and environmental exposure. The structure–property relations of epoxies (and thermosets in general), however, have received little attention compared to other commonly used polymer glasses. Because of the dependence of their chemical and physical structure on fabrication conditions and because of their infusible, insoluble nature, thermosets are more difficult to study than are noncrosslinked polymeric glasses.

This chapter reviews the fundamental areas necessary for making meaningful durability predictions for epoxies. These are: (1) their physical structure, which includes the crosslinked network morphology and the microvoid characteristics, (2) their chemical structure, (3) their modes of deformation and failure and (4) the effect of environmental factors on their structural and mechanical integrity. Although amine-cured epoxies are addressed primarily, the discussion also applies to the entire glassy thermoset family.

7.2. MATERIALS

Several specific epoxy systems are mentioned in this chapter. Most of them are amine-cured. Reference is made to a number of amine-cured bisphenol-A-diglycidyl ether (Dow; DER 332) (DGEBA) epoxies, to a diaminodiphenyl sulphone (Ciba-Geigy; Eporal)-cured tetraglycidyl 4,4′-diaminodiphenyl methane (Ciba-Geigy, MY 720) (TGDDM–DDS) epoxy, and to the following curing agents used in conjunction with DGEBA epoxides:

(1)   Jeffamine T-403, an aliphatic polyether triamine (Jefferson),
(2)   Diethylene triamine (DETA) (Eastman) and
(3)   2,5-Dimethyl,-2,5-hexane diamine (DMHDA) (Aldrich).

Another curing agent mentioned is the polyamide, Versamid 140 (General Mills). Some of the references to TGDDM–DDS systems are to 'in-house' preparations, while others are to the commercial systems Narmco 5208, which contains no catalyst, and Fiberite 934, which contains about 0·5 wt % BF_3 catalyst.[3]

7.3. CHEMICAL STRUCTURE

The chemical structure of an epoxy is often complex. The structure depends on specific cure conditions because more than one reaction can occur during cure and the kinetics of these reactions exhibit different temperature dependencies. In addition, the structure is affected by factors such as steric and diffusional restrictions of the reactants during cure,[4–9] the presence of impurities that act as catalysts,[10] the reactivities of the epoxide and curing agent,[11] isomerisation of epoxide groups,[11–13] nonhomogeneous mixing of the reactants[9,14] and cyclic polymerisation of the growing chains.[11] These factors can lead to heterogeneous network structures.

In amine-cured epoxides, networks are generally assumed to result from addition reactions of epoxide groups with primary and secondary amines,[11] as illustrated in Fig. 7.1. Epoxides and amines with more

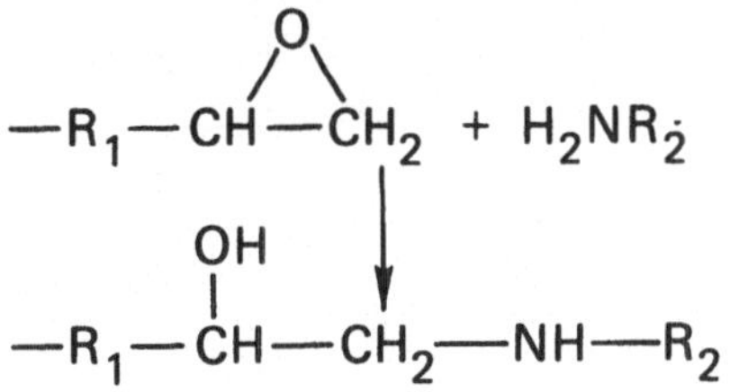

FIG. 7.1.   Epoxide–amine addition reaction.

than three functional groups can form highly crosslinked network structures. However, amine-cured epoxies often exhibit considerable ductility[8,9,14–17] and microscopic flow,[8,9,14–18] and they exhibit yield stresses that show thermal-history and strain–rate dependencies very similar to those of noncrosslinked glasses.[17] Such observations suggest either that the epoxies are not as highly crosslinked as expected if only simple epoxide–amine addition reactions had taken place or that the reactions themselves did not proceed completely. (Further possibilities discussed in later sections are that (1) crosslinks are broken under stress and (2) the microscopic flow processes are controlled by regions of low crosslink density where the network morphology is heterogeneous.)

In addition to the steric and diffusional restrictions that limit the cure reactions and the crosslink density, a ring-opening homopolymerisation of the epoxide groups can also occur,[10-12,19] as illustrated in Fig. 7.2.

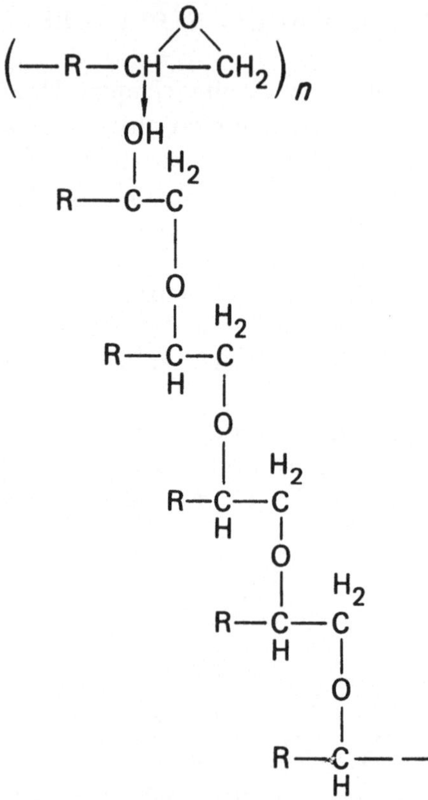

FIG. 7.2. Homopolymerisation of epoxides.

As a consequence of this *trans*-etherification reaction to form polyether linkages, fewer epoxide groups are available during the later stages of cure to react with the secondary amine groups and, subsequently, to form crosslinks. Although the epoxide homopolymerisation reaction by itself can lead to crosslinked networks in many amine-cured epoxide systems, a lower crosslink density generally results in these systems when the *trans*-etherification reaction occurs. The epoxide–amine reactions are controlled by the presence of H-bond donors, such as OH groups, that are necessary to open the epoxide rings.[5,6,10] The *trans*-etherification reaction, however, requires a tertiary amine as a catalyst and an H-bond donor as a cocatalyst.[10,12,19] Hence, the final chemical structure of the epoxy network can be complex because it depends on such parameters as (1) the relative rates of the chemical reactions at room temperature, (2) the final postcure temperature, (3) the concentrations of catalysts such as sorbed moisture in the system and (4) steric restrictions that inhibit reactions at secondary-amine sites. Fourier-transform infrared spectroscopy studies using spectral stripping, which reveals differences in the spectra recorded at different stages of cure, should enhance our knowledge of the chemical reactions involved in the network formation of amine-cured epoxy systems.

In the case of the TGDDM–DDS epoxy systems, there is significant evidence that these networks do not form exclusively from epoxide–amine addition reactions. The structure of the unreacted TGDDM epoxide and DDS amine monomers are illustrated in Fig. 7.3. Table 7.1 illustrates the wt % of DDS required for reaction of (1) all primary and secondary

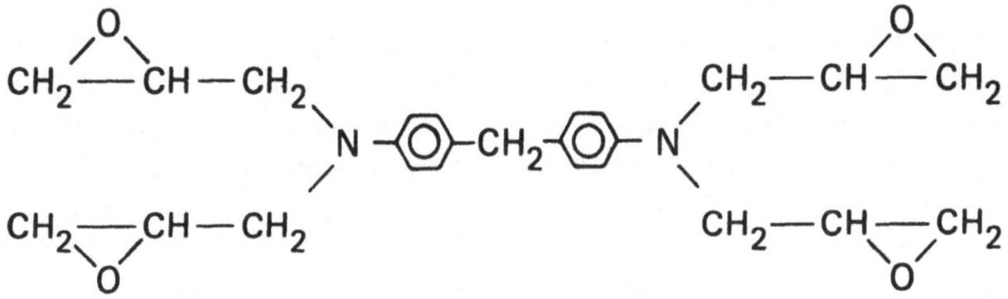

tetraglycidyl 4, 4′–diaminodiphenyl methane
TGDDM

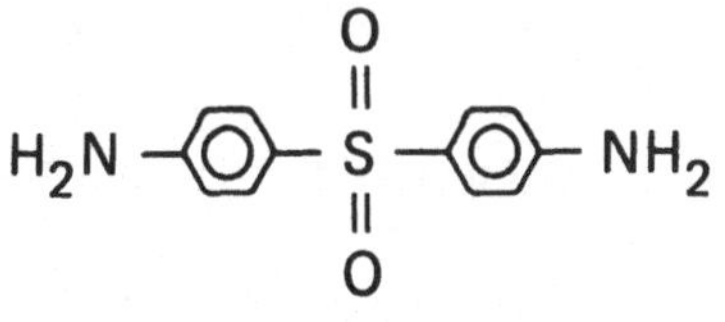

4, 4′-diaminodiphenyl sulfone
DDS

FIG. 7.3.   The TGDDM–DDS epoxy system.

TABLE 7.1
THEORETICAL REACTION MIXTURES FOR TGDDM–DDS EPOXY SYSTEM

	50% TGDDM epoxide groups react	100% TGDDM epoxide groups react
100% DDS primary and secondary amines react	23 wt % DDS	37 wt % DDS
100% DDS primary amines react	37 wt % DDS	54 wt % DDS

amines in the DDS and (2) only the primary amines in the DDS with 50% and 100% of the epoxide groups in the tetrafunctional TGDDM molecules. The glass transition temperature (T_g) of the TGDDM–DDS epoxies rises with increasing DDS concentration (up to 25 wt % DDS) because of corresponding increases in molecular weight and/or crosslink density.[9] The T_g, however, reaches a maximum of about 250 °C at about 30 wt % DDS and decreases, subsequently, for higher DDS concentrations because of plasticisation by unreacted DDS molecules. However, 37 wt % DDS is required to consume half the TGDDM epoxide groups when only epoxide–primary amine reactions occur (Table 7.1). Hence, the T_g maximum value at about 30 wt % DDS suggests that, assuming only epoxide–primary amine reactions occur, less than half of the TGDDM epoxide groups react because steric and diffusional restrictions inhibit further reaction. It seems doubtful whether networks in which only half of the epoxide groups have reacted would exhibit the respectable mechanical properties shown by the TGDDM–DDS epoxies (20–35 wt % DDS).[9] Evidently, other cure reactions, in addition to the epoxide–primary amine reactions, are occurring and possibly involve (1) epoxide homopolymerisation, (2) epoxide–secondary amine reactions and (3) internal cyclisation within the TGGDM epoxide as a result of hydroxyl and/or secondary amino groups reacting with adjacent unreacted epoxides.

Infrared spectroscopy also suggests that TGDDM–DDS networks do not form exclusively from epoxide–amine addition reactions. In Fig. 7.4 the percentage of epoxide groups consumed during cure, as determined by the disappearance of the epoxide band at $910 \, cm^{-1}$ in the infrared spectrum, is plotted as a function of cure conditions for both a BF_3-catalysed TGDDM–DDS epoxy (Fiberite 934; 25 wt % DDS) and a noncatalysed TGDDM–DDS epoxy (Narmco 5208; 20 wt % DDS). For a standard 177 °C cure lasting 2·5 h, all the epoxide groups are consumed (within experimental error) for the BF_3-catalysed system despite the fact

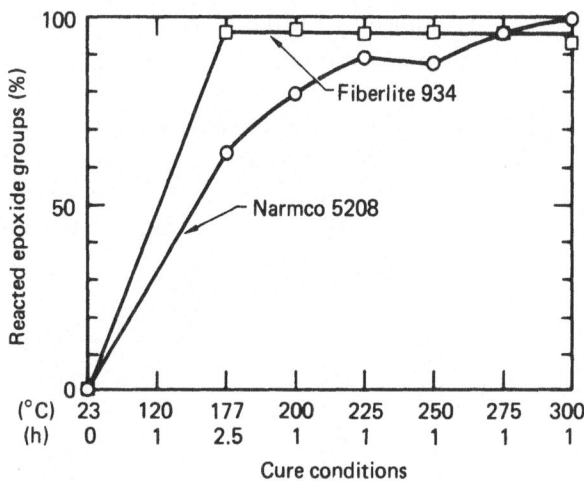

FIG. 7.4.   Percentage of reacted ẽpoxide groups as a function of cure conditions for TGDDM–DDS-based epoxies.

that the wt % of DDS in this system is well below the stoichiometric quantity necessary to consume all the epoxide groups by normal epoxide–amine addition reactions. In the case of the noncatalysed system about 35 % of the epoxide groups remain unreacted after curing at 177 °C for 2·5 h and all such groups reacted only after exposure at 300 °C for 1 h.

7.4.   PHYSICAL STRUCTURE

The primary physical and structural parameters that control the modes of deformation and failure as well as the mechanical response of epoxies are the crosslinked network structure and microvoid characteristics.[8,9,14–16,18]

The chemistry of the cure processes and the final network structure of epoxies have been deduced from the chemistry of the system. This knowledge is based on the assumptions that the cure reactions are known and that they go to completion. The experimental techniques used to determine the cure processes and epoxy network structure include IR and ^{13}C NMR spectroscopy and swelling, ultrasonic, thermal conductivity, dynamic mechanical, and differential scanning calorimetry measurements (see references in Ref. 8). However, as discussed in the previous section, in many epoxy systems the chemical reactions are diffusion controlled and do not proceed to completion. Furthermore, a heterogeneous distribution in the crosslink density can occur. Hence, a variety of crosslinked network structures can be produced.

Figure 7.5 schematically illustrates possible network topographies of epoxies.[8,20–22] An ideal uniform crosslinked network structure is illustrated in Fig. 7.5(a). In reality, however, networks contain loops and dangling chain-ends as illustrated in Fig. 7.5(b). Such networks can exhibit

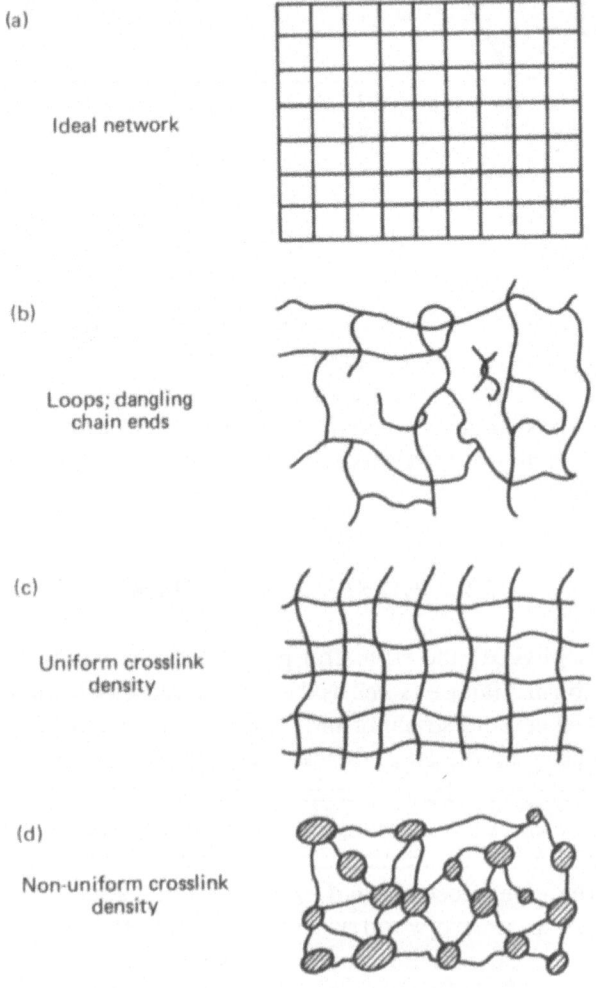

(a)

Ideal network

(b)

Loops; dangling
chain ends

(c)

Uniform crosslink
density

(d)

Non-uniform crosslink
density

FIG. 7.5.   Epoxy network morphologies.

an essentially uniform crosslink density in a low molecular weight or crosslink density matrix (Fig. 7.5(c)). Also, nonuniform crosslink density networks in which regions of high crosslink density form either a continuous (Fig. 7.5(d)) or a discontinuous phase in a low molecular weight or crosslink density matrix are other possible network morphologies.

High crosslink density regions, from 6000 to 10 000 nm in diameter, have been observed in crosslinked resins.[3,8,15,16,18,23 – 49] The formation of a heterogeneous rather than a homogeneous system depends on polymerisation conditions, i.e. temperature, solvent, and chemical composition. The high crosslink density regions have been described as agglomerates of colloidal particles[28,29] or floccules[31] in a lower molecular weight interstitial fluid. Funke[50,51] suggested a number of factors that can be responsible for heterogeneous network formation: (1) difference between the reactivities of different functional groups, (2) unreacted functional groups, (3) intramolecular cyclisation reactions and (4) phase separation. Phase separation in the form of microgelation is generally believed to be the primary mechanism in the formation of heterogeneous crosslinked networks. Solomon et al.[30] originally suggested that a two-phase system is produced by microgelation prior to the formation of a macrogel. Kenyon and Nielsen[34] indicated that the highly crosslinked microgel regions are loosely connected during the latter stages of the curing process. Karyakina et al.[41] suggested that microgel regions originate in the initial stages of polymerisation from the formation of microregions of aggregates of primary polymer chains. More recently, Luettgert and Bonart[47] discussed the morphology of epoxies in terms of the relative rates of microgel formation and the subsequent growth rate of these gel particles. At low cure temperatures, only a small number of gel particles are nucleated and, hence, large nonhomogeneities are produced; at higher cure temperatures, the rate of nucleation of microgel particles is faster, and larger numbers are produced, which, therefore, limits their growth in size. Finally, the high crosslink density regions have been reported to be only weakly attached to the surrounding matrix,[28,29,31] and their size also varies with the proximity of surfaces[31,45] and the presence of solvents.[30,52]

Most of the evidence for heterogeneous regions of crosslink density in epoxies is derived from electron microscope investigations. These microscopy studies involve carbon–platinum replication of etched and nonetched free surfaces and fracture surfaces. However, artefacts can often result from replication techniques. Scepticism of evidence of nodular morphology based on these techniques arises for the following reasons: (1) similar nodular structures that are observed by replication of epoxy surfaces can also be produced by replication of inorganic glass slides; (2) blisters that are produced as a result of the etching of epoxy surfaces can be interpreted as nodular regions of high crosslink density, (3) the fracture topographies can exhibit a nodular appearance in the initiation region as a result of fractured craze fibrils,[8,9,14,18] and (4) the surface structure implied by carbon–platinum replication techniques is not always observed by scanning electron microscopy studies of gold-coated surfaces.

More confidence can be placed in bright-field transmission electron microscopy studies of the morphology of thin epoxy films strained directly in the electron microscope. For example, in DGEBA–DETA epoxies, it was observed that particles 6 to 9 nm in diameter remain intact and flow past one another during flow processes. Figure 7.6 illustrates a network of these particles. It was suggested that the 6 to 9 nm diameter

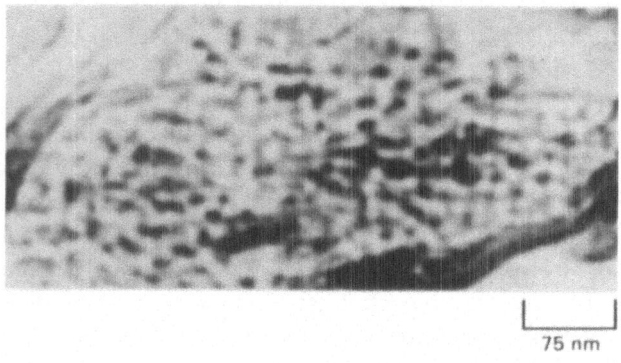

75 nm

FIG. 7.6. Bright-field transmission electron micrograph of strained network structure of particles of 6–9 nm in diameter in DGEBA–DETA epoxy.

particles were molecular domains that were intramolecularly crosslinked and that formed during the initial stage of polymerisation.[8,18] The interconnection of molecular domains by regions of either low or high crosslink density allows two types of network structure: (1) regions of high crosslink density embedded in a low or noncrosslinked matrix or, (2) noncrosslinked or low crosslink density regions embedded in a high crosslink density matrix. In the case of DGEBA–DETA epoxies, both types of network morphology were seen, with the first type more prevalent.[8,18] Deformation of the first type of network involves preferential deformation of the regions of low crosslink density without causing cleavage of the high crosslink regions. For the second type of network, the deformation process is more complex. Local affine deformation requires that network cleavage and flow occur simultaneously in the high crosslink density regions while at the same time flow occurs with little network cleavage in the neighbouring low crosslink density regions. This deformation process results in progressively larger regions that are poorly crosslinked.

In other amine-cured epoxies, such as TGDDM–DDS systems,[9] no evidence for heterogeneous crosslink density distributions was found from observation of strained films in the electron microscope. However, under polarised light, a network of larger 1 mm sized nodules was seen in BF_3-catalysed TGDDM–DDS epoxy systems (Fiberite 934).[3] These birefringent networks are permanently destroyed above T_g (or at 25 °C below T_g when stress is applied). The birefringence originates from the preferential alignment of the macromolecules within the birefringent network. The network alignment is caused by biaxial shrinkage stresses imposed on the epoxy sheet during cure. (The TGDDM–DDS epoxies are cured between glass plates, and the shrinkage stresses are not completely relieved by the release agent present on the surface of the glass plates.) The nonhomogeneous distribution of the catalyst within the TGDDM–DDS system produces a heterogeneous structure because regions of high catalyst concentration polymerise more rapidly than surrounding regions. The shrinkage stresses, in conjunction with the heterogeneous polymerisation conditions during gelation and glass formation, produce the birefringent network of aligned molecules; the more highly polymerised regions, which contained high concentrations of catalyst, align more readily under stress than do surrounding regions of lower molecular weight.

The microvoid characteristics of the epoxy are also important in controlling the mechanical response of the glass. Microvoids can have a deleterious effect on the mechanical properties of epoxies by acting as stress concentrators and by accumulating sorbed moisture. Microvoids can result when air is trapped in the system during cure and from trapped low molecular weight material that is subsequently eliminated from the glass during postcure. This low molecular weight material results from either nonhomogeneous mixing of epoxide and curing agent or from the aggregation of unreacted constituents. In polyamide-cured DGEBA epoxies, crystals of DGEBA epoxide monomer trapped in the partially cured resin at room temperature can produce microvoids by melting and volatilising under certain postcure conditions.[14] Thermal-anneal, moisture-sorption, and mechanical-property studies also indicate that in TGDDM–DDS epoxies, the melting and volatilisation of unreacted DDS crystallites during cure produces microvoids.[9]

7.5. MODES OF DEFORMATION AND FAILURE

Localised plastic flow has been reported to occur during the deformation and failure of epoxies; in a number of cases, the fracture energies were two to three times greater than the expected theoretical estimates for purely brittle fracture.[8] However, no systematic studies have been made to elucidate the microscopic flow processes occurring during the deformation of epoxies and to determine the relation of such flow processes to the network structure.

Recent investigations revealed that both DGEBA–DETA and TGDDM–DDS epoxies deform and fail by a crazing process.[8,9,14-16,18] Crazes were observed in films either strained directly in the electron microscope or strained on a metal substrate. The fracture topographies of these epoxies, fractured as a function of temperature and strain rate, were interpreted in terms of a crazing process. The TGDDM–DDS epoxies also deformed to a limited extent by shear banding, as indicated by multiple right-angle steps present in the fracture-topography initiation region (Fig. 7.7). Shear-band propagation in these partially crosslinked glasses produces structurally weak planes because of bond cleavage caused during molecular flow. Hull[53] and Mills[54] both noted that the intersection of shear bands, which occurs at right-angles, causes a stress concentration. This stress concentration is sufficient to cause a crack to propagate through the structurally weak planes caused by shear-band propagation. These phenomena can produce the multiple right-angle steps observed in the fracture topography. Mixed modes of deformation that involve both crazing and shear banding were also observed in the fracture topography of TGDDM–DDS epoxies. More recently, in the case of DGEBA–DMHDA epoxies, macroscopic shear bands were observed under polarised light.

The ability of amine-cured epoxies to undergo considerable microscopic flow can be explained by either, (1) incomplete cure reactions and/or epoxide homopolymerisation resulting in a lower crosslink density network (see Section 7.3), (2) regions of low crosslink density controlling the flow processes (see Section 7.4) or, (3) bond breakage under stress, which results in a lower overall crosslink density. Two pieces of evidence were recently published which suggest that bond breakage readily occurs

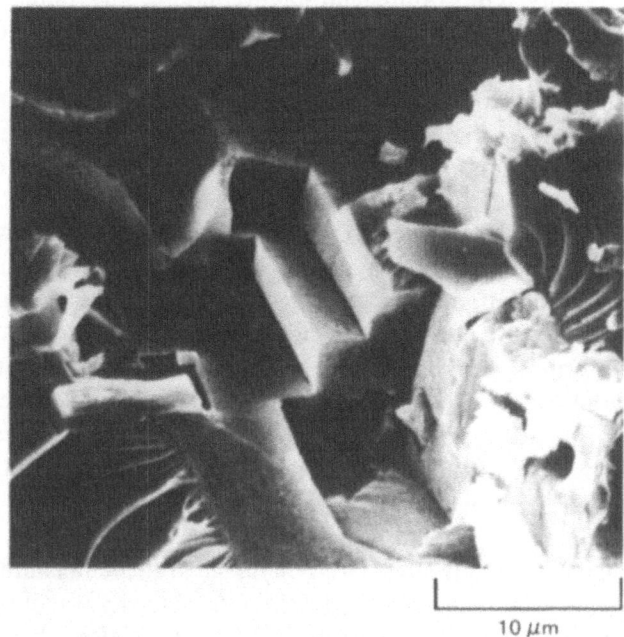

10 μm

FIG. 7.7. Scanning electron micrograph illustrating multiple right-angle steps in the fracture-topography initiation region of a TGDDM–DDS epoxy.

in epoxy networks under stress. Firstly, Levy and Fanter[55] reported enhanced chemiluminescence of TGDDM–DDS epoxies under stress. This chemiluminescence is associated with bond fracture and subsequent reaction of the macro-radicals with oxygen. Secondly, Gledhill et al.[56] reported that an amine-cured DGEBA epoxide is appreciably toughened after it is subjected to an applied load. This observation suggests that stress causes bond breakage, which lowers the crosslink density; this decreased crosslink density allows an increase in molecular flow at the crack tip, which results in increased toughness.

7.6. DURABILITY

The durability of epoxy matrices depends on many complex interacting phenomena. The factors that control the critical path to ultimate failure or unacceptable damage depend specifically on the particular prevailing environmental conditions. These environmental factors include service stresses, humidity, temperature, and solar radiation. The combined effects of thermal history, moisture exposure, and stress have a deleterious effect on the physical and mechanical integrity of epoxies. Morgan *et al.*[57] have recently studied the effect of specific combinations of moisture, heat, and stress on the physical structure, failure modes, and tensile mechanical properties of TGDDM–DDS epoxies. The main findings from these studies are outlined in the following paragraph.

Sorbed moisture plasticises TGDDM–DDS epoxies and lowers their tensile strengths, ultimate elongations, and moduli. The fracture topographies of the initiation cavity and mirror regions of these epoxies indicate that sorbed moisture enhances the craze initiation and propagation processes. The crazing process is more susceptible to sorbed moisture than is the T_g; this can be explained in terms of local moisture concentrations enhancing the local cavitation and flow processes. Hence, modification of T_g by sorbed moisture cannot be utilised alone as a sensitive guide to predict deterioration in the mechanical response and, hence, the durability of epoxies.

The effect of increasing stress levels on the subsequent moisture-sorption characteristics of initially dry TGDDM–DDS epoxies was investigated. In Fig. 7.8 the equilibrium moisture-sorption levels after about 40 days exposure to 100% relative humidity at room temperature are plotted versus the stress levels that were applied to the epoxies before moisture sorption. All data points fall within the shaded areas. Stresses in the

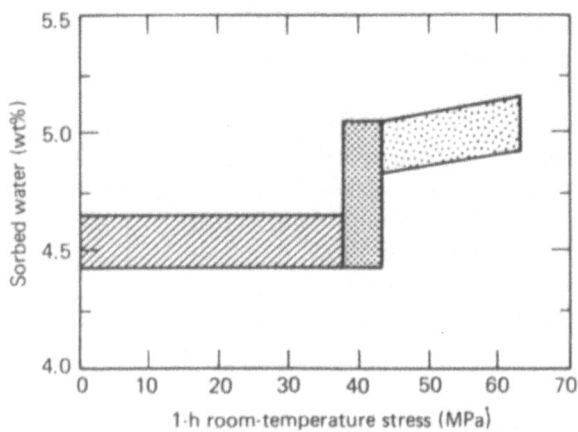

FIG. 7.8. Equilibrium wt % moisture sorbed by a TGDDM–DDS epoxy at 100% relative humidity and 23°C versus 1 h constant-stress levels that were applied before exposure to moisture.

0 to 38 MPa range had no detectable influence on the subsequent moisture-sorption levels. However, moisture-sorption levels increase sharply by up to about 11 % in the 38 to 43 MPa stress range. At higher stress levels in the 43 to 65 MPa range, in which a few specimens actually broke, there is only a slight trend towards higher moisture-sorption levels with increasing stress.

The data in Fig. 7.8 indicate that the initial stages of craze–crack growth enhanced the accessibility of moisture to sorption sites to a greater extent than the later stages of growth. (The primary sorption sites within the TGDDM–DDS epoxy are the hydroxyl, sulphonyl, and primary and secondary amine groups—all of which are capable of forming hydrogen bonds with water molecules.) The TGDDM–DDS epoxy specimens that fractured under constant load were found to exhibit fracture topographies similar to specimens previously studied that fractured in shorter times in the 10^{-2} to 10^1 min^{-1} strain–rate region.[9] Such topographies have been interpreted in terms of a craze–crack growth process[9] in which crazing followed by crack propagation predominates in the initial stages of failure, and crack propagation alone predominates during the later stages of failure. The dilatational changes produced in the epoxy glass by the crazing process enhance the accessibility of moisture to sorption sites within the epoxy to a greater extent than does crack propagation alone. Hence, the initial stages of failure in TGDDM–DDS epoxies enhance the accessibility of moisture to sorption sites to a greater extent than do the later stages of failure.

One of the more extreme environmental conditions experienced by an epoxy composite matrix occurs during a supersonic dash of a fighter aircraft. The aircraft dives from high altitudes (where outer surface temperature is -20 to $-55\,°C$) into a supersonic, low-altitude run during which aerodynamic heating raises the surface temperature to 100–$150\,°C$, in a matter of minutes. On reduction of speed, the outer surface temperature drops extremely rapidly at rates up to about $500\,°C/min$, thus exposing the epoxy composite to a thermal spike. Simulation of such thermal spikes has been shown to increase the amount of moisture sorbed by the epoxy or epoxy composite.[58–64] However, after a certain number of consecutive thermal spikes, the amount of moisture sorbed ceases to increase. Browning[60,63] suggested that such increases result from microcracks caused by the moisture and temperature gradients present during the thermal spike. McKague[62] recently noted that damage does not occur unless the thermal-spike maximum temperature exceeds the T_g of the moist epoxy.

The amount of moisture sorbed by TGDDM–DDS epoxies was enhanced by about 1·6 wt % after exposure to a 150 °C thermal spike.[57] The surfaces of the thermally-spiked epoxies were examined for the presence of surface microcracks using scanning electron microscopy. No significant areas of microcracking were observed in any of the specimens when examined under magnifications of up to 30 000 times. Hence, the sorbtion of additional moisture by the epoxies after exposure to thermal spikes is not primarily caused by microcracking.

The primary mechanism by which thermally spiked epoxies sorb additional moisture can be explained in terms of moisture-induced free volume changes. The molecular mobility of the epoxy is enhanced at the high temperatures experienced during the thermal spike as the T_g of the epoxy–moisture system is approached. This molecular mobility is sufficient to enhance the dissociation of H-bonds between the water molecules and active sites within the epoxy. Although the ruptured H-bonds can re-form at active sites, there is an overall decrease in the amount of hydrogen bonding and a corresponding increase in the mobility of the water molecules. The more mobile water molecules with fewer H-bonds require a greater free volume because H-bonding generally causes a volume decrease. The molecular mobility of the epoxy–moisture system during a thermal spike is sufficient to allow configurational changes to occur within the epoxy network, which accommodates the greater free volume required by both the more mobile water molecules and the normal moisture-induced swelling stresses imposed on the epoxy. Such free volume increases, which involve permanent rotational isomeric population changes within the epoxy network, are frozen into the epoxy glass during the rapid cool down portion of the thermal spike. The additional free volume allows water molecules access to previously inaccessible active sites within the epoxy.

To a lesser extent, the rupture of crosslinks, crazing and/or cracking, and the loss of unreacted material can also contribute to enhanced moisture sorption after thermal spike exposure. Thermal spike exposure can cause surface crazing and/or cracking of epoxies if the moisture-induced swelling stresses, together with those stresses that result from temperature gradients and relaxation of fabrication stresses, exceed the craze-initiation stress at the maximum thermal spike temperature. Thicker epoxy specimens are more susceptible to the growth of permanent damage regions during thermal spike exposures because they are exposed to larger temperature gradients and shrinkage stresses during cure that, in turn, produce larger fabrication stresses and strains.

7.7. ACKNOWLEDGEMENT

This chapter refers to work performed by the author under the auspices of the US Department of Energy, by the Lawrence Livermore Laboratory, under contract number W-7405-ENG-48. Neither the US nor the US Department of Energy, nor any of their employees, nor any of their contractors, subcontractors, or their employees, makes any warranty, express or implied, or assumes any legal liability or responsibility for the accuracy, completeness or usefulness of any information, apparatus, product or process disclosed, or represents that its use would not infringe privately-owned rights.

Reference to a company or product name does not imply approval or recommendation of the product by the University of California or the US Department of Energy to the exclusion of others that may be suitable.

REFERENCES

1. *Air Force durability workshop*, September 1975, Battelle Columbus Laboratories, Columbus, Ohio.
2. *Air Force conference on the effects of relative humidity and temperature on composite structures*, March 1976, University of Delaware, AFOSR-TR-77-0030 (1977).
3. MORGAN, R. J. and MONES, E. T. *Composites Tech. Rev.*, 1979, **1**(4), 18.
4. FRENCH, D. M., STRECKER, R. A. H. and TOMPA, A. S. *J. Appl. Polym. Sci.*, 1970, **14**, 599.
5. HORIE, K., HIURA, H., SAWADA, M., MITA, I. and KAMBE, H. *J. Polym. Sci.*, 1970, **8**, Part A1, 1357.
6. ACITELLI, M. A., PRIME, R. B. and SACHER, E. *Polymer*, 1971, **12**, 335.
7. PRIME, R. B. and SACHER, E. *Polymer*, 1972, **13**, 455.
8. MORGAN, R. J. and O'NEAL, J. E. *Polym. Plast. Technol. Eng.*, 1978, **10**, 49.
9. MORGAN, R. J., O'NEAL, J. E. and MILLER, D. B. *J. Mater. Sci.*, 1979, **14**, 109.
10. WHITING, D. A. and KLINE, D. E. *J. Appl. Polym. Sci.*, 1974, **18**, 1043.
11. LEE, H. and NEVILLE, K. *Handbook of Epoxy Resins*, 1967, McGraw-Hill, New York.
12. SIDYAKIN, P. V. *Vysokomol. Soedin.*, 1972, A14, 979.
13. BELL, J. P. and MCCAVILL, W. T. *J. Appl. Polym. Sci.*, 1974, **18**, 2243.
14. MORGAN, R. J. and O'NEAL, J. E. *J. Macromol. Sci. Phys.*, 1978, B15(1), 139.
15. MORGAN, R. J. and O'NEAL, J. E. *Structural parameters affecting the brittleness of polymer glasses and composites* in *Advances in Chemistry No. 154: Toughness and Brittleness of Plastics*, eds. Deanin, R. D. and Crugnola, A. M., 1976, American Chemical Society, Washington, D.C.

16. MORGAN, R. J. and O'NEAL, J. E. in *Chemistry and Properties of Crosslinked Polymers*, ed. Labana, S. S., 1977, Academic Press, New York, p. 289.
17. MORGAN, R. J. *J. Appl. Polym. Sci.*, 1979, **23**, 2711.
18. MORGAN, R. J. and O'NEAL, J. E. *J. Mater. Sci.*, 1977, **12**, 1966.
19. NARRACOTT, E. *Brit. Plast.*, 1953, **26**, 120.
20. NIELSEN, L. E. *J. Macromol. Sci., Rev. Macromol. Chem.*, 1969, C3(1), 69.
21. NIELSEN, L. E. *Crosslinking-effect on physical properties of polymers*, 1968, Monsanto/Washington University/ONR/ARPA Association, Report HPC-68-57.
22. MANSON, J. A., KIM, S. L. and SPERLING, L. H. *Influence of crosslinking on the mechanical properties of high T_g polymers*, 1976, Technical Report AFML-TR-76-124, for US Air Force Materials Laboratory, Wright–Patterson Air Force Base, Ohio, by Materials Research Center, Lehigh University, Bethlehem, Pa. Available from National Technical Information Service, Arlington, Va., Document AD-A033078.
23. CARSWELL, T. S. *Phenoplasts*, 1947, Interscience, New York.
24. ROCHOW, T. G. and ROWE, F. G. *Anal. Chem.*, 1949, **21**, 261.
25. SPURR, R. A., ERATH, E. H., MYERS, H. and PEASE, D. C. *Ind. Eng. Chem.*, 1957, **49**, 1839.
26. ERATH, E. H. and SPURR, R. A. *J. Polym. Sci.*, 1959, **35**, 391.
27. ROCHOW, T. G. *Anal. Chem.*, 1961, **33**, 1810.
28. ERATH, E. H. and ROBINSON, M. *J. Polym. Sci.*, 1963, **3**, Part C, 65.
29. WOHNSIEDLER, H. P., *J. Polym. Sci.*, 1963, **3**, Part C, 77.
30. SOLOMON, D. H., LOFT, B. C. and SWIFT, J. D. *J. Appl. Polym. Sci.*, 1967, **11**, 1593.
31. CUTHRELL, R. E. *J. Appl. Polym. Sci.*, 1967, **11**, 949.
32. NEVEROV, A. N., BIRKINA, N. A., ZHERDEV, YU. V. and KOZLOV, V. A. *Vysokomol. Soedin.*, 1968, **A10**, 463.
33. NENKOV, G. and MIKHAILOV, M. *Makromol. Chem.*, 1969, **129**, 137.
34. KENYON, A. S. and NIELSEN, L. E. *J. Macromol. Sci., Chem.*, 1969, A3(2), 275.
35. TURNER, D. T. and NELSON, B. E. *J. Polym. Sci., Polym. Phys. ed.*, 1972, **10**, 2461.
36. BASIN, V. YE., KORUNSKII, L. M., SHOKALSKAYA, O. Y. and ALEKSANDROV, N. V. *Poly. Sci. USSR*, 1972, **14**, 2339.
37. KESSENIKH, R. M., KORSHUNOVA, L. A. and PETROV, A. V. *Polym. Sci. USSR*, 1972, **14**, 466.
38. BOZVELIEV, L. G. and MIHAJLOV, M. G. *J. Appl. Polym. Sci.*, 1973, **17**, 1963; *J. Appl. Polym. Sci.*, 1973, **17**, 1973.
39. MORGAN, R. J. and O'NEAL, J. E. *Polym. Preprints*, 1975, **16**(2), 610.
40. SELBY, K. and MILLER, L. E. *J. Mater. Sci.*, 1975, **10**, 12.
41. KARYAKINA, M. I., MOGILEVICH, M. M., MAIOROVA, N. V. and UDALOVA, A. V. *Vysokomol. Soedin.*, 1975, **A17**, 466.
42. MAIOROVA, M. V., MOGILEVICH, M. M., KARYAKINA, M. I. and UDALOVA, A. V. *Vysokomol. Soedin.*, 1975, **A17**, 471.
43. SMARTSEV, V. M., CHALYKH, A. YE., NENAKHOV, S. A. and SANZHAROVSKII, A. T. *Vysokomol. Soedin.*, 1975, **A17**, 836.
44. MORGAN, R. J. and O'NEAL, J. E. *Polym. Plast. Technol. Eng.*, 1975, **5**(2), 173.
45. RACICH, J. L. and KOUTSKY, J. A. *J. Appl. Polym. Sci.*, 1976, **20**, 2111.

46. MANSON, J. A., SPERLING, L. H. and KIM, S. L. *Influence of crosslinking on the mechanical properties of high* T_g *polymers*, 1977, Technical Report, AFML-TR-77-109, for US Air Force Materials Laboratory, Wright–Patterson Air Force Base, Ohio by Materials Research Center, Lehigh University, Bethlehem, Pa. Available from National Technical Information Service, Arlington, Va., Document AD-A054431.
47. LUETTGERT, K. E. and BONART, R. *Prog. Colloid. Polym. Sci.*, 1978, **64**, 38.
48. DUSEK, K., PLESTIL, J., LEDNICKY, F. and LUNAK, S. *Polymer*, 1978, **19**, 393.
49. SCHMID, R. *Prog. Colloid. Polym. Sci.*, 1978, **64**, 17.
50. FUNKE, W. *Chimia*, 1968, **22**, 111.
51. FUNKE, W., BEER, W. and SEITZ, U. *Prog. Colloid. Polym. Sci.*, 1975, **57**, 48.
52. BELL, J. P. *J. Polym. Sci.*, 1970, **8**, Part A2, 417.
53. HULL, D. *Acta. Met.*, 1960, **8**, 11.
54. MILLS, N. J. *J. Mater. Sci.*, 1976, **11**, 363.
55. LEVY, R. L. and FANTER, D. L. *Polymer Preprints*, 1979, **20**(2), 543.
56. GLEDHILL, R. A., KINLOCH, A. J. and SHAW, S. J. *J. Mater. Sci.*, 1969, **14**, 1769.
57. MORGAN, R. J., O'NEAL, J. E. and FANTER, D. L. *J. Mater. Sci.*, (in press).
58. VERETTE, R. M. *Temperature/humidity effects on the strength of graphite/epoxy laminates*, 1975, AIAA Paper No. 75-1011.
59. McKAGUE, E. L., JR., HALKIAS, J. E. and REYNOLDS, J. D. *J. Compos. Mater.*, 1975, **9**, 2.
60. BROWNING, C. E. Ph.D. thesis, University of Dayton, Dayton, Ohio, 1976.
61. HEDRICK, I. G. and WHITESIDE, J. B. Effects of Environment on Advanced Composite Structures in *AIAA Conference on aircraft composites. the emerging methodology of structural assurance*, 1977, San Diego, Calif., Paper No. 77-463.
62. McKAGUE, E. L. Chapter 5 in *Proceedings of Conference on Environmental Degradation of Engineering Materials*, eds. Louthan, M. R. and McNitt, R. P., 1977, Virginia Polytechnic Inst. Printing Dept., Blacksburg, Virginia, p. 353.
63. BROWNING, C. E., The mechanism of elevated temperature property losses in high performance structural epoxy resin matrix materials after exposure to high humidity environments, *22nd National SAMPE Symposium and Exhibition*, San Diego, Calif., 1977, **22**, 365.
64. *Advanced Composite Materials—Environmental Effects*, ASTM STP 658, ed. Vinson, J. P., 1978, American Society for Testing and Materials, Philadelphia, Pa.

Chapter 8

SOME MECHANICAL PROPERTIES OF CROSSLINKED POLYESTER RESINS

W. E. Douglas and G. Pritchard

Kingston Polytechnic, Surrey, UK

SUMMARY

This chapter is concerned with (a) the dynamic mechanical behaviour and (b) the fracture toughness of crosslinked unsaturated polyester resins.

Four dynamic mechanical relaxations are normally seen. The reasons for assigning these relaxations to particular motions of the network structure are given, although there is still some uncertainty about the origin of the β-transition.

The fracture behaviour of crosslinked polyesters is also considered. The fracture surfaces are similar to those of glass in many respects. Fracture toughness is low, and only slightly improved by the addition of a dispersed elastomeric phase.

8.1. INTRODUCTION

This chapter is concerned chiefly with topics which, although not especially novel, have recently become of special interest to those concerned with unsaturated polyester resins. The main subjects discussed are:

(1) the relationship between polyester network structure and dynamic mechanical properties
(2) fracture toughness and fracture surface morphology.

It has already been established by those concerned with other mechanical properties how resin formulations affect the tensile strength, hardness,

231

elongation at break, heat distortion temperature, Izod impact strength, flexural modulus, etc., and the results have been reviewed by Boenig.[1] The effect of degree of cure on mechanical properties has also been given considerable attention.[2,3]

8.2. CROSSLINKED NETWORK STRUCTURES

Polyester networks may be described in chemical terms, by reference to the dibasic acids, diols, and crosslinking agents used, and also in physical terms, by considering the crosslink density, chain length, chain length distribution, average bridge length and sol fraction. The chemicals used in polyester formulations are discussed in Chapter 3, and the abbreviations for chemicals mentioned in this chapter are listed in Table 8.1.

TABLE 8.1
POLYESTER RESIN COMPONENTS: ABBREVIATIONS USED IN THIS CHAPTER

Acid	Code	Diol	Code	Crosslinking agent	Code
o-Phthalic	PA	1,2-Propane		Methyl	
Isophthalic	IPA	Diol	PG	Methacrylate	MMA
Terephthalic	TPA	Diethylene		Styrene	S
Maleic	MA	Glycol	DEG	Bromostyrene	BS
Fumaric	FA	1,6-Hexane			
Adipic	AA	Diol	HD		
Succinic	SA	Ethane			
Sebacic	SBA	Diol	ED		
Methyl succinic	MSA				

The importance of the correct selection of resin components is obvious, but many other considerations have to be taken into account. Some of these may be evident from Fig. 8.1, which shows a typical crosslinked network, from which the branched chain polyester molecules have been omitted. In particular, the following should be noted:

(1) the polycondensation chains (unbroken lines),
(2) the crosslinking bridges (–S–S–S),
(3) the polyester segments between bridges,
(4) unreacted crosslinking sites, ⊙(non-terminal maleic unsaturation is less reactive than the fumaric isomer),

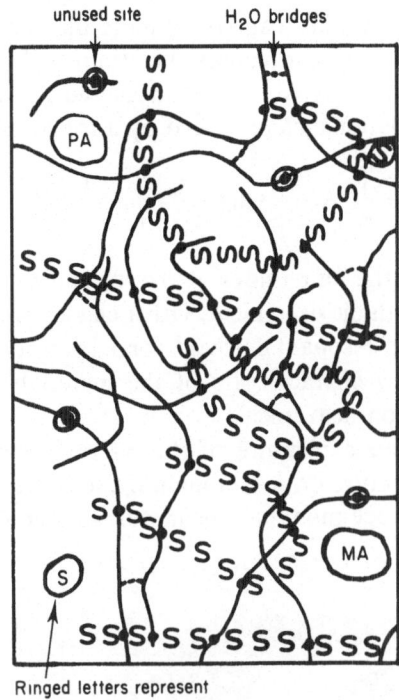

unused site H_2O bridges

Ringed letters represent
diluent monomer

FIG. 8.1.   Typical crosslinked polyester network.

(5)   uncrosslinked polyester molecules,
(6)   trace quantities of 'diluent' molecules, e.g. PG, and
(7)   water bridges (dotted lines, – – – –).

The polycondensation chains are usually much shorter than the cross-linking chains, which pass through several reactive sites. However, different crosslinking agents behave in very different ways. For instance, MMA forms a few very long bridges while S forms many short ones. It is partly for this reason that S is overwhelmingly preferred, and is always assumed to be the crosslinking agent, unless otherwise stated. Vinyl toluene is occasionally used; this resembles S in many ways.[4]

8.3. DYNAMIC MECHANICAL PROPERTIES

Dynamic mechanical testing involves subjecting a specimen of suitable geometry to regular periodic deformations, usually sinusoidal. The internal friction within the material (caused by molecular or other movements) dissipates energy as heat and causes the strain to lag behind the stress input. The phase lag angle (δ) can be measured by various means. For freely damped oscillations, as produced by a torsional pendulum,[5]

$$\tan \delta = \frac{1}{\pi} \cdot \Delta = \frac{1}{\pi} \cdot \ln\left(\frac{A_n}{A_{n+1}}\right) = \frac{1}{\pi k} \ln\left(\frac{A_n}{A_{n+k}}\right)$$

where A_n, A_{n+k} are the amplitudes of the nth and $(n+k)^{th}$ oscillations. Δ is called the logarithmic decrement. Other classes of instrument include forced vibration, non-resonant devices,[6] forced vibration devices utilising a resonant frequency characteristic of the material and its geometry[7] and pulse propagation instruments.[8]

These methods give a measure of the internal energy dissipation as a function of temperature or, less commonly, as a function of frequency. They also give a storage modulus (shear storage modulus G', or Young's

FIG. 8.2. Nonius torsional pendulum.

storage modulus E', in the appropriate case). A corresponding loss modulus G'', or E'' can be obtained when $\tan \delta < 0.3$, by putting

$$\tan \delta = G''/G' \quad \text{or} \quad E''/E'$$

or, more accurately,

$$\frac{G''}{G'} = \frac{4\pi\Delta}{4\pi^2 + \Delta^2}$$

Relaxation processes in polymers and the methods of studying them are discussed in several texts.[9,10] This chapter centres chiefly around studies carried out by means of the torsional pendulum, at frequencies of around 1 Hz, over a wide temperature range. This frequency is very suitable for the study of relaxations in polymers. A torsional pendulum is shown in Fig. 8.2.

8.4.   THE GLASS TRANSITION TEMPERATURE (T_g)

The temperature at which an amorphous glassy plastic material becomes rubberlike is classically determined by dilatometry, but dynamic mechanical methods also give an estimate of T_g. Crosslinked polymers have broad transitions, indicated by a fall in storage modulus (see Fig. 8.3). A peak in $\tan \delta$ or in Δ is associated with, but not always coincident with, the fall in storage modulus. As a result, different ways of identifying T_g have emerged. Cook and Delatycki[11] list three ways of determining T_g, by selecting either

(1)   the temperature at which $\log G'$ is midway between the glassy and rubbery plateaux, or
(2)   the temperature at which G' reaches 10^8 Pa, or
(3)   the temperature at which $\tan \tan \delta$ is a maximum.

These values do not differ very much at low frequencies, but high frequencies increase the T_g determined by the position of the damping peak.

The glass transition temperature is a major parameter, and frequently the relaxation causing the damping peak is known as the α-transition. It is believed to be caused by relaxation of substantial parts of the crosslinked network, and therefore depends on the inherent flexibility of both condensation and polyaddition chains as well as on the crosslink density. Lenk and Padget[12] studied the T_g of unsaturated polyesters as a function

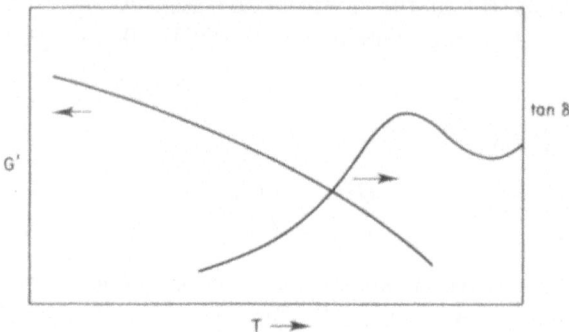

FIG. 8.3. G' and $\tan \delta$ in the region of the glass transition temperature of a crosslinked thermosetting resin.

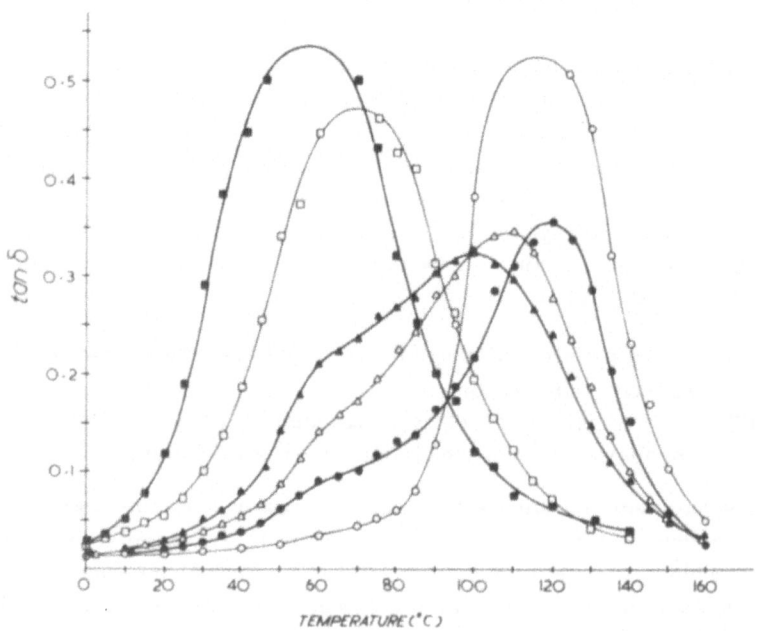

FIG. 8.4.   The effect of S concentration in the resin on the $\tan \delta$ peak, in the region of the α-transition (courtesy of John Wiley and Sons Inc. and the authors of Ref. 13).

of unsaturated:saturated acid radio, and found that increasing this ratio caused:

(1) a rise in T_g
(2) an increase in G' at T_g
(3) a decrease in tan δ (max.)
(4) a broadening of the transition range.

The temperature of the main transition (the α-relaxation) is dependent on polyester chain length, and falls with M_n (number-average molecular weight). It also depends on S bridge length, and was found to first rise, in the case of one resin (1:1 PA:FA), from 50 °C at 15% w/w S to 120 °C at 40% w/w S, and then decrease slightly to 115 °C at 70% w/w S. This is demonstrated by the tan δ peaks in Fig. 8.4.[13] The increasing S concentration produces an increase in the fumarate utilisation and, therefore, a higher crosslink density. This restricts the mobility of the polyester chains, and results in an increase in the temperature of the relaxation. Above 40% S, however, the fumarate unsaturation is probably already well used, and further increases simply augment average bridge length, thus reducing crosslink density.[11]

8.5.   CROSSLINK DENSITY AND T_g

Crosslink density can be determined from the stiffness of crosslinked polymers in the region of the rubbery plateau ($T > T_g + 50$ K). The dynamic shear storage modulus is given by

$$G' = 2\phi\rho dRT$$

where ϕ = a front factor
ρ = the molar concentration of crosslinks per unit mass
d = the density
R = the gas constant
T = the temperature (K).

According to Graessley,[14] the equation can be modified to the form

$$G' = \phi dRT(2C_4 + C_3)$$
$$= \phi dRT\rho_e$$

where C_3, C_4 are the concentrations in mole kg^{-1} of effective tri- and tetra-functional crosslinks and ρ_e is the concentration of elastically active network strands, in mole kg^{-1}. Blanchard and Wooton[15] proposed a further modification to allow for the space taken up by the strands themselves, so that

$$G' = \frac{\phi_1}{(1 - K\rho_e)} dRT\rho_e$$

where K is a steric hindrance term, and ϕ_1 is the ideal network front factor, i.e.

$$\phi_1 = \phi(1 - K\rho_e)$$

It should be remembered that the polyester chains are relatively short and, as indicated in Fig. 8.1, there is a sol fraction unconnected to the gel. Cook has pointed out[16] that in order to determine the effective crosslink density, ρ_e, a consideration of the varieties of crosslink found in the three-dimensional structure is required. Eight types are listed:

(1)   ordinary S–fumarate crosslinks in which the polyester chain is connected to other crosslinks on both sides,
(2)   those crosslinks in which the polyester is connected to the gel on one side but at the other side it is a 'loose end',
(3)   those crosslinks which form the only reacted site in a given polyester chain.

These crosslinks do not all make equivalent contributions to network properties. In low-S systems, where fumarate–fumarate homopolymerisation occurs, there may also be fumarate–fumarate links in which the two conjoined polyester chains have:

(4)   no other links with the gel, or
(5)   one link only and three loose ends, or
(6)   two links and two loose ends, or
(7)   three links and one loose end, or
(8)   four links and no loose ends.

It is found that G'/T passes through a maximum at a certain optimum concentration of S, and then falls towards zero as the polyester mole fraction diminishes. Cook and Delatycki[11] also express the effective crosslink density in terms of the parameters illustrated in Fig. 8.1, i.e.

$$\rho = \rho'\left(1 - \frac{M_{c_1}}{M_1} - \frac{M_{c_2}}{M_2}\right)$$

where: M_1 = number-average molecular weight of the polyester condensation chain (typically between 700 and 2500),

M_2 = number-average molecular weight of the S chain (estimated
(1) to be of the order of 8000 to 14 000),

M_{c_1}, M_{c_2} are the molecular weights between crosslinks of polyester segments and of S segments, respectively,

ρ' = number of moles of reacted unsaturation per unit mass.

In practice, the crosslink density of thermosetting resins is non-uniform,
and evidence for this has been submitted not only in the form of electron
microscopic observations but also as a consequence of uneven swelling
observed when the resins are placed in suitable liquids.[17] The structure
of crosslinked resins may be conceived as approximating to one of the
models in Fig. 1.1. Maizel et al.[18] determined the crosslink density of
a series of polyesters, made from MA, PA and DEG (MA:PA molar
ratios 1:0, 4:1, 2:1, 1:1 and 1:2), from the theory of high elasticity
and from the measurement of initial Young's modulus in the region
$T \geq T_g + 50\,\text{K}$. The density of crosslinks was found to increase almost
linearly with mole fraction of MA until the proportion of PA became
very small, whereupon this effect was reduced. The T_g fell from 106 °C
for the 1:0 resin to 36 °C for the 1:2 resin. Since the polyester:S ratio
was constant, a reduction in the MA:PA ratio should have given longer
S bridges (i.e. higher M_{c_2}).

Cook[19] also studied the effect of polyester crosslink density on T_g,
using DEG/SA/FA resins crosslinked with S. An expression was obtained
by combining the Di Marzio equation[20] i.e.

$$\frac{1}{T_g} = \frac{1}{T_g'} - \frac{1}{T_g'} K_{DM} \left(\frac{M_0}{10^3 \alpha} \right) \rho$$

where T_g' = the T_g of the uncrosslinked, infinite molecular weight polymer

K_{DM} = a crosslinking constant

M_0 = molecular weight in g mole^{-1} of a structural unit

α = number of rotatable bonds/unit

ρ = crosslink density

with a Fox–Loshaek expression.

This gave:

$$\frac{1}{T_g} = \frac{1}{T_g''} - K_1 \left(\frac{100 - C}{100} \right) \frac{10^3}{M_1} - K_2 \left(\frac{10^3}{M_2} \right) \left(\frac{C}{100} \right) + K_1' \rho$$

where M_1, M_2 are the molecular weights of polyester and S chains respectively, C is the weight % of S in the network, T_g'' is the T_g of a hypothetical blend of the two kinds of chain, both infinitely long. The other terms are constants. In practice, the S-fumarate crosslinking chains were considered long enough for the term

$$\left(\frac{10^3}{M_2}\right)\left(\frac{C}{100}\right)$$

to be neglected, and a plot of $1/T_g$ against $1/M_1$ was found to be linear. The slope of the lines was similar to that derived from the work of Grieveson.[22] Values obtained for the crosslinking constant K_1' were in the region of 4.8 to $6.7 \times 10^{-4}\,\mathrm{kg\ mole^{-1}\ K^{-1}}$.

The following are some of the factors which may give rise to the variability of crosslinked polyester properties. Crosslinks can be irregularly spaced, especially in IPA resins. Also, fumaric unsaturation is more easily utilised than the non-terminal maleic form, and the proportion of maleic to fumaric depends partly on the nature of the diol used. Chain length depends both on reaction time and on diol: diacid stoichiometry. Finally, residual diluent substances such as excess diol can plasticise the resin with resultant lowering of modulus. As with all polymers, rate or frequency of loading can influence properties.

8.6. SECONDARY RELAXATIONS

Many polymers, including crosslinked unsaturated polyester resins, show secondary transitions below T_g. These are customarily denoted by β, γ, δ, etc., in order of decreasing temperature. Sometimes the transitions detected by dynamic mechanical methods are also observed by analogous a.c. electrical measurements.[23] The existence of secondary transitions gives the material some capacity for energy absorption below T_g and increases the impact strength. Figure 8.5 shows schematically the transitions found in unsaturated polyester resins.

Perepechko et al.[24] studied the viscoelastic properties of polyesters made from 1 mole DEG to 0.17 mole FA and 0.83 mole saturated acid. The saturated acids consisted of the series $HOOC(CH_2)_nCOOH$ where n ranged from 2 (SA) up to 8 (SBA). In addition, MSA and thiovaleric acid were tried. An acoustic technique allowed measurements to be made of modulus E', tan δ, and velocity of sound, V, by means of a forced resonance cantilever beam. Frequencies were 150 and 900 Hz.

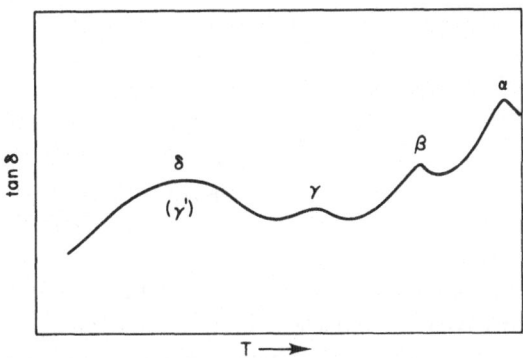

FIG. 8.5. Dynamic mechanical relaxations of a crosslinked unsaturated polyester resin.

Generally, V fell from about $2 \times 10^{-4} \, mm \, s^{-1}$ at $-100 \, ^\circ C$ to 1×10^{-4} $mm \, s^{-1}$ at about $0 \, ^\circ C$ and reached a plateau of less than $2 \times 10^{-5} \, mm \, s^{-1}$ above $+100 \, ^\circ C$. Four transitions were noticed in the case of the succinic resin: at $-46 \, ^\circ C$, $-25 \, ^\circ C$, $+12 \, ^\circ C$ and $+98 \, ^\circ C$. The transition at $-25 \, ^\circ C$ was assumed to be equivalent to T_g. The AA polyester resin was found to have an additional transition at $-16 \, ^\circ C$ and the SBA at $+12 \, ^\circ C$. The relationship between the various transitions and the acid chain length was observed but not found to be particularly simple. A liquid–liquid transition (T_{ll} transition) ($> T_g$) in amorphous homopolymers and block copolymers of S has been studied by torsional braid analysis, and is thought to arise from a molecular rather than a macroscopic relaxation.[25]

8.7. RELAXATIONS IN THE CONSTITUENT CHAIN

The polyester network is made up of two interlocking chain types, and it may be useful to consider what transitions are found in the linear chain structures when these are examined separately.

Homopolystyrene has a T_g of $100 \, ^\circ C$. The transitions in S-maleic anhydride alternating copolymers have been studied by Block et al.,[26] but these copolymers do not contain opened maleic rings, but rather have the structure

Measurements with a Nonius pendulum over the temperature range 140 K to 460 K showed a prominent β-transition at about 375 K, and a loss peak at about 170 K which could be correlated with an analogous dielectric relaxation observed at 240 K and 10^3 Hz (the frequency difference would be sufficient to shift the peak considerably). This latter peak was attributed to the twisting of the unopened ring in the amorphous region. A γ-transition, attributed to the same motion in the crystalline phase, could be found only by electrical measurements.

More information is available about transitions in linear polyesters. Four or more methylenic groups have been found to co-operate to produce a low-temperature, β-relaxation which is not very pronounced in semi-crystalline, linear polyesters such as poly(hexamethylene sebacate) but much more prominent in the completely amorphous isomer, poly(2-methyl,2-ethyl,1,3-propylene sebacate), i.e.

$$\left(\!\!-CO(CH_2)_8CO\cdot O\cdot CH_2\cdot \overset{\displaystyle CH_3}{\underset{\displaystyle CH_2CH_3}{\overset{\textstyle |}{\underset{\textstyle |}{C}}}}\cdot CH_2\cdot O\!\!-\right)_{\!n}$$

The methylenic groups were believed to rotate by 180° or 360° about the chain axis.[27] The α-relaxation for this polyester occurred at $-60\,°C$.

The β-transition temperature falls as the diol length increases through the series of linear poly(alkyl terephthalates).[28] Recently, the results of a study of the effect on the β-relaxation of structural variation of several aromatic polyesters have been published.[29] It was found that the relaxation was controlled by the structure of the diol. When Cook and Delatycki[30] measured T_β for a series of crosslinked FA/DEG/SA resins of differing FA:SA ratios, the values obtained were found to approach the temperature of the T_g of poly(diethylene succinate) ($-30\,°C$) as the FA proportion approached zero. Therefore, these investigators concluded that the β-transition temperature in crosslinked polyesters was closely related to the T_g of the corresponding linear polyesters.

8.8. THE β-TRANSITION IN CROSSLINKED RESINS

The β-transition temperature has been studied systematically as a function of polyester molecular weight, of diol, of diacid, and of water content. It was found that this relaxation could be related to the motion of the polycondensation segments; its temperature decreased as diol length increased.[30]

The polycondensation chains are made more flexible by addition of ether-containing diols such as DEG, but this is not very effective in lowering T_β because, although the ether groups have low rotational energy barriers, polar interactions reduce the mobility of the segments. There was also very little change in the value of G_β'' (max.) as the diol was progressively lengthened from ED to HD. This suggests that the phthalate group is mainly responsible for the β-relaxation. All three phthalic isomers produce a β-transition, with the order of increasing T_β being PA < IPA < TPA. However, the groups causing the relaxation must be able to relax at temperatures below the T_g, but in some environments this is not possible. For instance, Isaoka et al.[31] found no relaxation at all below 219 °C for crosslinked poly(diallyl phthalate), but according to Roshchupkin et al.,[32] the more open structure of the crosslinked polymer of bis-methacrylato-triethylene glycol phthalate

$$CH_2{=}C{-}C{-}O{-}(CH_2{-}CH_2{-}O)_3{-}C \qquad C{-}O{-}(CH_2{-}CH_2{-}O)_3{-}C{-}C{=}CH_2$$

gave a transition at 66 °C and 1 Hz.

The ether-type diols had little effect in lowering T_β but it was found that increasing the length of such diols did increase the specific β loss (i.e. G_β'' (max.) per mole of saturated dibasic acid).

A β-transition is also found when there is no phthalic isomer present; T_β increases as the SA:FA ratio increases in the series of phthalic-free resins.

Lenk and Padget[12] drew a different conclusion from investigations of a series of maleic/bis-acid A2 resins: that the β-transition was associated with S crosslink relaxation. This was because the loss tangent is increased by a reduction in maleic proportion (and hence by an increase in S bridge length). However, Cook and Delatycki pointed out[30] that this proposal is not consistent with the observed dependence of the β-relaxation on the S concentration. For FA/DEG/SA resins with various SA mole fractions, these authors explained results similar to those obtained by Lenk and Padget, by suggesting that the β-relaxation is the result of motion of the polyester segments between crosslinks.

8.9. γ AND γ' TRANSITIONS IN CROSSLINKED RESINS

Transitions have also been found at much lower temperatures than those of the β-relaxation. The fourth transition for crosslinked polyesters is called γ',[33] perhaps to avoid confusion with the loss angle δ. It was found that the low temperature region is affected by water contained in the resins, and since all polyesters contain water when first made, desiccation is necessary.

A low-temperature relaxation zone was found by Witort et al.[34] in the range $-93\,°C$ to $-133\,°C$. The resin used was made by reacting MA (sometimes partly replaced by PA) with 2,2-di(4-hydroxypropoxyphenyl)propane and 1 to 2 mole % polypropylene glycol ($M_n = 3000$). The aromatic diol was sometimes partly replaced by PG.

The low-temperature, γ-transition was studied by Cook and Delatycki.[33] Conventional FA/PG resins with 40% w/w S were found to exhibit γ-transitions at about $-100\,°C$. On desiccation, the tan δ peak indicating this transition was progressively reduced, but a still lower temperature transition (γ' between $-190\,°C$ and $-140\,°C$) increased in prominence. Complete drying left hardly any sign of the γ peak; at the other extreme it approached a constant value as the water content reached 3·0%.[35] T_γ was found to decrease as S concentration increased from 25% to 70% (for FA/PA/PG resins). The size of the peak was linearly related to fumarate concentration. This led the investigators to the conclusion that the γ peak requires water to be present and to form bridge structures with the fumaric units, e.g.

These fumarate–water complexes can relax, provided that they are not sterically hindered. T_γ increases, for instance, when the ester groups are close enough to induce interchain interactions between fumarate groups. Insertion of PA results in random structures with greater opportunities for intrachain ester interactions. (The γ'-transition was attributed to motion of the diol section of the polyester chain.)

Addition of 4 % w/w PG (or xylene) to the resin lowered T_g by plasticisation, but did not affect the γ-relaxation. Polyester chain length was not found to affect either the γ- or γ'-transition significantly in the crosslinked resin.

The same authors examined the γ'-transition further[36] by using resins of various diol and diacid units. The γ'-transition was most prominent in plots of G'' versus T, especially with dry resins. However, drying did not affect $T_{\gamma'}$ so much as it affected peak height. The peak became less prominent as the S concentration increased and the polyester proportion fell.

The γ' peak is due to motions of diol segments. It becomes more prominent on drying, because water restricts these motions, although whether it does so by hydrogen bonding or simply by filling spaces is not known. The γ' peak also increases as diol length increases (PG to HD) and is shifted to lower temperatures.

8.10. THE FRACTURE TOUGHNESS OF CROSSLINKED POLYESTER RESINS

In this section some recent studies of the fracture of polyester resins are discussed and reference is made to experiments with pre-notched specimens such as the single edge notched tensile (SEN), centre-notched tensile (CN) and double cantilever beam (DCB) types. The propagation of cracks in thermosetting resins is discussed in Chapter 9 as a separate topic.

Extensive research has been carried out to determine the fracture behaviour of metals, ceramics, glass and linear organic glassy thermoplastics. Less has been achieved in the case of thermosetting resins, partly because of their variable and uncertain composition. Another factor is that thermosetting resins are generally toughened by fibre reinforcement, which completely alters their mode of failure. Nevertheless, there is growing interest in the relationship between chemical structure and toughness in crosslinked polymers.[37,38]

The fracture surface energy, γ, of a material containing either a through-thickness edge crack of length a, or a through-thickness central crack of length $2a$, is given by

$$\sigma_f = \left(\frac{G_c E}{\pi a}\right)^{\frac{1}{2}} = \left(\frac{2E\gamma}{\pi a}\right)^{\frac{1}{2}} \qquad \text{(plane stress)}$$

or

$$\sigma_f = \frac{G_c E}{\pi a (1 - \mu^2)^{\frac{1}{2}}} = \frac{2E\gamma}{\pi a (1 - \mu^2)^{\frac{1}{2}}} \qquad \text{(plane strain)}$$

where σ_f = stress at fracture

G_c = critical potential energy release rate with respect to crack length at fracture

E = Young's modulus

μ = Poisson's ratio

Berry[39] calculated that γ should be about $0.45\,\text{J}\,\text{m}^{-2}$ for linear poly-(methyl methacrylate), whereas experimental values were found to be several hundred times greater. (See Table 8.2.) The fracture surfaces showed coloration under reflected light. Crosslinking gradually reduces γ to a value very much smaller than before, which suggests that viscous flow is responsible for the additional work of fracture. The values given in Table 8.2 for unsaturated polyester resins are great enough to suggest that some flow still occurs, despite crosslinking.

TABLE 8.2

FRACTURE SURFACE ENERGY OF POLY(MMA) AND OF
UNSATURATED POLYESTER RESINS

Polymer	Fracture surface energy γ $(J m^{-2})$	Reference
Poly(MMA) [uncrosslinked]	300	39
Poly(MMA) [uncrosslinked]	188	40
Poly(MMA) [uncrosslinked]	500	41
Poly(MMA) [lightly crosslinked]	80	42
Poly(MMA) [crosslinked]	46	43
Poly(MMA) [densely crosslinked]	23	42
Unsaturated polyester	42	44
Unsaturated polyester	10	42
Unsaturated polyester	47	45

The flow occurs at the tip of the propagating crack, but it is uncertain whether this flow involves movement of uncrosslinked (or lightly cross-linked) regions with respect to denser micelles, or whether it involves more extensive movement within a single, dense and highly interconnected structure. (See Fig. 1.1.)

Owen and Rose[46] determined several fracture toughness parameters for mixtures of a conventional low reactivity polyester resin (molar ratio $MA = 1$, $PA = 2$, $PG = 3$) with a flexibilising polyester resin of identical alkyd:S ratio (the alkyd was made from MA, AA and PG). Measurements were made in tension using CN plates. It was found that G_c was hardly changed by the addition of 15% w/w flexibilising resin, but it increased tenfold on addition of 30%. Young's modulus declined steadily, but Poisson's ratio passed through a minimum at 15% addition. As the resin blend became more flexible, the difference between values of the critical stress intensity factor, K_{Ic}, obtained using thick specimens, and the corresponding values (K_c) obtained from thin specimens, increased. This was attributed to the increasing tendency for the thin specimens to fail by a mixture of modes rather than by 'mode I' opening.

The poly(propylene maleate adipate) flexibiliser was reactive and capable of being incorporated into the network, so apart from unspecified differences in chain length, acid number, etc., the main difference between the various blends was the ratio of aromatic phthalic to aliphatic adipic sequences in the polyester condensation chains. (The ratio of MA to AA in the flexibiliser was not stated.)

This indicates that increasing the mobility of polyester segments greatly increases the work of fracture. A similar increase in mobility can be achieved by large increases in crosslink spacing, leading to a change from brittle to ductile behaviour.[47]

Christiansen and Shortall[48] carried out rather similar investigations using a PA-based flexibiliser. This additive did not lower the modulus so drastically and was probably more akin to the base resin than was the flexibiliser used by Owen and Rose. Fracture energy increased from $22 \cdot 8 \, J \, m^{-2}$ to $28 \cdot 5 \, J \, m^{-2}$ at 20% addition of flexibiliser. These workers also used the expression,[49,50]

$$D = \frac{\pi}{8} \left(\frac{K_c}{\sigma_y} \right)^2$$

to obtain the value of the Dugdale plastic zone length D in terms of the compressive yield stress σ_y. Assuming $\mu = 0 \cdot 35$, the value of D for the medium reactivity orthophthalic base resin was calculated to be

$5.26\,\mu$m. An estimate of the radius of the assumed plastic zone r_y, obtained from the expression,[51]

$$r_y = \frac{1}{6\pi}\left(\frac{K_c}{\sigma_y}\right)^2$$

was $0.8\,\mu$m. These values can be compared with the considerably larger estimated sizes of dense globular micelles commonly reported to be contained in thermosetting resins. The critical flaw size (c) obtained by applying the Griffith equation

$$\sigma_f^2 = \frac{2E\gamma}{\pi c(1 - \mu^2)}$$

to polished tensile specimens was about $22\,\mu$m, which is similar to the size of the smallest globular aggregates mentioned above. Linear polymeric glasses have higher critical flaw sizes; for example, polystyrene has $c = 1000\,\mu$m, and polymethyl methacrylate has $c = 50\,\mu$m. Crosslinking reduces these values considerably. Abeysinghe et al.[52] obtained the value $c = 38\,\mu$m for a polyester resin formulated from equimolar quantities of MA, PA, PG and DEG.

In practice, flaws are often holes within materials and not through-cracks. Polyester resins sometimes develop internal disc-shaped cracks after prolonged immersion in hot water. The distribution of stress in the neighbourhood of a disc-shaped internal crack, of diameter a, within an elastic solid is given by,[53]

$$K_{1c} = \frac{2\sigma_f a^{\frac{1}{2}}}{\pi^{\frac{1}{2}}}$$

By use of this expression, a critical diameter for these osmotically generated disc cracks was found to be $90\,\mu$m for the resin mentioned above.[52]

The fracture toughness of unsaturated polyester resins depends on several structural factors. Rhoades[54] found that K_{1c} falls progressively as the maleic:phthalic ratio is increased from 1:1 to 1:0. It had previously been found that K_{1c} falls after prolonged immersion in hot water[55] but Rhoades concluded that the extent of this susceptibility to water is itself structure-dependent, being more noticeable at high MA:PA ratios. The cure temperature affects the crosslink density, and it is not surprising that a low cure temperature favours a high fracture energy.[48] Some typical values of the fracture toughness of unsaturated polyester resins are given in terms of K_{1c} in Table 8.3.

TABLE 8.3

CRITICAL STRESS INTENSITY FACTORS OF UNSATURATED POLYESTER
RESINS

K_{Ic} $(MN\,m^{-3/2})$	Specimen geometry	MA:PA mole ratio	Reference
0·49	DCB	(probably) 1:1	48
0·72	CN	1:2	46
0·84	SEN	3:1	55
0·71	CN	1:1 (contained PG and DEG)	52
0·71	SEN	1:1	54

8.11. TOUGHENING WITH ELASTOMERIC PHASES

Attempts have been made to increase the fracture toughness of unsaturated polyester resins by the same method as that used for polystyrene, i.e. by addition of an elastomeric phase. The use of carboxyl-tipped butadiene acrylonitrile rubbers has been successfully established for the toughening of thermosetting epoxide resins, but compatibility problems arose when these elastomers were first blended with unsaturated polyester resins. Whereas epoxy resins can be toughened fifty-fold, the improvements achieved with polyesters are more commonly three-fold or less. Tetlow et al.[56] added butadiene–acrylonitrile rubbers with various end-groups (carboxyl, vinyl, amine) to a highly reactive IPA polyester resin and measured the fracture toughness by using DCB fracture toughness specimens. In general, addition of vinyl terminated elastomers led to the greatest increase in fracture toughness, although even in these cases the enhancement was less than 100% at 8% rubber addition. The combination of both rubber and added flexibilising resin had no greater effect. The theory of rubber toughening described by these investigators is that rubber particles of at least 1 μm in diameter induce crazing with void formation, whilst particles of size $\leq 0\cdot1$ μm induce deformation by shear banding. The optimum size distribution is therefore bimodal, with the two mechanisms operating jointly, so that shear banding prevents crazes from developing into cracks. According to this theory, very highly crosslinked resins are the ones with least molecular mobility and are thus the least able to be toughened by such mechanisms. Therefore, the resins most in need of toughening are those for which the improvement is most difficult to achieve.

Butadiene–S block copolymers have been used to toughen polyester resins, with better effects than those obtained by use of uncured S–butadiene rubber.[57]

8.12. FRACTURE SURFACES

The fracture morphology of crosslinked polyester resins resembles that of glass in many respects.[48] The fracture surfaces of simple tensile specimens show small regions of smooth, mirror-like appearance and small mist regions, followed by hackle and, eventually, crack branching. The initiation site is often at the surface and, from this site, long narrow 'river lines' are occasionally found, passing through the mirror into the mist region. These river lines are caused by crack propagation at different levels.

Different features are seen according to whether the specimen geometry favours stable, i.e. not self-propagating, or unstable crack growth. Owen and Rose[58] examined the surfaces of CN fracture toughness specimens, which contained only a small region of river markings adjacent to the crack tip, followed at a distance by a smooth surface characteristic of slow, unstable propagation and, eventually, by a region of greater roughness. This region contained conic markings, caused by the spread of secondary fractures, in a plane close to, and parallel to, the main fracture

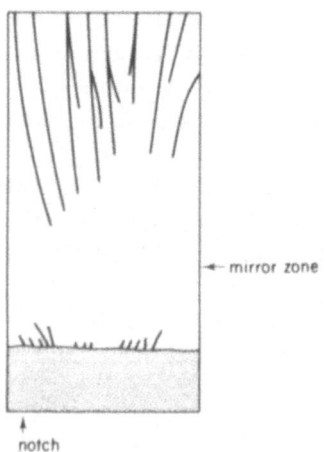

FIG. 8.6.   The position of the mirror zone in the fracture surface of a SEN sample of an unsaturated polyester resin, broken in tension.

plane. With the addition of increasing proportions of flexibiliser, the river markings became more widespread and regular.

The surfaces of specimens broken in tension after crack tip production by cyclic loading were similar to those of CN fracture toughness specimens.

Some of the features found on polyester resin fracture surfaces are indicated schematically in Fig. 8.6. The rough region is in most cases related to increased propagation velocity. Flexible resins were found to have larger smooth regions of slow, unstable crack propagation.[58] Christiansen and Shortall[48] found that the rough hackle region did not appear in tapered DCB specimens.

Rhoades[54] examined the fracture surfaces of three specimens of a 2:1 MA:PA resin (Fig. 8.7). The top, relatively uncomplicated surface

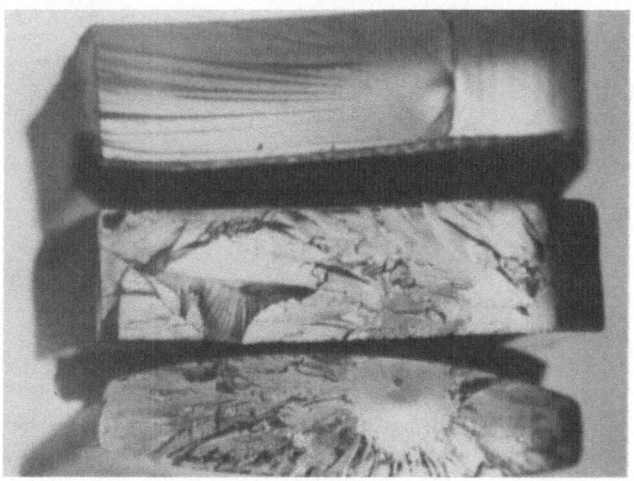

FIG 8.7.   Tensile fracture surfaces of samples of a 2:1 MA:PA unsaturated polyester resin (top) SEN; (middle) unnotched; (bottom) unnotched but containing a prominent flaw.[54]

was that of an SEN specimen broken in tension, whilst the middle surface was of a similar specimen without the notch; this surface appears much more complex. The bottom surface was also an unnotched tensile specimen, but the fracture initiation site can be seen to be a void inadvertently introduced during casting.

As mentioned earlier, polyester resins immersed in water for long periods frequently develop disc cracks. Figure 8.8 shows the fracture surface

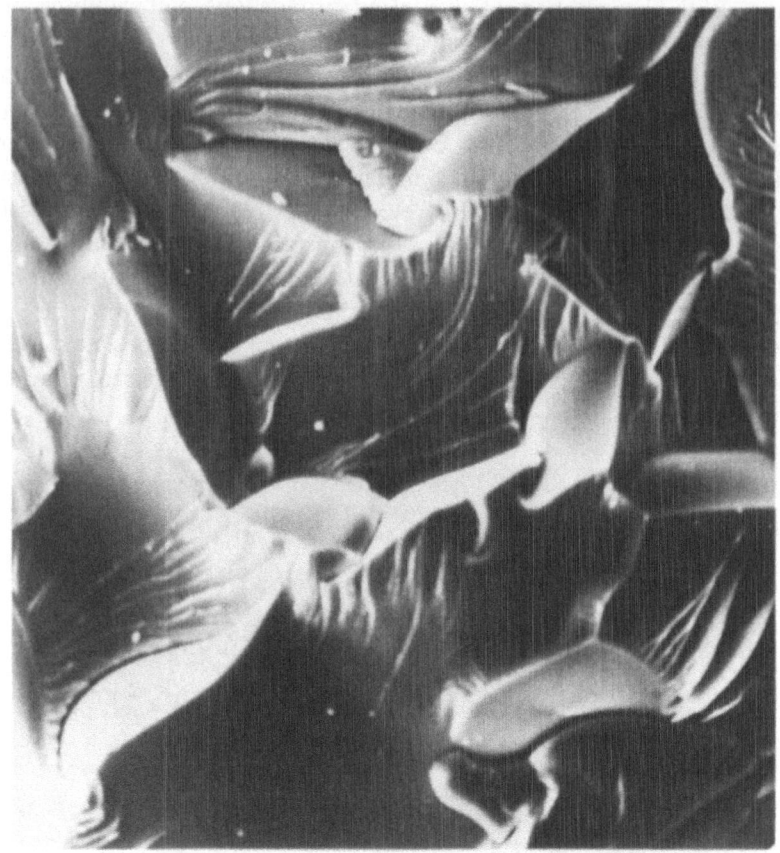

FIG. 8.8.   Fracture surface of half of a CN unsaturated polyester resin specimen,
broken in tension after 6000 h immersion in distilled water at 65 °C. The figure shows
the central region (× 215).

of a CN specimen, broken in tension after 6000 h immersion in distilled
water at 65 °C. The figure shows a close-up of one part of the region
midway between the notch and the far edge of the specimen. The development of disc cracks releases strain energy and is accompanied, at least
initially, by a small increase in the fracture toughness.

Mushkatel and Marom[59] found that boiling of fracture toughness
specimens in water before fracture was carried out had the effect of
converting the surface morphology to a smooth, mirror-like form, but

if part of the S was replaced by ring-substituted BS, the water-boiling treatment no longer had this effect. BS appeared to suppress the development of water-induced microcracks about 0·02 mm wide. Scanning electron micrographs showed a little crazing at low BS concentrations but none at all at higher levels despite 107 h boiling. Also, the retention of fracture initiation energy (determined by three-point bending of notched bars) was improved.

This was explained by the theory that large crosslinking monomers such as BS, chlorostyrene and vinyl toluene produce a relatively open structure, allowing water molecules to penetrate without generating excessive internal stresses. Craze resistance has long been thought to be inversely proportional to density.[60]

Improved resistance to microcracking was also shown by the BS resins after combined humidity and ultraviolet treatment, but photo-oxidative degradation was increased.

Fracture surfaces of polyester resins broken in the three-point bending mode are also discussed by Kitoh *et al.*[61] Resin formulations ranged from FA:DEG molar ratios of 1:1 to FA:AA:DEG of 1:4:5. Again, mirror and rough regions were observed. A very wide range of crosshead speeds were employed (3 to 120 000 mm/min).

8.13.   CONCLUSIONS

Crosslinked unsaturated polyester resins exhibit four dynamic mechanical relaxations. The α-transition is believed to involve the whole network, while the β-transition is probably caused by polyester mobility although present views are not unanimous on this point. The γ-transition is caused by water-bridges linking ester groups in neighbouring polyester chains, and the lowest temperature, γ'-transition is caused by movement of diol segments.

Most polyester resins, except those containing high proportions of linear aliphatic acid, or high proportions of diols containing ether groups, undergo brittle fracture under normal conditions. The fracture toughness falls as the degree of crosslinking increases. Characteristic fracture surfaces are glass-like. The effect of water immersion on these surfaces has been studied. Addition of an elastomeric phase has very little beneficial effect on the toughness of polyester resins.

REFERENCES

1. BOENIG, H. V. *Unsaturated Polyesters*, 1964, Elsevier, Amsterdam.
2. IMAI, T. *J. Appl. Polym. Sci.*, 1967, **11**, 575.
3. PRITCHARD, G., Ph.D. thesis, *The crosslinking of polyesters: a study of network formation by physical methods*, 1968, University of Aston in Birmingham, England.
4. HEBERT, N. T. *Plast. World*, 1975, **33**, 47.
5. SEEFRIED, C. G., JR. and KOLESKE, J. V. *J. Test. and Eval.*, 1976, **4**, 220.
6. DEKKING, P. *Determination of dynamic mechanical properties of high polymers at low frequencies*, 1961, Report from TNO Laboratory, Delft, Netherlands.
7. LEARMONTH, G. S. and PRITCHARD, G. *Soc. Plast. Eng. J.*, 1968, **24**, 47.
8. SOFER, G. A., DIETZ, A. G. H. and HAUSER, E. A. *Ind. Eng. Chem.*, 1953, **45**, 2743.
9. WARD, I. M. *Mechanical Properties of Solid Polymers*, 1971, Wiley, London.
10. HAWARD, R. N., ed. *The Physics of Glassy Polymers*, 1973, Applied Science Publishers Ltd, London.
11. COOK, W. D. and DELATYCKI, O. *J. Polym. Sci., Polym. Phys. Ed.*, 1974, **12**, 1925.
12. LENK, R. S. and PADGET, J. C. *Europ. Polym. J.*, 1975, **11**, 327.
13. COOK, W. D. and DELATYCKI, O. *J. Polym. Sci., Polym. Phys. Ed.*, 1974, **12**, 2111.
14. GRAESSLEY, W. W. *Macromolecules*, 1975, **8**, 865.
15. BLANCHARD, A. F. and WOOTON, P. M. *J. Polym. Sci.*, 1959, **34**, 627.
16. COOK, W. D. *Europ. Polym. J.*, 1978, **14**, 721.
17. FUNKE, W. *J. Polym. Sci.*, 1967, Part C3, 1497.
18. MAIZEL, N. S., MOZZHECHKOVA, N. I., TSVETKOVA, O S., GURINOVICH, L. N. and MIKHAILOVA, Z. V. *Vysokomol. Soedin.*, 1978, A**20**, 636.
19. COOK, W. D. *Europ. Polym. J.*, 1978, **14**, 715.
20. DI MARZIO, E. A. *J. Res. Nat. Bur. Stds.*, 1964, **68A**, 611.
21. FOX, T. G. and LOSHAEK, S. *J. Polym. Sci.*, 1955, **15**, 371.
22. GRIEVESON, B. M. *Polymer*, 1960, **1**, 499.
23. POCHAN, J. M. and HINMAN, D. F. *J. Polym. Sci., Polym. Phys. Ed.*, 1975, **13**, 1365.
24. PEREPECHKO, I. I., SVETOV, A. YA., SEDOV, L. N. and SAVICHEVA, O. I. *Plasticheskie Massy*, 1974, **1**, 42.
25. GILLHAM, J. K. *Polym. Eng. Sci.*, 1979, **19**, 749.
26. BLOCK, H., COLLINSON, M. E. and WALKER, S. M. *Polymer*, 1973, **14**, 68.
27. FROIX, M. F. and GOEDDE, A. O. *J. Macromol. Sci., Phys.*, 1975, **B11**, 345.
28. FARROW, G., MCINTOSH, J. and WARD, I. M. *Makromol. Chem.*, 1960, **38**, 147.
29. URALIL, F., SEDEREL, W., ANDERSON, J. M. and HILTNER, A. *Polymer*, 1979, **20**, 51.
30. COOK, W. D. and DELATYCKI, O. *Europ. Polym. J.*, 1978, **14**, 369.
31. ISAOKA, S., MORI, M., MORI, A. and KUMANOTANI, J. *J. Polym. Sci.*, 1970, **8**, Part A1, 3000.
32. ROSHCHUPKIN, V. P., KOCHERVINSKII, V. V, SEL'SKAYA, O. G. and KOROLEV, G. V. *Vysokomol. Soedin.*, 1971, **B13**, 497.

33. COOK, W. D. and DELATYCKI, O. *J. Polym. Sci., Polym. Phys. Ed.*, 1975, **13**, 1049.
34. WITORT, I., SLUPKOWSKI, T., CHOMKA, W. and JACHYM, B. *Polimery Tworzywa Wielkoczasteczkowe*, 1976, **21**, 155.
35. COOK, W. D. and DELATYCKI, O. *J. Polym. Sci., Polym. Phys. Ed.*, 1977, **15**, 1967.
36. COOK, W. D. and DELATYCKI, O. *J. Polym. Sci., Polym. Phys. Ed.*, 1977, **15**, 1953.
37. PRITCHARD, G. and RHOADES, G. V. *Mater. Sci. Eng.*, 1976, **26**, 1.
38. YOUNG, R. J. Chapter 6 in *Developments in Polymer Fracture—1*, ed. Andrews, E. H., 1979, Applied Science Publishers Ltd, London.
39. BERRY, J. P. *J. Polym. Sci.*, 1961, **50**, 107.
40. WEICHERS, W., M.Sc. thesis, *The fracture properties of thermosetting resins*, 1969, Loughborough University, England.
41. DAVIDGE, R. W. and TAPPIN, G. *J. Mater. Sci.*, 1968, **3**, 165.
42. BROUTMAN, L. J. and McGARRY, F. J. *J. Appl. Polym. Sci.*, 1965, **9**, 609.
43. BERRY, J. P. *J. Polym. Sci.*, 1963, **1**, Part A1, 993.
44. DIGGWA, A. D. S. *Polymer*, 1974, **15**, 101.
45. PRITCHARD, G. and RHOADES, G. V., unpublished data, Kingston Polytechnic, England.
46. OWEN, M. J. and ROSE, R. G. *J. Phys. D. Appl. Phys.*, 1973, **6**, 42.
47 MATVEYEVA, N. G., ZEMSKOVA, Z. G., SIVERGIN, YU. M. and BERLIN, A. A. *Vysokomol. Soedin.*, 1974, A**16**, 588.
48. CHRISTIANSEN, A. and SHORTALL, J. B. *J. Mater. Sci.*, 1976, **11**, 1113.
49. WILLIAMS, J. G. *Int. J. Fracture Mechs*, 1972, **8**, 393.
50. MARSHALL, G. P., COUTTS, L. G. and WILLIAMS, J. G. *J. Mater. Sci.*, 1974, **9**, 1409.
51. KNOTT, J. F. *Mater. Sci. Eng.*, 1971, **7**, 1.
52. ABEYSINGHE, H. P., PRITCHARD, G. and ROSE, R. G. unpublished data, Kingston Polytechnic, England.
53. SNEDDON, I. N. *Proc Roy. Soc. Lond.*, 1946, A**187**, 229.
54. RHOADES, G. V., Ph.D. thesis, *The effect of the structural parameters of polyester resins on the mechanical properties of polyester moulding compounds*, 1979, Kingston Polytechnic, England.
55. PRITCHARD, G., ROSE, R. G. and TANEJA, N. *J. Mater Sci.*, 1976, **11**, 718.
56. TETLOW, P. D., MANDELL, J. F. and McGARRY, F. J. *34th SPI Reinforced Plastics/Composites Conf.*, New Orleans, La., USA, 1979, Paper 23-F.
57. HAYASHI, Y., IBATA, J., TOYOMOTO, K. and UTA, B. Japan. Kokai 74 30,480., *Chem. Abst.*, **81**: 107084e.
58. OWEN, M. J. and ROSE, R. G. *J. Mater. Sci.*, 1975, **10**, 1711.
59. MUSHKATEL, M. and MAROM, G. *J. Appl Polym. Sci.*, 1976, **20**, 2979.
60. HAWARD, J. B. *Soc Plast. Eng. J.*, 1959, **15**, 379
61. KITOH, M., MIYANO, Y. and SUZUKI, K. *Kobunshi Ronbunshu Eng. Ed.*, 1976, **5**, 122.

Chapter 9

CRACK PROPAGATION IN THERMOSETTING POLYMERS

ROBERT J. YOUNG

Queen Mary College, University of London, UK

SUMMARY

Thermosetting polymers are widely used as matrix materials in fibre-reinforced composites. They are generally brittle and their fracture properties are of considerable interest. Recent developments in the study of crack propagation in thermosetting polymers have been reviewed using a linear elastic fracture mechanics (LEFM) approach. It has been shown that considerable advances have recently been made in understanding the criteria controlling crack propagation. It appears that for both continuous and stick/slip propagation the criterion for failure is that a critical stress of the order of 3–4 times the yield stress of the material must be achieved at a critical distance ahead of the crack. The possibility of crazing and the effect of microstructure upon crack propagation have also been discussed.

9.1. INTRODUCTION

Over recent years the subject of crack propagation in thermosetting resins has received considerable interest.[1-17] Thermosetting polymers are being used increasingly as adhesives and matrix materials in reinforced plastics. Because of this, knowledge of their mechanical properties is extremely important. In particular it is desirable to know how, and under what conditions, cracks propagate in these materials so that the properties of the adhesive joint or composite can be predicted. Although the ultimate properties of reinforced polymer composites are dominated by the strength of the reinforcing medium (usually fibres), it is highly likely that the long-term strength and performance of these materials may be controlled by crack propagation in the matrix.

The general subject of fracture in thermosetting resins has been reviewed recently by the author in another book in this series.[1] It was shown that considerable advances had been made in the understanding of fracture in these materials. The fracture of typical thermosets such as epoxy resins and polyester resins was reviewed and it was shown that although a considerable amount of experimental data had been collected there was a notable lack of understanding of the mechanisms of crack propagation. This chapter is concerned with advances that have recently been made in explaining how and why cracks propagate in thermosetting polymers and, in particular, understanding the mechanisms of crack propagation in epoxy resins.

9.2. FACTORS AFFECTING CRACK PROPAGATION

9.2.1. Fracture Mechanics Testing

The most convenient way of studying crack propagation in thermosetting resins is through the use of linear elastic fracture mechanics (LEFM).[18] For LEFM to be applied to a particular material it is necessary that the stress–strain curve of the material is linear to fracture, and that any plastic deformation is confined to regions close to the crack tip. Since thermosetting polymers are brittle materials showing only very limited plastic deformation during tensile loading their mechanical properties normally meet the requirements of LEFM. It has been shown[1] that a variety of test pieces can be used to monitor crack propagation in brittle polymers. However, the most versatile specimens are those for which the stress intensity factor[1.8] is independent of crack length. Such specimens are the double torsion (DT) and tapered double cantilever beam (TDCB) and they have been used widely for the study of crack propagation in thermosets.[15–17] They are shown schematically in Fig. 9.1.

One of the most characteristic aspects of crack propagation in thermosets is that, under certain circumstances, cracks propagate in a stable, continuous manner and that at other times they propagate in a stick/slip, discontinuous way. These types of behaviour are shown in Fig. 9.2 for cracks propagating in DT or TDCB specimens. In these test pieces a constant cross-head speed leads to propagation at a fixed load (or stress

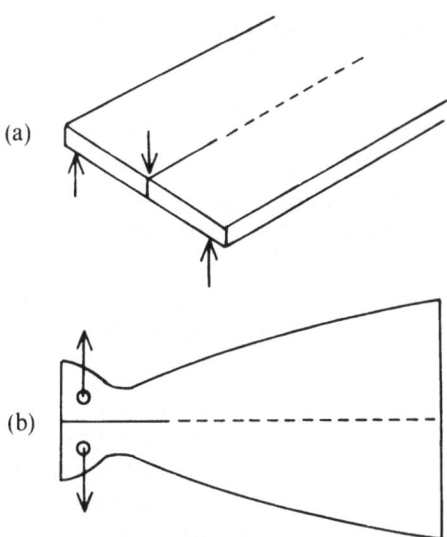

FIG. 9.1.   Schematic diagrams of two of the various fracture mechanics specimens that have been used with thermosetting polymers. (a) Double torsion (DT). (b) Tapered double cantilever beam (TDCB).

intensity factor, K_{Ic}) when the specimen is undergoing continuous propagation. They allow excellent control of the way in which cracks propagate. For example, if the cross-head speed is reduced the crack slows down and if the cross-head speed is increased the crack speeds up. Also slow crack growth can be followed in some materials when the cross-head is held at a fixed displacement. This has been done for PMMA (polymethyl-methacrylate)[19] and it has enabled the crack velocity to be measured as a function of the stress intensity factor (K_{Ic}), over a wide range of velocity, using only one specimen. If the material is

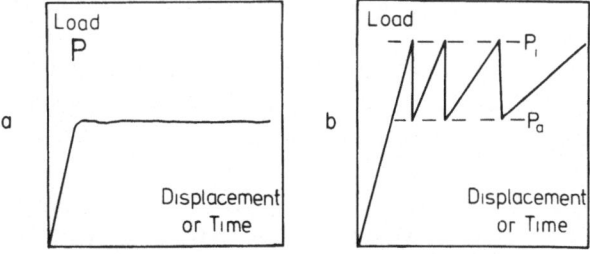

FIG. 9.2.   Schematic load/displacement or load/time curves for crack propagation in thermosetting resins. (a) Continuous propagation. (b) Stick/slip propagation.

exhibiting stick/slip propagation the DT and TDCB specimens are again very useful. In this case the load/displacement curves have characteristic saw-tooth shape. Propagation takes place at the peak load and ceases at the minimum load as the crack grows in a stick/slip manner. This explanation has recently been shown[1,20] to be slightly incorrect and it is now thought that propagation ceases not at the minimum load but at some point up to the reloading part of the curve. The use of stable test pieces has greatly improved our understanding of crack propagation in brittle polymers and most of the results discussed in this chapter were obtained using DT or TDCB specimens.

9.2.2. Material Variables

It is now well established[1] that there are a variety of material variables which affect the stability of crack propagation and the stress levels at which it occurs in thermosetting resins. These include the type of resin and curing agent used, the amount of curing agent, the temperature of the curing treatment and time of curing. The effect of the type and quantity (in parts per hundred (phr)) of curing agent used upon the fracture energies (G_{Ic}) of a series of epoxy resins is shown in Table 9.1. The values of G_{Ic} quoted are the initiation values when the mode of propagation is stick/slip and all the measurements have been made at room temperature. The details of cure schedule, testing geometry and rate can be found by consulting the original references. It is difficult to make generalisations from the data in Table 9.1 since they have been obtained from such a wide variety of resins and curing agents, using different testing geometries and rates. It can be seen that the values of G_{Ic} are typically of the order of $100 \, \text{J m}^{-2}$ and in some cases much larger. They are considerably higher than values of G_{Ic} obtained for other thermosetting polymers.[1,11] This shows that epoxy resins are some of the toughest thermosets available and hence are used in the most critical applications such as in the matrices of high-performance composites.

The detailed effect of changing the amount of curing agent and cure schedule upon crack propagation in a given resin can only be found by altering one particular variable at a time. The effects of changing curing agent content and post-cure period upon K_{Ic} (critical stress intensity factor for initiation) for a DGEBA (bis phenol-A-diglycidyl ether) epoxy resin cured with triethylene tetramine (TETA) are shown in Fig. 9.3. It can be seen that the value of K_{Ic} increases as the amount of curing agent is increased. A similar effect has been found with an

TABLE 9.1

G_{Ic} VALUES MEASURED FOR CRACK PROPAGATION IN DGEBA EPOXY RESINS CURED WITH VARIOUS HARDENERS (AFTER REFS. 1 AND 21)

Resin	Hardener (phr)	G_{Ic} (Jm^{-2})	Ref.
Epikote 828	10 DETA	86	22
	95 MNA + 0·5 BDMA	154	3
	27 DDM	340	4
	14·6 MPD	110	4
	4 DMP	180	4
	27 DDM	312	5
CT200	13 PA	220	6
DER332	5 PIP	121	23
	Various TEPA	52–227	24
	Various HHPA	158–262	25
MY750	8·3 EDA	329	9
	12·2 TDA	489	9
	16·1 HDA	575	9
	11·5 DETA	130	9
	11·0 TETA	141	9
	15·0 TEPA	136	9

Key to hardeners
BDMA = benzyl dimethylamine
DDM = diphenyl diaminomethane
DETA = diethylene triamine
DMP = tris(dimethylaminomethyl)phenol
EDA = ethylene diamine
HDA = hexamethylene diamine
HHPA = hexahydrophthalic anhydride
MNA = methyl nadic anhydride
MPD = *m*-phenylene diamine
PA = phthalic anhydride
PIP = piperidine
TDA = tetramethylene diamine
TEPA = tetraethylene pentamine
TETA = triethylene tetramine.

increase in the post-cure temperature.[17] It appears, therefore, that, at least for this particular system, K_{Ic} increases as the amount of crosslinking in the resin (characterised by the degree of cure) is increased. At first sight these observations would appear to be in conflict with those of other workers[21] who studied crack propagation in similar DETA-cured

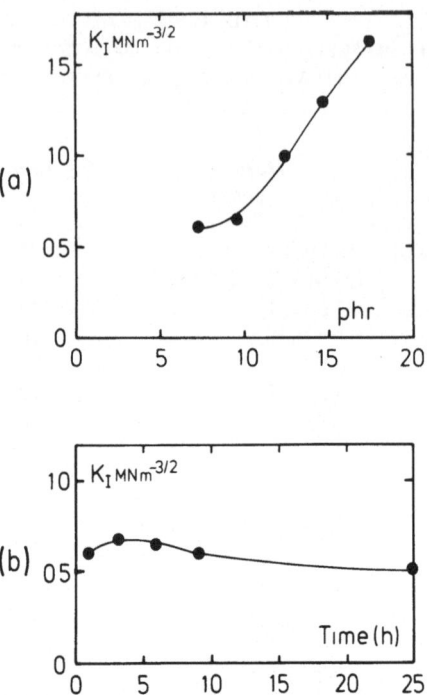

FIG. 9.3. (a) Variation of the critical stress intensity factor for initiation, K_{Ic_i}, with parts per hundred (phr) of TETA curing agent for a DGEBA epoxy resin (data taken from Ref. 26). (b) Variation of K_{Ic_i} with post-cure period for a TETA cured epoxy resin post-cured at $100\,^{\circ}C$ (data taken from Ref. 17).

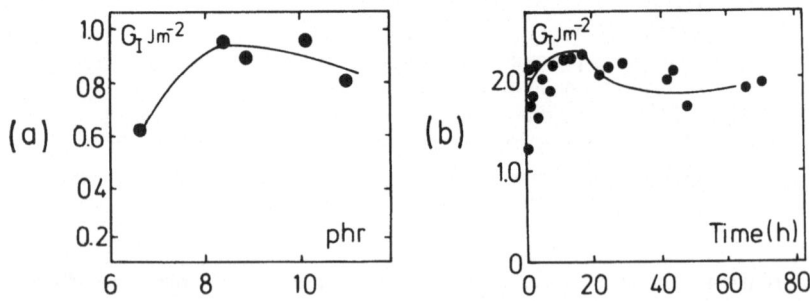

FIG. 9.4. (a) Variation of critical strain energy release rate for crack initiation, G_{Ic_i}, with phr of DETA curing agent for a DGEBA epoxy resin (after Ref. 21). (b) Variation of G_{Ic_i} with post-cure time for a DETA cured epoxy resin post-cured at $106\,^{\circ}C$ (after Ref. 21).

systems and found that G_{Ic} peaked at particular curing agent contents or post-cure periods. The results of Mijovic and Koutsky[21] are shown in Fig. 9.4. They have employed considerably longer cure schedules than those for the specimens used in Fig. 9.3 and it must be remembered that the relationship[18] between K_{Ic} and G_{Ic} ($K^2 \simeq EG$) involves the Young's modulus, E, which can also vary with cure.[26] These factors may help to reconcile the apparent discrepancies between the data in Figs. 9.3 and 9.4.

9.2.3. Testing Variables
Although it is clear from the previous section that crack propagation behaviour in different systems varies with the composition of the resin, the most significant advances in understanding the mechanisms of crack propagation have been made by following the effect of changing testing

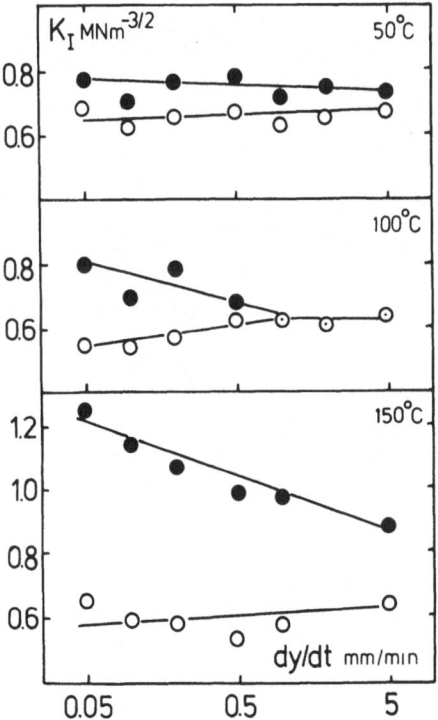

FIG. 9.5.   Variation of K_{Ici} (●) and K_{Icd} (○) with cross-head speed for a TETA cured epoxy resin post-cured for 3 h at different stated temperatures, all tested at room temperature (after Ref. 17).

variables such as rate[2,15] and temperature[15-17] and keeping the composition of the resin and curing conditions constant. The variation of the critical stress intensity factors for initiation and arrest, K_{Ici} and K_{Ica}, for the same composition of resin, cured at different temperatures for the same length of time and tested at different cross-head speeds (i.e. rates), is given in Fig. 9.5. It can be seen that there is a tendency for the size of the jumps to decrease as the cross-head speed is reduced, and for propagation to be continuous at high cross-head speeds and stick/slip at low rates of testing. This effect has been shown to be similar for a wide variety of thermosetting resins when they are tested at different rates.[1,2,15-17] This behaviour is clearly unusual and any theoretical explanations of crack propagation in thermosetting polymers must be able to account for this effect.

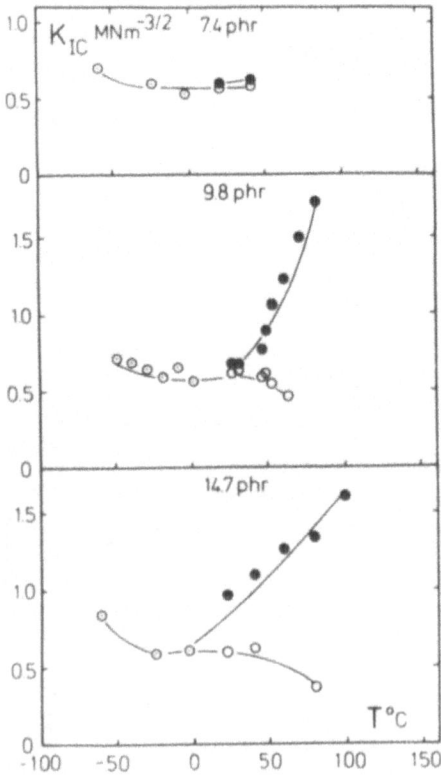

FIG. 9.6. Variation of K_{Ic} with testing temperature for epoxy resins containing different amounts of TETA. ● K_{Ici}, ○ K_{Ica}, ☉ K_{Ic} continuous. Data obtained using a DT specimen at a cross-head speed of 0·5 mm min⁻¹.

It is possible to shed more light upon the mechanisms of crack propagation by looking at the effect of changing the temperature of testing. Figure 9.6 shows the variation of K_{Ici} and K_{Ica} for three different formulations of an epoxy resin cured under identical conditions but tested at different temperatures. It can be seen that in all cases propagation is continuous at low temperatures but stick/slip at higher temperatures. This behaviour has been shown to be applicable to other formulations of epoxy resins[9,16,17,27] and is consistent with the effect of testing rate. Since epoxy resins are viscoelastic solids it would be expected that reducing the rate of testing would be similar to increasing the testing temperature. Both of these factors appear to promote stick/slip behaviour.

Some recent observations by Scott et al.[27] have shown that the picture may be somewhat more complex than was previously thought. They extended measurements of K_{Ic}, as a function of testing temperature, down to liquid nitrogen temperature ($-196\,^\circ$C). They found that below about $0\,^\circ$C propagation became continuous, which is consistent with previous observations.[9,15-17] It remained continuous down to about $-100\,^\circ$C but became discontinuous (stick/slip) again below this temperature. Some of their results are given in Fig. 9.7. This low-temperature stick/slip propagation has been observed below $-100\,^\circ$C in a series of amine-cured epoxy resins.[27] It appears to be different from the stick/slip propagation at high temperatures as the size of the jump increases as the temperature is reduced. The origin of this low-temperature stick/slip behaviour is

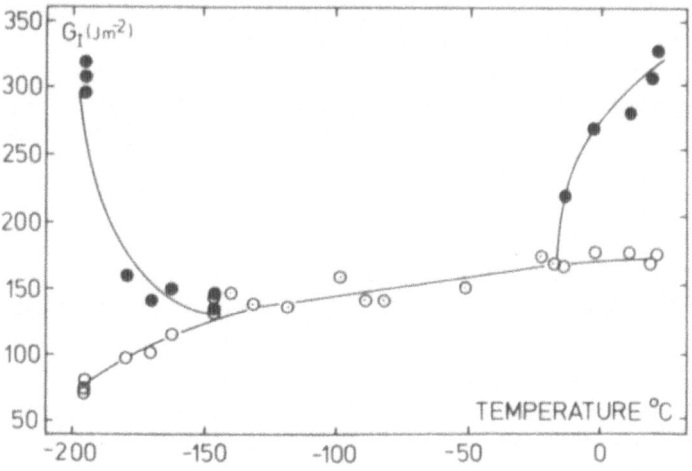

FIG. 9.7.   Variation of G_{Ic} with testing temperature for an EDA-cured epoxy resin tested at different temperatures (after Ref. 27).

not known. The cooling agent used by Scott et al.[27] was nitrogen, N_2, and this could be responsible for the observed effect. It is well known that gases such as N_2 can affect the crazing and fracture of thermoplastics at low temperatures[28] and it is possible that they have a similar effect upon the epoxy resins. A critical test to prove this is to look at the effect of cooling with helium gas upon the crack propagation. This gas does not affect thermoplastics and so testing epoxy resins in a helium environment would be of great interest.

It has been suggested[29] that the high-temperature stick/slip propagation in epoxy resins may also be due to an environmental effect such as the absorption of water vapour from the atmosphere. It appears that this is not the case[30] since stick/slip behaviour is found when cracks are propagated under vacuum.[15] It will be shown later that recent investigations have indicated that this type of stick/slip propagation is due to crack blunting which is controlled by the plastic deformation characteristics of the resin.

9.3. MECHANISMS OF CRACK PROPAGATION

It has been recognised for several years that crack propagation in thermoplastics, such as PMMA, takes place through the breakdown of crazes.[31] There have also been suggestions that crazing occurs in thermosetting resins such as epoxies.[32,33] The evidence for this has been discussed in a previous publication[1] and it seems unlikely that in fully cured resins the amount of tensile drawing required to form a craze would be possible. Recent experimental observations[21] have tended to suggest that crazing does not normally occur. It has also been suggested[21] that glassy polymers such as epoxy resins have a nodular structure and that this affects crack propagation.

9.3.1. Fractography

Examination of the fracture surfaces of epoxy resin samples using optical and electron microscopy has enabled both the possibility of crazing and the effect of the nodular structure to be investigated. Previous publications[9,17] have shown that there are several characteristic features that can be seen on the fracture surfaces of epoxy resins. When crack propagation is continuous the surfaces tend to be smooth and relatively featureless. On the other hand, when stick/slip propagation occurs there are features on the surface in the vicinity of the crack arrest lines. These features

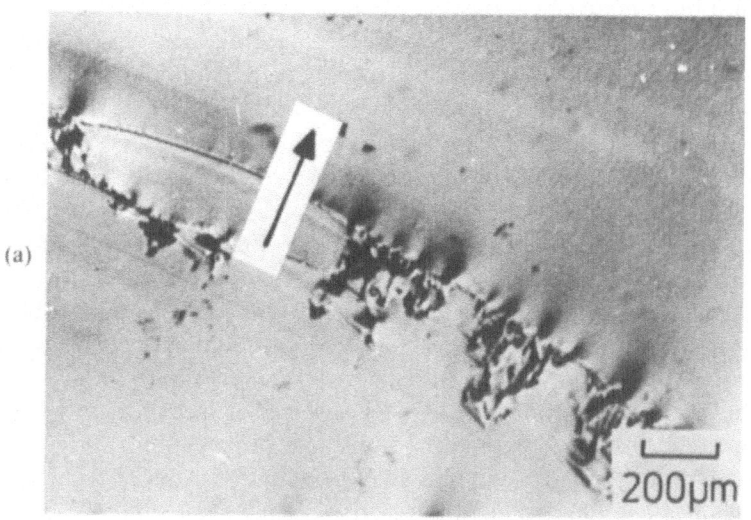

(a)

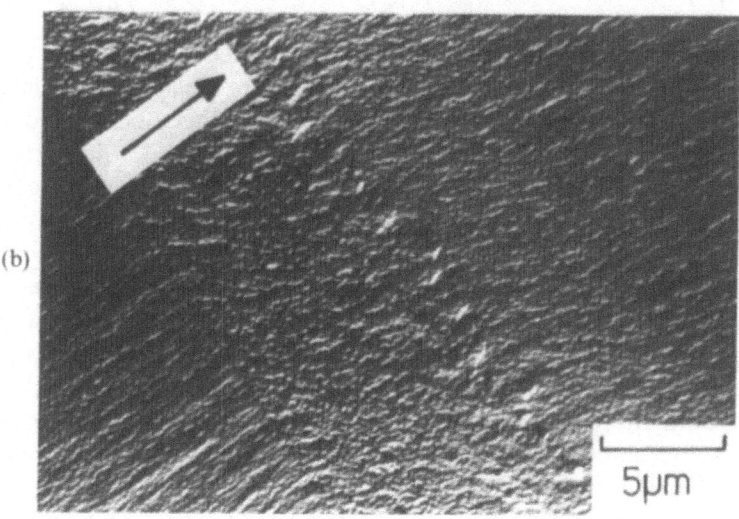

(b)

FIG. 9.8.   Fracture surfaces of an epoxy resin cured with 9·8 phr of TETA. The specimen was post-cured at 50 °C for 3 h and tested at 22 °C. (a) Optical micrograph of surface. (b) EM replica of crack arrest line. (Crack growth direction indicated by arrows.)

fall into three main categories: (1) triangular features, (2) fine arrest lines and (3) broader rough areas.[17] The occurrence of each type of feature depends upon the state of cure of the resin and on the temperature of testing.

The triangular markings are found typically in under-cured resins,[17] i.e. due to low curing temperature and/or small quantities of curing agent. They appear along rather indistinct arrest lines as shown in Fig. 9.8(a). The electron micrograph in Fig. 9.8(b) was obtained from the same type of area and it can be seen that the surface features are streaked in the direction of crack propagation. There is also a broad band along the arrest line and an underlying nodular structure on the scale of ~ 1000 Å.

Fine arrest lines are obtained on the surfaces of specimens just under-going the transition from stable to unstable propagation.[26] Figure 9.9 shows an optical micrograph and an electron micrograph of the fracture surface of a specimen at the transition. The fine arrest line can be seen in Fig. 9.9(a), and Fig. 9.9(b) shows the same line at a higher magnification. It can be seen that the structure is again streaked and the arrest line corresponds to an abrupt change in the direction of the streaks. An underlying nodular structure on the scale of ~ 500 Å can also be resolved.

A typical example of the broader type of crack arrest line is given in Fig. 9.10(a). This type of feature is typical of well-cured specimens fractured at high temperature.[17] The fracture surface up to the crack arrest line is relatively featureless. After crack arrest there is a slow-growth region[9] of closely spaced striations parallel to the crack growth direction. There is, then, a rough hackled area where the crack accelerates during the 'slip' process. Examination of the fracture surface through electron microscope replicas has shown that the smooth areas of such specimens are relatively featureless, in contrast to the slow growth region which is shown in Fig. 9.10(b). In this area there are V-shaped features which appear to be caused by the crack propagating on different levels. It is not normally possible to resolve any underlying nodular structure in well-cured specimens such as that used in Fig. 9.10.

It is clear that there is no evidence of any craze debris on the surfaces of the epoxy resins shown in Figs. 9.8–9.10. Mijovic and Koutsky[21] came to a similar conclusion for DETA-cured resins. It seems likely, therefore, that the crazing in epoxy resins suggested by Morgan and O'Neal[33] from the examination of thin films, may not occur in bulk samples.

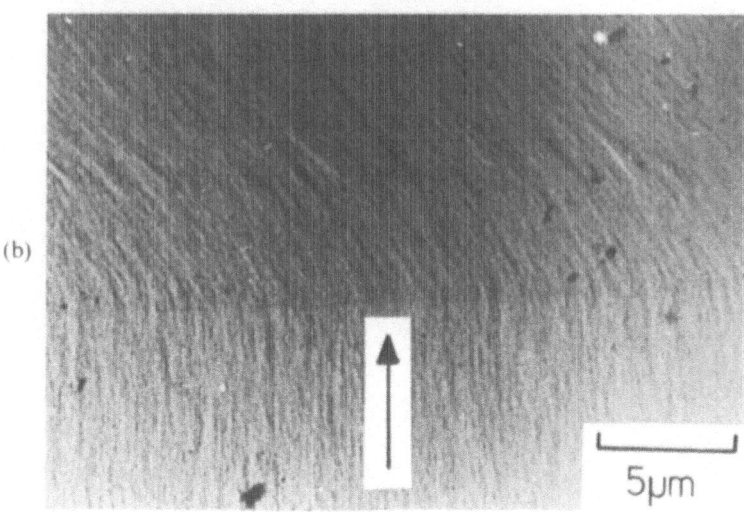

FIG. 9.9.   Fracture surface of an epoxy resin cured with 9·8 phr of TETA. The specimen was post-cured at 100 °C for 3 h and tested at 22 °C. (a) Optical micrograph of fracture surface. (b) EM replica of crack arrest line. (Crack growth direction indicated by arrows.)

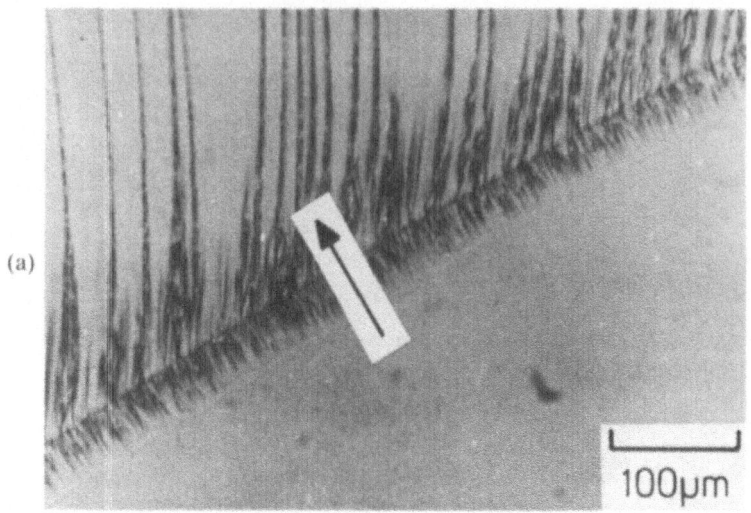

(a)

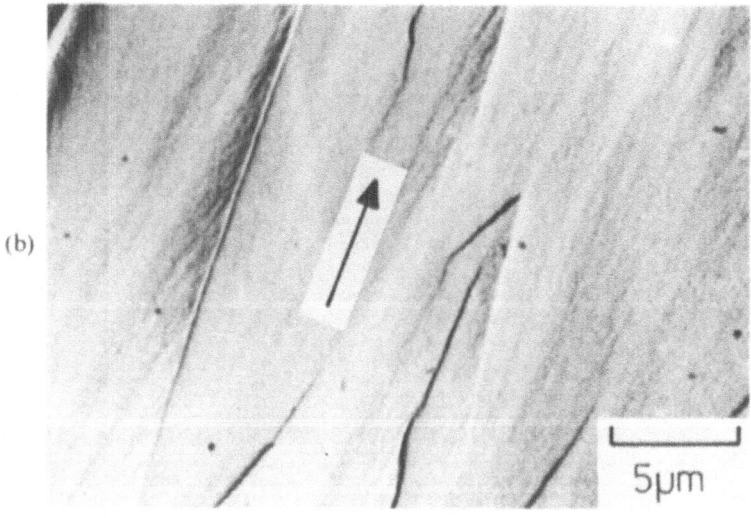

(b)

FIG. 9.10. Fracture surface of an epoxy resin cured with 9·8 phr of TETA. The specimen was post-cured at 150 °C for 3 h and tested at 22 °C. (a) Optical micrograph of fracture surface. (b) EM replica of slow-growth region following the crack arrest line. (Crack growth direction indicated by arrows.)

Koutsky and co-workers[21,34] have recently suggested that the mechanical properties of epoxy resins may be controlled by the nodular structure. This is an attractive proposition as there is a well established dependence of the properties of metals upon the microstructure. However, even the existence of nodules in glassy polymers is a matter of some controversy, with Uhlmann[35] suggesting that they are artefacts, on one hand, and Yeh[36] affirming that they represent the true structure, on the other. Supporters of the existence of nodular structure have problems explaining the lack of small-angle X-ray scattering that should occur if the nodules are present in glassy polymers, but even Uhlmann suggests[35] that the X-ray scattering evidence is ambiguous for epoxy resins. It would seem reasonable that thermosetting polymers, which are normally made by the addition of a curing agent to a pre-polymer, could have a non-uniform, two-phase structure. The micrographs in Figs. 9.8–9.10 and the work of Mijovic and Koutsky[21] suggest that the nodule size in a given resin curing agent system varies with the state of cure. Since the properties of the resin also depend upon the cure it is tempting to relate the mechanical properties to the nodular microstructure. Mijovic and Koutsky have recently done this and, although they find variations of properties with microstructure, they have not been able to explain why a particular nodule size gives rise to particular properties.

9.3.2. Plastic Deformation at the Crack Tip

It is clear that plastic deformation most occur at the crack tip during crack propagation in thermosetting resins. However, it is extremely difficult to show directly that plastic deformation has taken place. It has been possible to see blunt cracks with tip radii of the order of several microns[37] but observation of plastic zones, which must be present, has proved extremely difficult. Phillips and co-workers[9] have shown that when stick/slip propagation occurs in epoxy resins there is a characteristic slow-growth region after the crack arrest line. They observed the growth of a crack in an epoxy sample and found that after the 'slip' process the crack became stationary at the arrest line. They found that prior to the next 'slip' step it grew slowly through a small region similar to that shown in Fig. 9.10 before bursting through and jumping ahead. It is found that the length of the slow-growth region (l_r) increases as the temperature of testing is raised. This effect can be clearly seen in Fig. 9.11 for a TETA-cured resin, where micrographs are given for specimens which have been fractured at two different temperatures. It was shown in Fig. 9.6 that for this formulation the value of K_{Ic} also increases

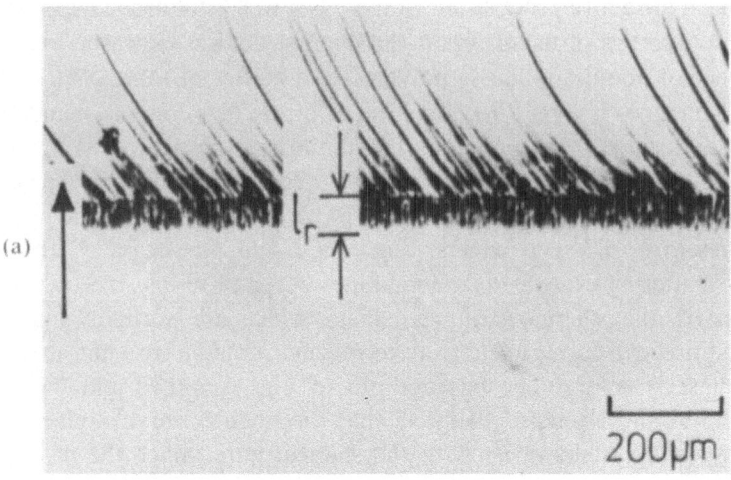

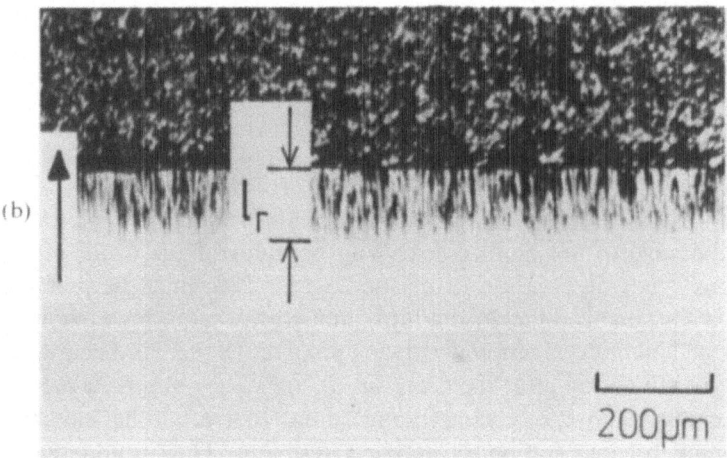

FIG. 9.11. Optical micrographs of the crack arrest region on fracture surfaces of an epoxy resin hardened with 14·7 phr of TETA and post-cured for 3 h at 100 °C. The length of the slow-growth region is given by l_r. (a) Specimen fractured at 22 °C. (b) Specimen fractured at 60 °C.

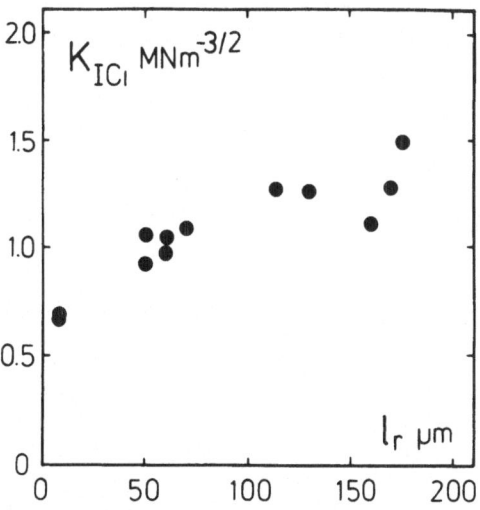

FIG. 9.12.   Plot of K_{Ic_1} against l_r for the specimens used in Fig. 9.11 and other formulations of resin tested at different rates and temperatures.

with increasing temperature. Figure 9.12 shows that there is a unique correlation between the length of the slow-growth region l_r and K_{Ic_1} for different epoxy formulations tested under a variety of experimental conditions. In fact, it is found that l_r is approximately the same as the radius of a Dugdale plastic zone (r_p) calculated using the equation:[18]

$$r_p = \frac{\pi}{8}\left(\frac{K_{Ic}}{\sigma_y}\right)^2 \qquad (1)$$

where σ_y is the yield stress of the resin. Figure 9.13 is a plot of $(K_{Ic_1}/\sigma_y)^2$ against l_r using some of the data from Fig. 9.12 and values of σ_y from earlier publications.[16,26] The straight line has a slope of $8/\pi$ (~ 2.55) and represents the relationship between r_p and $(K_{Ic_1}/\sigma_y)^2$ if a Dugdale plastic zone is present at the crack tip. The proximity of the experimentally determined points to the theoretical line strongly suggests that l_r is closely related to r_p.

A clear picture of stick/slip propagation is now emerging. It appears that during loading after crack arrest a plastic zone forms at the tip of the crack. Propagation then takes place by slow growth through the plastic zone followed by rapid propagation through virgin material. The slow-growth region therefore defines the plastic zone at the crack tip. Although this is only indirect evidence of the presence of the plastic

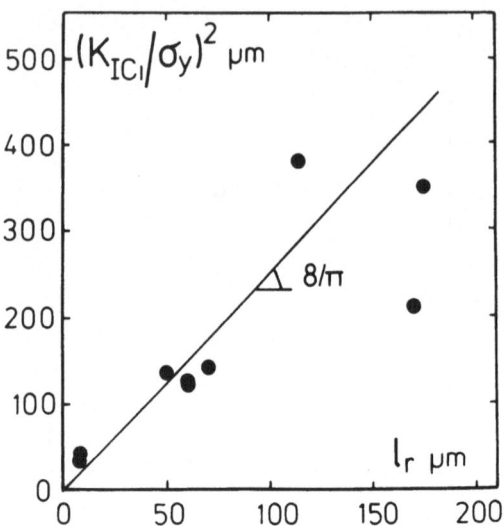

FIG. 9.13.   Plot of $(K_{Ic}/\sigma_y)^2$ against l_r, using some of the data from Fig. 9.12. The straight line is drawn with a slope of $8/\pi$ according to eqn. (1), assuming that $l_r \simeq r_p$.

zone it gives an important new insight into processes taking place at the crack tip.

9.4.   CRACK GROWTH CRITERIA

Criteria controlling the propagation of cracks in metals and thermoplastic polymers[18] have been developed over the past few years and recent experimental and theoretical investigations have allowed criteria to be developed to account for crack propagation in thermosets, particularly in epoxy resins. Slightly different theories must be applied to the two different types of crack growth encountered in thermosets (continuous or stick/slip propagation) although it will be shown that the basic criterion controlling fracture is the same in both cases.

9.4.1.   Continuous Propagation

The continuous propagation of cracks in epoxy resins that occurs at low temperature is somewhat similar to the type of propagation that takes place in brittle thermoplastics such as polymethyl-methacrylate, PMMA.[18,38] It has been clearly demonstrated that in PMMA propagation takes place through a constant crack-opening displacement (δ) criterion.[18.38] There is now strong evidence that a similar criterion can be applied to continuous propagation in thermosetting polymers such as epoxy resins.[16] The value of δ is given by:[16,18,38]

$$\delta = \frac{K_{Ic}^2}{\sigma_y E} \simeq e_y \left(\frac{K_{Ic}}{\sigma_y}\right)^2 \qquad (2)$$

where σ_y is the yield stress, e_y the yield strain and E the Young's modulus of the material. The criterion for continuous propagation is that crack growth takes place when δ reaches a critical value, δ_c. Calculations have shown that δ_c for continuous propagation in epoxy resins is remarkably constant,[16] but as soon as stick/slip propagation ensues, the value of δ_c rises rapidly and the constant δ criterion no longer applies. The effect of this criterion upon K_{Ic} for continuous propagation can be seen in Fig. 9.6. In the continuous regions K_{Ic} decreases slightly as the temperature is increased. This is because the yield stress and modulus σ_y and E are also falling but δ_c is remaining constant (eqn. (2)).

In thermoplastics such as PMMA the constant δ criterion is associated with the growth of a crack through a single craze at the tip of the crack, which can be accurately modelled by a Dugdale plastic zone.[18,38] The evidence for the presence of crazes at the tips of moving cracks in fully cured thermosets is not strong[1,21] and it has been shown in Section 9.3.1 that examination of the fracture surfaces implies that crazing has not occurred.

9.4.2. Stick/Slip Propagation
There have been recent important developments in the understanding of stick/slip propagation in thermosetting polymers and it is now thought that this type of behaviour can be explained in terms of blunting at the crack tip and a quantitative theory has been developed very recently[37,39] to explain stick/slip propagation. It has been known for several years[16] that if the resin has a high yield stress then crack propagation is continuous, whereas if the material has a low yield stress, due to changes in composition or testing variables, propagation tends to be stick/slip in nature. Since epoxy resins are brittle, measurements of yield stress can normally only be made in compression. This behaviour can be quantified as shown in Fig. 9.14 where K_{Ic} is plotted against σ_y for a whole range of TETA-cured epoxy resins tested at different rates and temperatures. The values of K_{Ic} and σ_y have been taken from this and previous publications.[16,26] It can be seen that all the data fall

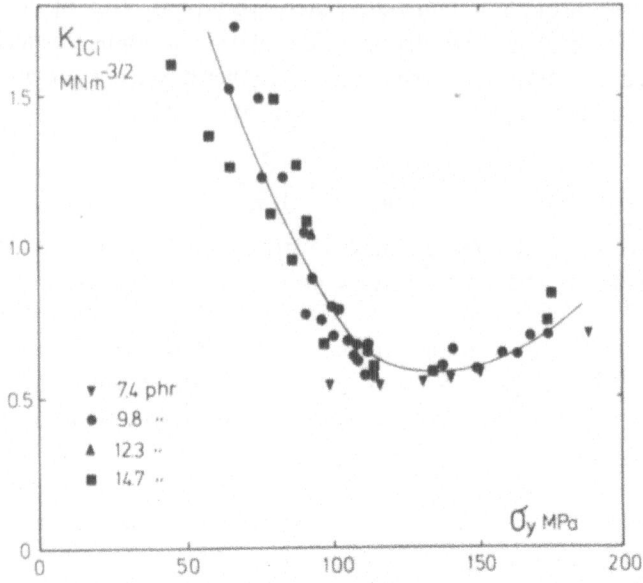

FIG. 9.14.   Plot of K_{Ic_i} against σ_y for different formulations of resin tested at a variety of rates and temperatures. The values of σ_y have been taken from previous publications.[16,25]

approximately upon a master curve indicating a unique relationship between K_{Ic_i} and σ_y. Kinloch and Williams[37] have obtained a similar correlation between K_{Ic_i} and σ_y for a series of resins hardened with different curing agents. This correlation has allowed stick/slip propagation to be explained quantitatively.[39]

It can be shown[40] that for a crack under an applied stress of σ_0, the stress (σ_{yy}) normal to the axis of the crack at a small distance (r) ahead of the crack is given by:

$$\sigma_{yy} = \frac{\sigma_0\sqrt{a}}{\sqrt{2r}}\frac{(1+\rho/r)}{(1+\rho/2r)^{3/2}} \tag{3}$$

where ρ is the radius of the crack tip and a is the crack length. If it is postulated that the failure criterion is that fracture occurs when a critical stress (σ_c) is reached at a distance $r = c$, then eqn. (3) can be rewritten as:

$$\frac{\sigma\sqrt{\pi a}}{\sigma_c\sqrt{2\pi c}} = \frac{(1+\rho/2c)^{3/2}}{(1+\rho/c)} \tag{4}$$

The term $\sigma_c\sqrt{2\pi c}$ can be considered to be the critical stress intensity factor for a 'sharp' crack, K_{Ic},[20] and $\sigma\sqrt{\pi a}$ as the stress intensity factor for a blunt crack, K_{IB}.[20] Hence it follows that:

$$\frac{K_{IB}}{K_{Ic}} = \frac{(1 + \rho/2c)^{3/2}}{(1 + \rho/c)} \tag{5}$$

This equation relates K_{IB} to the radius of a blunt crack and the theory can be checked by measuring the variation of K_{IB} with ρ. However, direct measurement of ρ is difficult as it tends to be small ($\sim 10\,\mu m$) for natural cracks in thermosetting polymers. Kinloch and Williams[37] have overcome this problem by measuring K_{IB} as a function of ρ for a series of epoxy resin samples containing pre-drilled holes of large, known diameter. They showed that a relationship of the form of eqn. (5) held for these materials.

However, there is still the problem of determining ρ for natural cracks in specimens undergoing stick/slip propagation. Kinloch and Williams[37]

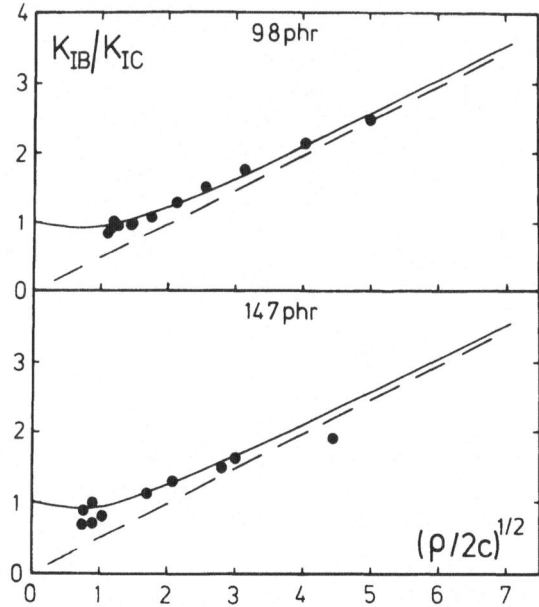

FIG. 9.15. (K_{IB}/K_{Ic}) as a function of $(\rho/2c)^{1/2}$. The value of ρ has been taken as δ as given by eqn. (2). The experimental points have been fitted to the theoretical (solid) curve by assuming the values of c given in Table 9.2. The curve is asymptotic to the dashed line of slope $= \frac{1}{2}$.

have shown that extrapolation of the relationship between K_{IB} and ρ for artificially drilled holes to natural cracks indicates that ρ is approximately the same as the crack opening displacement, δ. Moreover, it has been shown in Section 9.3.2 that there is strong evidence for a Dugdale plastic zone at the crack tip during stick/slip propagation, and so ρ can be estimated from δ in eqn. (2) which can then be used to predict K_{IB}/K_{Ic} using eqn. (5). The theory has been tested by measuring the ratio K_{IB}/K_{Ic} as a function of $(\rho/2c)^{1\ 2}$ as shown in Fig. 9.15. The theoretical line is fitted to the experimental points by choosing suitable values of the critical distance, c, which is the only fitting parameter. K_{Ic} has been taken arbitrarily as the value of K_{Ic} at the lowest temperature of measurement and K_{IB} as the value of K_{Ici} at higher temperatures. The agreement has been found to be equally good for epoxy resins cured with different curing agents[37] and even for rubber-modified epoxy resins.[37] The values of critical distance (c) for these materials and the data in Fig. 9.15 are given in Table 9.2. The values of critical stress (σ_c) can be determined from the relationship $K_{Ic} = \sigma_c\sqrt{2\pi c}$ and they are also given in

TABLE 9.2

DERIVED VALUES OF CRITICAL STRESS σ_c AND CRITICAL DISTANCE c FOR DGEBA EPOXY RESINS OF DIFFERENT FORMULATIONS

Formulation	σ_c (MPa)	c (μm)	σ_y (25°C) (MPa)	σ_c/σ_y	Ref.
9·8 phr TETA	360	0·60	112	3·2	45
12·3 phr TETA	300	1·10	91	3·3	45
14·7 phr TETA	270	1·60	86	3·2	45
9·4 phr 3° Amine	360	0·38	83	4·3	37
10 phr TEPA	495	0·12	117	4·2	37
Rubber-modified	220	8·30	70	3·1	37

Table 9.1. The yield stress σ_y of each formulation at 25°C is also given and it can be seen that the ratio σ_c/σ_y for each system is approximately the same (~ 3–4). The temperature of 25°C is only an arbitrary reference, but the constant ratio implies that the failure criterion is that a stress of the order of three or four times the yield stress must be reached in the plastic zone, regardless of the resin composition or yield stress. The critical distance, c, varies from 0·1 to 10 μm and its significance is not yet clear.

It is important to mention at this stage that the constant δ criterion

(Section 9.4.1) for continuous propagation can also be considered as requring a critical stress at a critical distance. In this case σ_c will be virtually constant as K_{Ic} falls only slightly as the temperature is increased. Since the yield stress also drops it is also likely that σ_c/σ_y will, again, be invariant.

9.5. TIME-DEPENDENT FAILURE

Many materials fail under prolonged loading at stresses which they can sustain, without failure, for short periods of time. This phenomenon is known as static fatigue and is exhibited by ceramics, glasses[41,42] and thermoplastics such as PMMA.[19] It is thought that static fatigue in ceramics and glasses is due to environmental attack[41,42] whereas in PMMA it is known to be caused by the slow propagation of cracks which takes place because of the time dependence of the Young's modulus of the material.[38] It is known that, under certain conditions, thermosetting polymers are prone to failure at constant load by static fatigue, but the results of different groups of workers appear to be in conflict. Some people have observed static fatigue in epoxy resins[43] and others have not.[9,44] It seems that this conflict may be resolved by examination of the careful work of Gledhill and Kinloch[43,44] who have recently shown that, at least at room temperature, certain formulations of epoxy resins appear to show static fatigue while others do not. They looked at the failure of TDCB aluminium adhesive joints bonded with a thin layer of epoxy (0·5 mm). This particular test piece is useful because similar ones made from the pure resin are prone to creep over long periods of loading which complicates the stress distribution.

Gledhill and Kinloch[43] found that, if the epoxy was cured with a tertiary amine, the critical strain energy release rate at failure, G_{Ic}, dropped dramatically as the loading time was increased, as shown in Fig. 9.16. The G_{Ic} measured after 10^8 s (~ 3 years) was only 25 % of that needed to propagate a crack during short-term loading (over a few seconds). Using exactly the same testing geometry and testing procedure they found that when the same epoxy resin cured with tetraethylenepentamine (TEPA), was used in the adhesive joint, static fatigue was not observed.[44] Moreover, they found that when specimens had been held under a load corresponding to 86 % of the failure load required to cause crack growth during rapid loading, there was an increase in failure load measured on subsequent rapid loading of the specimens. This effect is shown in

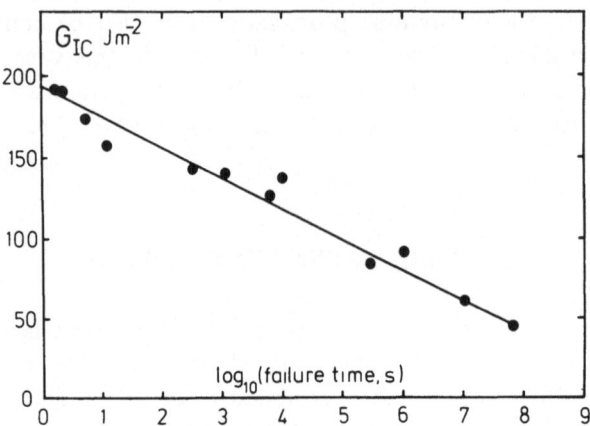

FIG. 9.16.   Relationship between G_{Ic} and failure time for TDCB adhesive joints consisting of an epoxy resin cured with a tertiary amine (after Ref. 43).

Fig. 9.17 where G_{Ici} (measured on retesting the specimen following static loading) is plotted against the static loading period. This means that the epoxy resin is actually becoming tougher on static loading and this effect has been termed a 'self-toughening mechanism'.[44]

The question that must be answered is, why does the same resin hardened with different curing agents show such a difference in behaviour? The answer almost certainly lies in the behaviour of the two systems during testing at a constant loading rate. The tertiary amine-cured resin shows stick/slip propagation at room temperature,[43] whereas, there is continuous

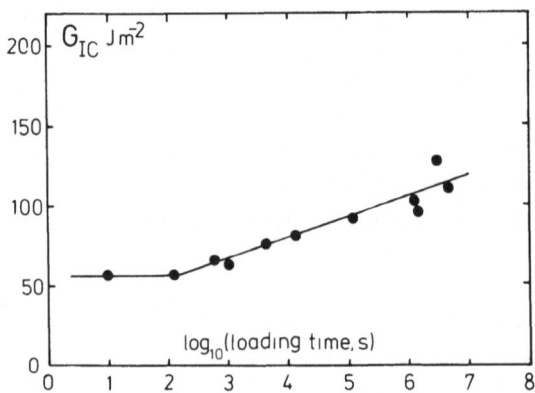

FIG. 9.17.   Increase in the value of G_{Ic} with testing time for TDCB adhesive joints consisting of an epoxy resin cured with TEPA (after Ref. 44).

propagation in the TEPA cured resin.[43] This means that cracks are relatively blunt even during short-term loading in the tertiary amine-cured material, and this is reflected in the significantly higher values of G_{Ic} that this material can sustain over short periods of time. On the other hand since the TEPA-cured resin undergoes continuous propagation at room temperature the cracks are relatively sharp leading to low values of G_{Ic} during short-term loading. However, as the loading period increases blunting starts to take place and so the value of G_{Ic} increases. Indeed, it is thought that the blunting only occurs for the initial crack since after the initial 'jump', propagation takes place in a continuous way as for an unprestressed specimen. Increasing the period of loading is equivalent to either reducing the testing rate or increasing the temperature, both of which have the effect of causing stick/slip propagation and hence, promote blunting. The drop in G_{Ic} observed for the tertiary amine cured resin has been explained in terms of propagation taking place when a critical plastic zone size or crack opening displacement is achieved. The yield stress and modulus of the resin will drop as the loading time increases, and so the critical conditions can be reached at a lower value of G_{Ic}.

It is worth considering at this stage if the two types of behaviour can be reconciled. Examination of Figs. 9.16 and 9.17 shows that the lowest value of G_{Ic} found for both systems is about the same ($\sim 50\,\mathrm{J\,m^{-2}}$), and it may be that the two plots are representing similar types of behaviour. Clearly measurements over longer periods of time are required to see if, eventually, a limiting value of G_{Ic} will be reached for the tertiary amine-cured resin and if static fatigue will occur in the TEPA-cured material after a sufficiently long loading period.

9.6. CONCLUSIONS

It is clear that there has been a considerable advance in the understanding of the factors controlling crack propagation in thermosetting resins. The most important recent advance has been in developing a theory which quantitatively explains stick/slip propagation in terms of a crack blunting process. It also follows from this theory that for both stick/slip and continuous propagation, the criterion for crack propagation is that a critical stress (typically three or four times the yield stress) must be reached at a critical distance ahead of the crack.

It has been shown that there is no strong evidence for crazing in fully cured thermosetting resins and the crack propagation appears to take place without the involvement of crazes. Attempts have been made to relate the nodular microstructure of thermosetting polymers to their mechanical properties. It has been shown that although both the microstructure and mechanical properties vary with resin formulation and cure schedule there is no strong evidence for any direct correlation between the two.

It is evident that more work must be done before there can be a full understanding of crack propagation in thermosetting polymers. By far the greatest proportion of the work discussed in this chapter has been concerned with epoxy resins and it is important that similar investigations should be carried out on other thermosets to see how far the conclusions that have been drawn concerning epoxy resins, can be extended to other thermosetting polymers. Also, more work on epoxy resins is required, particularly concerning the effect of microstructure upon mechanical properties. The long-term aims of such investigations should be to enable the properties of thermosetting polymers to be tailored by controlling their chemical and physical microstructure.

REFERENCES

1. YOUNG, R. J. Chapter 6 in *Developments in Polymer Fracture—1*, ed. Andrews, E. H., 1979, Applied Science Publishers Ltd, London.
2. YOUNG, R. J. and BEAUMONT, P. W. R. *J. Mater. Sci.*, 1976, **11**, 766.
3. DIGGWA, A. D. S. *Polymer*, 1974, **15**, 101.
4. MEEKS, A. C. *Polymer*, 1974, **15**, 675.
5. SELBY, K. and MILLER, L. E. *J. Mater. Sci.*, 1975, **10**, 12.
6. GRIFFITHS, R. and HOLLOWAY, D. G. *J. Mater. Sci.*, 1970, **5**, 302.
7. EVANS, W. T. and BARR, B. I. G. *J. Strain Anal.*, 1974, **9**, 166.
8. PHILLIPS, D. C. and SCOTT, J. M. *J. Mater. Sci.*, 1974, **9**, 1202.
9. PHILLIPS, D. C., SCOTT, J. M. and JONES, M. *J. Mater. Sci.*, 1978, **13**, 311.
10. YOUNG, R. J. and BEAUMONT, P. W. R. *J. Mater. Sci.*, 1977, **12**, 684.
11. PRITCHARD, G. and RHOADES, G. V. *Mater. Sci. Eng.*, 1976, **26**, 1.
12. OWEN, M. J. and ROSE, R. G. *J. Phys. D. Appl. Phys.*, 1973, **6**, 42.
13. CHRISTIANSEN, A. and SHORTALL, J. B. *J. Mater. Sci.*, 1976, **11**, 1113.
14. PRITCHARD, G., ROSE, R. G. and TANEJA, N. *J. Mater. Sci.*, **11**, 1976, 718.
15. YAMINI, S. and YOUNG, R. J. *Polymer*, 1977, **18**, 1075.
16. GLEDHILL, R. A., KINLOCH, A. J., YAMINI, S. and YOUNG, R. J. *Polymer*, 1978, **19**, 574.
17. YAMINI, S. and YOUNG, R. J. *J. Mater. Sci.*, 1979, **14**, 1609.
18. WILLIAMS, J. G. *Adv. Polym. Sci.*, 1978, **27**, 69.
19. BEAUMONT, P. W. R. and YOUNG, R. J. *J. Mater. Sci.*, 1975, **10**, 1334.

20. HAKEEM, M. and PHILLIPS, M. G., Private communication.
21. MIJOVIC, J. and KOUTSKY, J. A. *Polymer*, 1979, **20**, 1095.
22. BROUTMAN, L. J. and McGARRY, F. J. *J. Appl. Polym. Sci.*, 1965, **9**, 609.
23. BASCOM, W. D., COTTINGTON, R. L., JONES, R. L. and PEYSER, P. J. *J. Appl. Polym. Sci.*, 1975, **19**, 2545.
24. MOSTOVOY, S. and RIPLING, E. J. *J. Appl. Polym. Sci.*, 1966, **10**, 1351.
25. MOSTOVOY, S. and RIPLING, E. J. *J. Appl. Polym. Sci.*, 1971, **15**, 641.
26. YAMINI, S., Ph.D. thesis, *Crack propagation in epoxy resins*, 1979, University of London.
27. SCOTT, J. M., WELLS, G. and PHILLIPS, D. C. *J. Mater. Sci.* (to be published).
28. IMAI, Y. and BROWN, N. *J. Mater. Sci.*, 1976, **11**, 417.
29. HAKEEM, M. I. and PHILLIPS, M. G. *J. Mater. Sci.*, 1978, **13**, 2284.
30. YAMINI, S. and YOUNG, R. J. *J. Mater. Sci.*, 1978, **13**, 2287.
31. KAMBOUR, R. P. *J. Polym. Sci., Macromol. Rev.*, 1973, **7**, 1.
32. LILLEY, J. and HOLLOWAY, D. G. *Phil. Mag.*, 1973, **28**, 215.
33. MORGAN, R. J. and O'NEAL, J. E. *J. Mater. Sci.*, 1977, **12**, 1966.
34. RACICH, J. L. and KOUTSKY, J. A. *J. Appl. Polym. Sci.*, 1976, **20**, 2111.
35. UHLMANN, D. R., *Disc. Farad. Soc.*, 1979, **68**, (to be published).
36. YEH, G. S. *Crit. Rev. Macromol. Chem.*, 1972, **1**, 173.
37. KINLOCH, A. J. and WILLIAMS, J. G. *J. Mater. Sci.* (to be published).
38. YOUNG, R. J. and BEAUMONT, P. W. R., *Polymer*, 1976, **17**, 717.
39. YOUNG, R. J. and YAMINI, A. *Paper presented at IUPAC Symposium on Macromolecules, Mainz*, 1979.
40. WILLIAMS, J. G. *Stress Analysis of Polymers*, 1973, Longmans, London.
41. WIEDERHORN, S. M. *J. Amer. Ceram. Soc.*, 1967, **50**, 407.
42. EVANS, A. G. *J. Mater. Sci.*, 1972, **7**, 1137.
43. GLEDHILL, R. A. and KINLOCH, A. J. *Polym. Eng. Sci.*, 1979, **19**, 82.
44. GLEDHILL, R. A., KINLOCH, A. J. and SHAW, S. J. *J. Mater. Sci.*, 1979, **14**, 1769.
45. YAMINI, S. and YOUNG, R. J. *J. Mater. Sci.* (to be published).

INDEX

Activation energy, 134, 156
Alkylene oxide, 60–3
Alumina trihydrate, 75–6
Aluminium trihydrate, 22
Amino resins, 11–13
Ammonium
  dimolybdate, 76
  persulphate, 80
Antimony trioxide, 71
Arrhenius equation, 134, 147
Asbestos-filled mouldings, 108
Astrel-360, 153
Atlac-580, 36
Azo compounds
  decomposition rate, 132–3
  half-life, 132–3

Benzaphenone tetracarboxylic acid
    (BTDE), 188
Benzoin ethers, 35
Benzophenone, 35
Benzoyl peroxide (BPO), 39, 132,
    136, 141
β-transition in crosslinked resins,
    242–3
Bis(4-t-butyl cyclohexyl)
    peroxydicarbonate, 136
Bis-diene resins, 20
Bis(2-hydroxyethyl ether), 70

Bismaleimide polyimides, 184
Bisphenol A, 14, 70, 118
  epoxy, 32
  fumaric acid condensation
    polyester, 36
Bisphenol-A-diglycidyl ether. *See*
    DGEBA
Bromostyrene, 71
Bulk moulding compounds, 15, 127,
    137, 142
Butadiene–acrylonitrile rubbers, 21
Butadiene-S block copolymers, 250
t-Butyl perbenzoate, 39
t-Butyl peroxybenzoate, 136, 142
t-Butyl peroxyoctoate, 142

^{13}C NMR
  neutron scattering, 24
  spectroscopy, 217
Carbon–carbon initiators, 142–3
Carbon fibre reinforced composites,
    99, 118
Carbonic acid ester anhydrides, 82
Cerium compounds, 142
Chemical resistance of phenol–aralkyl
    resins, 112–14
Chemiluminescence, 223
Chlorine content, 22
Cobalt naphthenate, 128

Cold curing, 9
Condensation reactions, 13
Contact moulding processes, 126
Corrosion resistance of vinyl ester
  resins, 32, 44, 48–51
Costs, 26
Crack
  opening displacement, 278
  propagation
    continuous, 259, 274, 275
    criteria controlling, 274–9
    curing and cure schedule, effect
      of, 260
    epoxy resins, in, 258
    factors affecting, 258–66
    mechanisms of, 266–74
    stick/slip, 259, 265–6, 273, 275–9
    testing variables, 263–6
    thermosetting polymers, in,
      257–83
    unstable, 251
  tip', plastic deformation at, 271–4
Craze–crack growth process, 225
Crazing in epoxy resins, 268
Crosslink density, 214, 219–22,
  237–40, 248
Crosslinked networks, 214, 218
Crosslinked polyester resins, 231–55
Crosslinking, 9, 11, 14, 18–20, 25, 27,
  89, 116, 149–50, 186
Curing
  agents, 212, 260
  vinyl ester resins, of, 39
Cyclisation reactions, 14
1:4 Cyclohexane dimethanol
  (CHDM), 69
Cyclohexyl peroxides, 137

Decachlordiphenyl, 73
Decomposition temperature, 148
DEG/SA/FA resins, 239
Derakane-411, 32
Derakane-470, 42, 44
Derakane-510-A-40, 34
DGEBA, 212, 221, 260, 261
DGEBA–DETA system, 220, 222–3

DGEBA–DMHDA system, 222
Diacyl peroxides, 132
$\alpha\alpha'$-Dialkoxy-p-xylenes, 95
Diaminodiphenylmethane, 177
4,4'-Diaminodiphenylmethane, 187,
  188
4,4'-Diaminodiphenylsulphone (DDS),
  187, 212, 215–17
Dibromo neopentyl glycol (DBNPG),
  70, 73
4,4'-Di(chloromethyl)diphenylether, 89
Dichlorosilanes, 13
1,2-Dichloro-1,2,2-tetraphenyl ethane
  (DCTPE), 143
$\alpha\alpha'$-Dichloro-p-xylene, 91
Dicyclopentadiene (DCPD), 64–7
  acrylate, 44
Diels–Alder
  addition, 65
  reaction, 186
Diethylaniline (DEA), 128
Diethylene glycol, 67
Diethylene triamine (DETA), 212, 268
4,4'-Dihydroxy-2,2'-diphenyl propane,
  15
Dɪ Marzio equation, 239
1,2-Dimethoxy-1,1,2,2-tetraphenyl
  ethane (DMTPE), 143
$\alpha\alpha'$-Dimethoxy-p-xylene, 95–101
Dimethylaniline (DMA), 128, 132
N,N-Dimethylaniline, 39
2,5-Dimethyl-2,5-di(2-
  ethylhexanoylperoxy)
  hexane, 136
2,5-Dimethyl-2,5-hexane diamine
  (DMHDA), 212
Diols, 21
Di-2-phenoxyethyl peroxydicarbonate,
  136
Diphenyl ether, 97
Disc cracks, 251–2
Doryl resins, 91
Dough moulding compounds, 15, 23,
  76–80, 84
Dow XD-8084, 38
Dow XD-9002, 35
Dugdale plastic zone, 273
Durestos boards, 91

Electrical properties
  phenol–aralkyl resins, 112–14
  phenol–formaldehyde resin, 163
Electron-transfer oxidation–reduction
  reactions, 127
Energy
  conservation, 26
  dissipation, 234
  requirements, 26
Epichlorhydrin, 15
Epoxide–amine addition reaction, 213
Epoxide(s)
  cure of phenol–aralkyl resins, 115
  homopolymerisation of, 214
Epoxy Novolac vinyl ester resins, 34
Epoxy resins, 15–17
  amine-cured, 211
  chemical structure, 213–17
  composite, 99
  crack propagation, 258
  crazing, 268
  deformation modes and failure,
    222–3
  durability, 223–6
  fracture surfaces, 266
  mechanical properties, 271
  microvoid characteristics, 221
  network structure, 3, 217–18
  physical structure, 217–22
  stresses in, 25
  structure–property relationships
    and environmental
    sensitivity, 211–29
ESCA ^{13}C NMR, 24
Estercrete, 80–1
Ethanolamine, 73
2-Ethoxyethanol, 101
Ethylene glycol, 67

Filament wound pipe, 41
Fire retardancy, 15, 22, 71–6, 170,
  206
Flame retardant resins, 34–5
Flammability resistance of
  phenol–aralkyl resins, 114
Flexibilisers, 247
Foaming agent, 82

Fox–Loshaek expression, 239
Fractography, 266
Fracture
  mechanics testing, 258
  surface(s)
    energy, 245
    epoxy resins, 266
    polyester resins, 250–3
  toughness
    crosslinked polyester resins,
      245–8
    elastomeric phases, with, 249–50
Free radicals, sources of, 142–4
Friedel–Crafts
  catalyst, 97, 100–1
  condensation, 87
  reaction, 99, 103–5
  resins, 88, 90–9
Furan resins, 17–18
Furane resins, 169–70
Furfuryl alcohol, 17

γ and γ' transitions in crosslinked
  resins, 244–5
Gel permeation chromatography, 24
Glass transition temperature, 148,
  235–7
Glycols, 67–70
Griffith equation, 248

Halogenated additives, 22, 74
Halogenated formulations, 71
Halogen-containing derivatives, 21
Hardeners, 7, 261
Health and safety measures, 27
Heat distortion temperature of vinyl
  ester resins, 44
HET acid, 73, 74
Hexamethylene tetramine, 7, 9, 100
Hexamine, 115
Hexel F178, 184, 185
High-temperature properties, 145–210
  phenol–aralkyl resins, 110, 115
  thermoplastic resins, 150–60
  thermoset resins, 160
  vinyl ester resins, 34

HM-S carbon fibre/PMR-15
    composites, 188
Hydrogen peroxide, 130
Hydroperoxides, 128
Hydroquinone, 39
Hydroxyl content of thermosetting
    resins, 25

Impact properties of vinyl ester resins,
    41
Infra-red spectroscopy, 24
Inhibitors, 39–40
Initiator(s)
    blends, 139
    efficiency, temperature effect, 136
    half-life effect on cure
        characteristics, 135
    high temperature curing, for, 126,
        134–9
    organic, 123–7
    unsaturated polyester resins, for,
        121–44
Injection moulding, 10, 23, 26
Injection-compression, 11
Isophthalic acid, 63, 64

Jeffamine T-403, 212

Kapton film, 173, 175
Kerimid-353, 178–83
    autoclave lamination, 183
    processing and properties, 183
    thermal degradation, 181–3
Kerimid-601
    processing and properties, 178
    thermal degradation, 178
Kerimid-711, 184
Ketone peroxides, 128, 129, 132, 141

Laminated structures, 40–2
Laser–Raman spectroscopy, 24
Linear elastic fracture mechanics
    (LEFM), 256, 258

Maleic anhydrides, 63, 66
Matrix function, 2
Maturation agent, 77
Mechanical properties
    crosslinked polyester resins, 231–55
    epoxy resins, 271
Melamine, 11–13
Melamine–formaldehyde resins, 4, 5,
    163–7
    thermal stability, 166–7
Methyl ethyl ketone peroxide
    (MEKP), 39, 128–31
l-Methyl imidazole, 143
Methyl methacrylate, 70
Methylol derivatives, 165
Methylolmelamines, 165
N-Methylpyrrolidone, 185
Microcracking, 226
Moisture effects, 150
Moisture-sorption characteristics,
    224–6
Molecular weight distribution
    investigations, 101
Molybdenum
    compounds, 76
    trioxide, 76
Monochlorosilanes, 13
Moulding
    compounds, 76–80, 106, 108
    powders, 12

Neopentyl glycol (NPG), 67–8
Network structures, 3, 4
Nexus polyamide fibres, 44
Nodular structure, 271
Norbornene systems, 186–90
5-Norbornene-2,3-dicarboxylic acid
    (NE), 188
Novolaks, 7, 17, 51, 161
NR-150 polyimides, 194–200
Nuclear magnetic resonance studies,
    101

Organic initiators, 123–7
Organometallic resins, 94–5

Peroxides
  decomposition
    kinetics, 133
    rate, 132–3
  half-life, 132–3
Peroxyketals, 137–9, 142
Phase lag angle, 234
Phenol–aralkyl pre-polymer, 102
Phenol–aralkyl resins, 87–120
  chemical reactivity, 105
  chemical resistance, 112–14
  electrical properties, 112–14
  epoxide cured, 115–18
  examples of, 103
  flammability resistance, 114
  Friedel–Crafts reaction, 103–5
  hexamine cured, 115
  high-temperature
    properties, 115
    stability and strength, 110
  moulding compounds, 106
  physical properties, 116–18
  preparation and hexamine cure,
    100–2
  radiation resistance, 114
  range of, 103
  structure and properties, 105–8
  upgrading of phenolic novolacs,
    108
  wetting characteristics, 116
Phenol–formaldehyde
  composites, typical properties, 163
  condensates, oxidative degradation,
    106
  resins, 7–11, 24, 161–3
    electrical properties, 163
    thermal degradation, 162–3
Phenolic inhibitors, 39
Phenolic nuclei, substitution in, 102
Phenolic resins, 23, 105, 108–10
  injection moulding, 10
Phenolic resol resins, 108–10
Phenyl compounds, 21–2
Phosphorus based additives, 75
Phosphorus compounds, 72
Photodegradation, 22
Photoinitiators, 35, 143
Phthalic anhydrides, 63

Pipe, filament wound, 41
Plastic deformation at crack tip, 271–4
PMR-15, 188, 190
Pollution control, 27
Polyalkyl acrylate polymers, 30
Polyamide-imides, 202–4
Polyaminobismaleimides, 177–8
Poly(arylene ether sulphones), 151–3,
  157
Polybenzyl, oxidative degradation, 89
Polybenzyl polymers, 88
Polydiphenyl ether resins, 89–91
Polyester
  foam, 81–2
  resins, 21, 22, 24, 59–86
    alkylene oxide route, 60–3
    'built-in' halogen, 73
    bulk handling, 23
    crosslink density, 237–40
    crosslinked, mechanical
      properties, 231–55
    crosslinked network structures,
      232–3
    DCPD modified, 66–7
    dough moulding compounds, 80,
      84
    dynamic mechanical properties,
      233–5
    flame retardant, 71–6
    foams, 81–2
    fracture
      morphology, 250
      surfaces, 250–3
      toughness, 245–8
    future trends, 84–5
    glass transition temperature, 235–7
    glycols in, 67–70
    halogenated additives, 74
    high performance unsaturated, 64
    manufacturing processes, 60–3
    maturation rate, 76–8
    monomers, 70–1
    moulding compounds, 76–80
    refinement of existing processes,
      63
    room temperature curing, 129
    selecting initiators for curing at
      elevated temperatures, 134–9

Polyester—*contd.*
  resins—*contd.*
    sheet moulding compounds, 23, 80, 84
    shrinkage, 76, 78–80
    styrene emission, 83–4
    toughened DMC and SMC, 80
    unsaturated, 14–15, 23, 59
      initiator systems, 121–44
Polyesterification process, 63
Polyethersulphones, 21
Polyimide-2080, 200–2
Polyimides, 18–20, 24, 171–204, 206
  addition type, 177–8, 184
  aluminium-filled, 183
  condensation type, 171–3
  NR-150, 194–200
  preparation, 172–3
  processing and properties, 175–6
  thermal degradation, 173–5
  thermal stability, 153–6, 171
  thermoplastic, 194–204
Polymer cement, 80
Polymerisation *in situ*, 20
Poly(methyl methacrylate), 246, 259, 274, 275, 279
Poly(phenylene sulphide) resins, 158–60
  physical properties, 159
Poly(phenylene sulphone), 151
Poly(propylene maleate adipate), 247
Premix compounds, 15
Prepreg processing, 23
Price increases, 26
PRM-11, 188
Propylene glycol, 66
1,2-Propylene glycol, 67
Proton NMR, 24
PSP resin, 204–5
  press cure cycle for laminates, 205
  processing and properties, 204–5
  synthesis, 204
Pyrolysis gas chromatography, 24

Quality control
  finished products, 25
  thermosetting resins, 23–5

Radiation
  curing, 35, 53
  resistance of phenol–aralkyl resins, 114
Recycling, 23
Redox
  initiators for ambient temperature cure systems, 125
  mechanisms, 127–32
Relaxation(s)
  constituent chain, in, 241–2
  processes, 235
Resole formation, 9

Scrap additions, 23
Secondary relaxations, 240
Secondary transitions, 240
Self-toughening mechanism, 280
Sheet moulding compounds, 23, 33, 52, 55, 76–80, 84, 127, 137
Shrinkage and shrinkage stresses, 76, 78–80, 226
Silicone resins, 13–14, 167–9
  thermal stability, 168–9
Siloxane/phosphate composites, 95
Skybond-700, 173, 175
Smoke
  generation, 22, 170
  suppressants, 76
Stannic chloride, 101
Stick/slip propagation, 259, 260, 265, 266, 273, 275, 277–8
Storage modulus, 234
Strain energy release rate at failure, 279–81
Stress
  concentration, 222
  epoxy resins, in, 25
  intensity factor, 249, 258–61, 264, 275, 277–9
Styrene
  content of cured polyester resins, 139–42
  emission in polyester resins, 83–4
  monomer, 42–3
Supramolecular structure, 2–3
Swelling stresses, 226

Tensile elongation of resin castings, 40, 42
Terephthalic acid, 63–4
Tetrabromo phthalic anhydride, 73, 74
Tetrachlorophthalic anhydride, 73
Tetraethylenepentamine (TEPA), 279, 281
Tetraglycidyl 4,4′-diaminodiphenyl methane, 212, 215
TGDDM–DDS systems, 212, 213, 215, 221–6
  theoretical reaction mixtures, 216
Thermal degradation
  Kerimid-353, 181–3
  Kerimid-601, 178
  phenol–formaldehyde resins, 162–3
  polyimides, 173–5
Thermal spike exposure, 225, 226
Thermal stability, 146
  melamine–formaldehyde resins, 166–7
  polyimides, 171
  polysulphones, 153–6
  silicone resins, 168–9
Thermally stable resins, 145–210
Thermid-600, 190–3
Thermogravimetric analysis (TGA), 146, 148, 153, 163, 182, 192
Thermoplastic resins
  fabrication, 149
  high-temperature properties, 150–60
Thermosetting resins, 1–28
  crack propagation, 257–83
  fabrication, 149
  filled, 5
  future prospects, 25–7
  handling and processing, 22–3
  hardening reaction, 22
  high-temperature properties, 160
  hydroxyl content, 25
  limitations, 7
  matrix function, 2
  properties of, 3–7
  property improvements, 21–2
  quality control, 23–5
  reground, 23
  reinforced, 26
  supramolecular structure, 2–3
  unreinforced, 5

Thin layer chromatography (TLC), 24
Threshold limit value (TLV), 83
Time-dependent failure, 279
Toxicological aspects of vinyl ester resins, 55–6
Toxicological properties, 27
Triallylisocyanurate, 185
Trichlorosilanes, 13
Triethylene tetramine (TETA), 260, 271
2,2,4-Trimethyl-1,3-pentanediol (TMPD), 68–9

Urea, 11–13
Urea–formaldehyde, 22, 24

Vinyl ester resins, 17, 29–58
  aircraft industry applications, 53
  applications, 47–54
  ballistic armour, 54
  basic structure, 32–3
  brominated, 34
  corrosion resistance, 32, 44
  curing, 39
  dental filling material, 53
  dicyclopentadiene acrylate-diluted, 44
  early history, 30–1
  effects of cast resin high tensile elongation on performance of laminated structures, 40–2
  electrical insulation applications, 52–3
  filament wound laminate, 47
  filament wound pipe, 41
  flame retardant, 34–5
  generalised structure, 31
  glass fibre laminates, 44
  hand lay-up laminate, 47
  heat distortion temperature, 44
  high-temperature applications, 34
  HSMC, 52
  impact properties, 41
  inhibitors, 39–40
  laminate properties, 44–7

Vinyl ester resins—*contd.*
  land transportation applications,
    51–2
  marine applications, 53
  matched die laminates, 47
  mechanical properties, 44
  monomers, 42–4
  new developments, 54–5
  physical properties, 37
  radiation
    curable, 35
    curing applications, 53
  rubber-modified, 38
  SMC resins, 33, 55
  structure-imparted characteristics,
    31–2
  structures, 32–8
  synthesis, 31
  tensile elongation, 40, 42
  toughness, 32
  toxicological aspects, 55–6
  urethane-based, 36

Vinyl toluene, 44
Viscosity, 4
Viscosity–pressure characteristics, 21
Viscosity–temperature relationships, 4

X-ray
  diffraction, 25
  scattering, 271
Xylok-210, 110
Xylok-225, 108, 109
Xylok-237, 116

Yield stress, 278, 279
Young's modulus, 2

Zinc borate, 71